ADVANCES IN GRAPHENE SCIENCE

Edited by **Mahmood Aliofkhazraei**

Advances in Graphene Science
http://dx.doi.org/10.5772/51689
Edited by Mahmood Aliofkhazraei

Contributors

Nan-Fu Chiu, Teng-Yi Huang, Hsin-Chih Lai, Yuhai Hu, Xueliang Sun, Kai Wang, Ajay Kumar, Chee Huei Lee, Shannon Notley, Russell Crawford, Elena Ivanova, N.S. Sankeshwar, S. S. Kubakaddi, B.G. Mulimani, Artem Baskin, Petr Kral, Boyang Wang, Recep Zan, Ursel Bangert, Quentin Ramasse, Demie Kepaptsoglou, Rashid Jalil, Guocai Dong, Dirk Van Baarle, Joost Frenken

Published by InTech

Janeza Trdine 9, 51000 Rijeka, Croatia

Publishing Process Manager

Technical Editor InTech DTP team

Cover InTech Design team

First published August, 2013

Printed in Croatia

Additional hard copies can be obtained from orders@intechopen.com

Advances in Graphene Science, Edited by Mahmood Aliofkhazraei

p. cm.

ISBN 978-953-51-1182-5

For EU product safety concerns:
IN TECH d.o.o., Prolaz Marije Krucifikse Kozulić 3, 51000 Rijeka, Croatia,
info@intechopen.com or visit our website at intechopen.com.

Meet the editor

Mahmood Aliofkhazraei is a researcher in the corrosion and surface engineering group at the Tarbiat Modares University. Dr. Aliofkhazraei has received several honors, including the Khwarizmi award and the best young nanotechnologist award of Iran. He is a member of the National Association of Surface Sciences, Iranian Corrosion Association, and National Elite Foundation of Iran. His research focuses on nanotechnology and its use in surface and corrosion science.

Contents

Preface

Graphene is proving to be the magic material of the 21th century. It is widely accepted that it is the strongest material ever studied and can be an efficient substitute for silicon. Besides, fascinating properties of graphene, such as the highest electrical conductivity among the discovered substances, have dramatically shocked science and technology world. Graphene is a carbon based material with high atomic density. Its extraordinary characteristics such as extremely high mechanical strength, hardness, and adjustable thermal and electrical conductivity, as well as excellent surface and optical feature through chemical marking, have received great deal of attention by many researchers. The fact that chemists can barely find an alternative for graphene has caused development of various applications for it in nanoelectronics, solar cells, and storage tools such as batteries and super capacitors.

This book collects advances in graphene science. I like to appreciate all of contributors to this book and thank them for their high quality manuscripts. I wish open access publishing of this book help all researchers to benefit from this collection.

Dr. Mahmood Aliofkhazraei
Tarbiat Modares University
Iran

Section 1

Structure of Graphene

Chapter 1

Atomic Structure of Graphene and h-BN Layers and Their Interactions with Metals

Recep Zan, Quentin M. Ramasse, Rashid Jalil and
Ursel Bangert

Additional information is available at the end of the chapter

http://dx.doi.org/10.5772/56640

1. Introduction

Since the isolation of graphene and other two-dimensional (2D) materials, such as hexagonal boron nitride (h-BN), these just one atom layer thick materials have spurred a flurry of investigations into their structural properties, morphology and chemistry [1, 2]. Graphene is a 2D crystalline form of carbon and a basic building block for carbon allotropes such as fullerenes, carbon nanotubes and graphite [3]. In graphene, carbon atoms are packed in a planar honeycomb network. The unit cell of single-layer graphene consists of two carbon atoms, separated by 1.42 Å, with a lattice constant of 2.46 Å. Each atom has s, p_x and p_y orbitals and is bonded to three neighbor atoms in the lattice, forming an sp^2 atomic network. The p_z orbitals overlap between neighboring atoms resulting in so-called filled π and empty π^* states, which respectively form the valence and the conduction bands in graphene. However, so-called Bernal (AB-) stacked double layer graphene has four electrons in the unit cell: in this configuration, the individual layers are stacked but shifted with respect to one-another, so that the carbon atoms in one layer sit on top of the empty hexagon centers in the underlying layer. Due to the absence of any measurable bandgap in both single and double layer graphene these materials are often called zero-gap semiconductors. The lack of a bandgap is a serious limitation to the use of graphene in electronics as switching off any graphene-based device would be extremely difficult. Additionally, researchers have shown that the substrate on which graphene layers are placed has dramatic effects on its transport properties. SiO_2/Si substrates were used for initial graphene devices [1]. However, due to their surface roughness and to the presence of charged impurities graphene devices built directly on SiO_2 substrates could not demonstrate any marked improvement over more traditional designs: they were seemingly never able to utilize the intrinsic properties of graphene [4]. Other substrates (2D materials) such as hexagonal boron nitride (h-BN) [5] and, recently, dichalcogenides [6] have therefore been tried as support.

One of the members of 2D materials is h-BN, which consists of boron and nitrogen atoms arranged in a hexagonal structure such that each boron atom is bonded to three neighboring nitrogen atoms and *vice versa*. The resulting structure resembles graphene, with a similar lattice constant of 2.52Å. When stacked, h-BN adopts a so-called AA′ configuration whereby each boron atom sits atop a nitrogen atom, and *vice versa* [7]. By contrast, graphene is most frequently observed in an AB stacking arrangement. Despite the presence of structural similarity between the graphene and h-BN structure, the electronic characteristics are totally different; graphene is a conductor (semi-metal) whereas h-BN is an insulator or a wide gap semiconductor, with a gap of 5.9 eV, which makes it the thinnest possible insulator. In addition, h-BN is mostly inert, its surfaces are missing dangling bonds, and it has large optical phonon modes, making it an ideal substrate (dielectric support) for high-quality graphene electronic devices [5]. h-BN has strong covalent bonds in the plane between the atoms and weak Van der Waals bonds between different planes, which is similar to graphene, but, the bond is slightly ionic in comparison to graphene.

Due to its unique properties, graphene can be employed in a variety of applications. It is thought it may eventually replace many of the materials currently used in today's technology and result in higher performance, energy efficiency, flexibility and durability [8]. Graphene is being used in electronics [9], energy storage [10, 11], photonics [12], composite materials [13], sensor technology [14] and as a sample support for TEM applications [15]. One of the most promising electronic applications of graphene is the manufacture of transistors [1]. However, transistors, as any other graphene-based electronic device, must involve the incorporation of metal contacts, which link the graphene to the other device components. The choice of the metal used as a contact has been shown to dramatically affect the performance of the resulting devices [16], with little understanding hitherto of why this is the case. While the macroscopic properties of graphene devices are readily measured and characterized, understanding the metal-graphene interactions necessitates an investigation and direct 'visualization' at the atomic level, which is a remarkably challenging task. Indeed, there are only few reported observations at the atomic scale of the metal-graphene system, especially for suspended (or 'free-standing') graphene. A large effort has in fact been recently devoted to suspended devices [17, 18]. Using suspended graphene (rather than supported on a different material) eliminates substrate effects, which are known as a limiting factor for the mobility in graphene based devices [19]. A fully suspended geometry also allows exploiting directly the intrinsic, exceptional properties of the material. It is thus essential to carry out detailed systematic studies of the nucleation following deposition, and of the resulting coverage of metals on suspended graphene, in order to determine what the optimal contact might be and ultimately to improve device performance.

Transmission Electron Microscopy (TEM) or Scanning Transmission Electron Microscopy (STEM) are an ideal, perhaps the only, tool for such studies. Indeed, the technique's ability to image and identify directly each and every atom in 2D materials has already played a significant role in improving our understanding of graphene properties [20, 21]. Rapid progress in microscope components in recent years, in particular the successful implementation of aberration correctors for electron optical lenses, has heralded a new era in materials- and nano- science research. These instrumental advances have substantially improved the overall performance of electron microscopes: much higher resolutions are now attainable in both structural and spectroscopic data, with very clear benefits to the understanding of the investigated material's structure. More crucially, they enable the use of much lower acceleration voltages, e.g., <80 keV, below the carbon knock-on threshold (voltages, typically ranging from 100 to 300kV, have conventionally been used to accelerate the electron beam towards the specimen inside the microscope column) whilst maintaining atomic resolution. Structural changes and damage to the specimen due to its interactions with the high energy electron beam can thus be reduced without losing any of the structural details in the micrographs. Most of the results presented in this chapter were obtained on a Nion UltraSTEM microscope. The microscope was operated at 60 kV acceleration voltage to prevent knock-on damage to the graphene [22]. High angle annular dark field (HAADF) imaging was employed to produce micrographs. This chemically-sensitive 'Z-contrast' mode is ideally suited to directly identify the nature of individual atoms and it is usually complemented by further chemical fingerprinting through Electron Energy Loss Spectroscopy (EELS).

In the following in turn, we will summarize fabrication of graphene and h-BN flakes, their preparation for microscopy measurements, pristine graphene and h-BN S/TEM characterization, metal doped graphene and h-BN characterization and finish by conclusions.

2. Graphene and h-BN fabrication

Two different techniques were used to produce the graphene membranes for the experiments: mechanical cleavage of highly ordered pyrolytic graphite (HOPG) and chemical vapor deposition growth on copper substrates. For h-BN fabrication only the former method was used. Obtaining graphene and h-BN by exfoliation (mechanical cleavage) is simply done by pressing a piece of HOPG or bulk h-BN crystal against some adhesive tape and peeling the tape off [1, 23, 24]. Due to the layered structure of graphite and h-BN and the weak interlayer bonding (Van der Waals), the layers are easily separated, so that after repeating the peeling process a few times extremely thin layers are left on the tape. The thin graphite and h-BN flakes obtained in this fashion are then placed onto an oxidized silicon substrate. The obtained layers are initially observed under the optical microscope to identify single layer regions, whose specific color and contrast against the oxidized silicon wafer background provide a quick and convenient screening criterion. Identification of graphene flakes on the wafer is straightforward. An optical micrograph of exfoliated graphene layers on an oxidized silicon substrate is presented in figure 1. The flakes on the wafer vary in thickness from single layer to much thicker regions: a single layer area, 100x100μm in size, is magnified. This size of the single layer graphene flake is adequate not only for electron microscopy but also for many other characterization techniques. However, due to their weak contrast on the wafer and their small size compared to graphene, identification of h-BN flakes under the optical microscope is a quite challenging task.

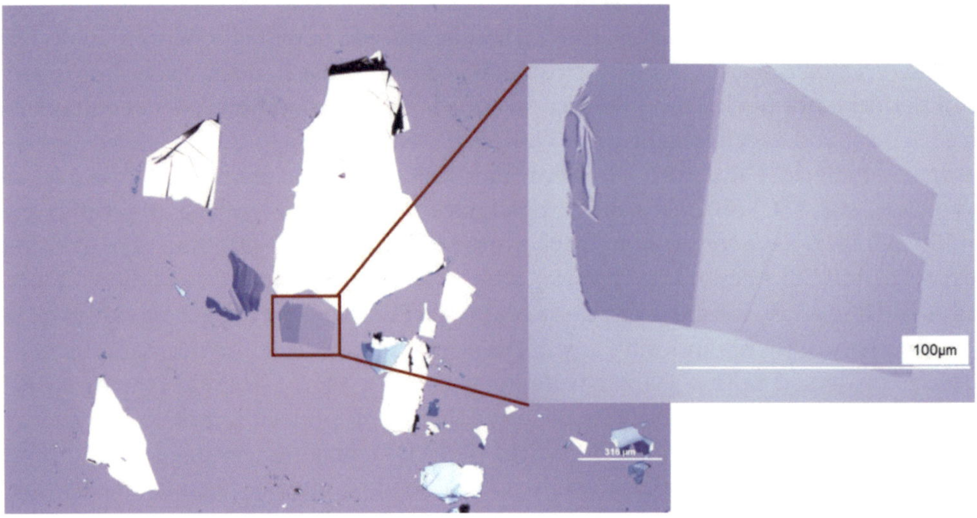

Figure 1. Optical image of the exfoliated graphene layers on 290 nm thick oxidized silicon wafers and enlarged view of single layer region, shown as inset.

Graphene growth by the chemical vapour deposition (CVD) technique is based on the decomposition of hydrocarbon gases at elevated temperature and the diffusion of carbon atoms ona metal surface, followed by rapid cooling to 'freeze in' any graphene layers formed on the surface of the substrates. The most widely used combination of gas and substrate at present consists of methane gas and a copper substrate; the low solubility of carbon in copper provides a means of restricting the graphene growth to a single layer [25, 26]. In our case, graphene was obtained on Cu substrates and methane gas was used as precursor in a quartz tube at pressure 600mTorr.

2.1. Fabrication of suspended sample for microscopy

A so-called *wet transfer* technique was used to prepare suspended samples for the microscopy experiments from both exfoliated and CVD-grown flakes, with only small variation in the methodology depending on how the flakes were initially grown [27, 28].

In the case of exfoliated samples for both graphene and h-BN, following the initial identification of regions of interest on the oxidised silicon wafer under an optical microscope, a layer of ~300 nm of poly-methyl-methacrylate (PMMA) was 'spun' onto the graphene flakes using a spinner, followed by a 130°C bake of the samples for 5mins. This additional layer is designed to protect the underlying graphene layers, providing mechanical support and facilitating the handling of the flakes during the transfer procedure. The substrate, or more precisely the oxidised silicon layer (topmost layer of the wafer), is then etched away using a 3% potassium hydroxide (KOH) solution, detaching the PMMA and the flakes from the silicon wafer. Once almost detached, the wafers are dipped into de-ionized (DI) water, to fully detach the PMMA and the flakes and to rid them of chemical residues from the etching process as well as of any possible contamination from the silicon wafer. Rinsing in DI water is repeated to ensure the

http://dx.doi.org/10.5772/56640

best sample cleanliness. Once the flakes are transferred onto a TEM grid by simply 'fishing them out' of the DI water bath, the sample is dried at 50°C for 10 mins, followed by a further 120°C bake for 10 mins in clean room conditions to improve the adhesion of the flakes to the grid, get rid of any remaining water content, and importantly, to flatten out (anneal) any possible wrinkles, which may have been created during the transfer. The PMMA protective layer is then dissolved by dipping the sample into acetone for 15 mins. Finally, the sample is dried in a Critical Point Dryer (CPD) in order to protect the delicate, now suspended, graphene layers against any rupture, damage or deformation due to surface tension. The transfer process to TEM grid is schematically summarised in figure 2.

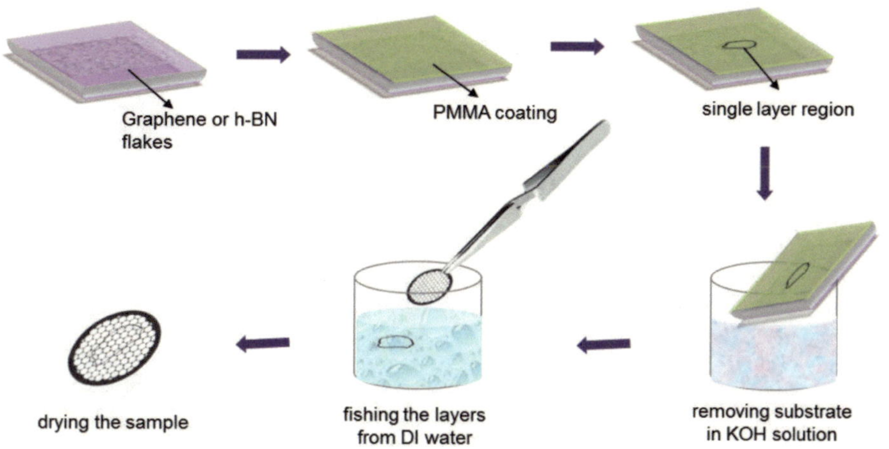

Figure 2. Schematic representation of the transfer of exfoliated graphene layers from an oxidized silicon wafer to a TEM grid by wet-etching using a KOH solution.

Optical images of exfoliated graphene flakes on the substrate (figure 3a) and after transfer onto the grid are shown in figures 3b and c, along with a low magnification TEM image in figure 3d.

In the case of graphene obtained by CVD on Cu substrate, the graphene growth takes place on both sides of the Cu surface. An additional step is therefore necessary to remove the flakes present on one of the two sides to avoid interference of layers during the transfer. The side being prepared for transfer is covered with PMMA, in order to protect the underlying graphene and to ease the transfer process as was the case for exfoliated samples, while the other side is exposed to a 20 mTorr oxygen plasma for a few minutes to remove the graphene layer. Then an aqueous ammonium persulphate ($(NH_4)_2 S_2O_8$) solution (0.1 molar) was used to etch the Cu substrate to produce the material investigated here. After the etching procedure, the PMMA/graphene layers are repeatedly rinsed in DI water, to remove further possible contaminants. The flakes are again directly transferred onto TEM grids by 'fishing them out' of the DI water bath. As for the exfoliated graphene transfer procedure previously described, the grids are then baked, the PMMA layers are dissolved in acetone and finally dried in a CPD.

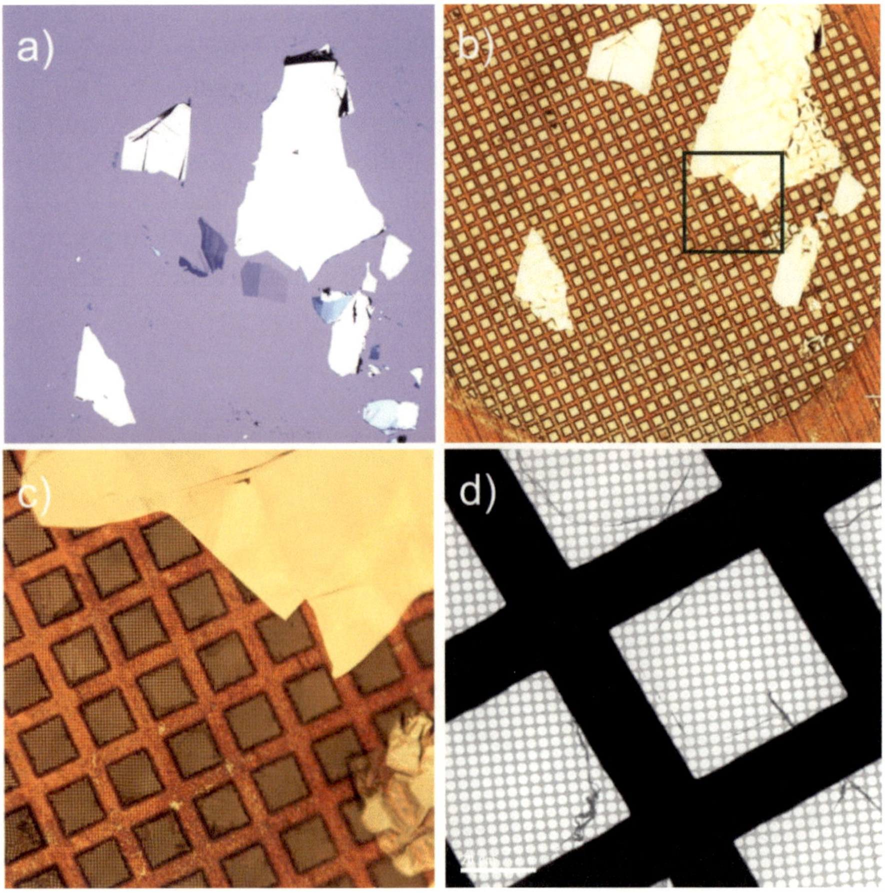

Figure 3. Graphene flakes a) as prepared on an oxidized silicon wafer, b) after transfer onto a TEM grid; c) magnified optical image of the green square indicated in (b) illustrating how suspended single layer regions are not easily visible; d) low magnification TEM image of a single layer area. Broken flakes and rolled up regions produce some visible contrast and hint at the presence of the single layer graphene sheet.

Thermal and electron-beam evaporators were used to deposit metal impurities onto the suspended graphene membranes. A fixed amount of material was deposited in fixed conditions for a large variety of metallic species, allowing a thorough comparison between metals, of the distibution, size and shape of the clusters. A total amount of 2Å of Au, Ti, Ni, Al and Pd was evaporated onto different suspended graphene and h-BN layers on TEM grids at a rate of 0.1 nm/s, in vacuum better than 10^{-6} mbar (up to ~10^{-8} mbar).

3. Transmission electron microscopy

Electron microscopy is based on the interactions of matter with a high energy beam of electrons (from a few keV to several 100 keV). These interactions provide insights about the structure,

http://dx.doi.org/10.5772/56640

topology, morphology and composition of a material. The interactions of an electron beam with matter can be divided into two categories: *elastic* and *inelastic* interactions, which are used in the TEM and STEM to obtain structural and compositional information about the specimen.

One of the main imaging modes of a TEM is known as bright field (BF) in which only the direct, unscattered and small-angle scattered electrons are allowed to contribute to the image formation. Phase contrast is the main contrast mechanism in BF imaging and constitutes the basis for so called high-resolution TEM (HRTEM). Phase contrast occurs due to the difference in the phase of the electron waves inferred by interaction with the atomic potential of the atoms in the thin specimen.

3.1. Scanning transmission electron microscope

In contrast to employing a broad and stationary beam in conventional TEM, in the STEM a small, focussed probe is scanned over the specimen. Transmitted or scattered electrons are collected to build up images serially from a very large number of image points over a reasonably long time (for high signal/noise ratio), whereas in a TEM image recording is parallel (simultaneous for all image points) and thus usually much shorter.

A STEM is more powerful to obtain Annular Dark Field (ADF) images than a conventional TEM. One of the most important characteristics of ADF imaging is using incoherent, elastically scattered electrons in contrast to BF imaging which relies mostly on phase contrast and therefore interference phenomena. STEM ADF signals are predominantly produced by electrons that have been scattered by the atomic nucleus (Rutherford scattering). This makes the ADF intensity of individual atoms depend on the charge of the atom's nucleus. Different Z atoms therefore give different ADF image intensities and this imaging technique is called Z-contrast imaging [29]. Only when sufficiently high angles are used in STEM ADF imaging does it become predominantly incoherent. This imaging (collecting very high angle, incoherently scattered electrons) is also often called HAADF imaging as the intensity recorded in this mode by positioning the electron beam on an atomic site is approximately proportional to the square of the average atomic number Z of this site [30, 31]. The HAADF-STEM imaging mode is particularly useful to visualize the structure and defects of materials, as its chemical sensitivity provides a direct visual guide to identify ad-atoms and impurities of a different weight from the surrounding material [22, 32].

As a result of inelastic interactions between the incident beam and the specimen, the scattered electrons can lose a part of their energy. This loss of energy can be measured by a spectrometer to yield elemental, chemical and dielectric information about the specimen. In addition to HAADF imaging, EELS was employed to identify foreign species on the graphene and h-BN surfaces in our investigations.

3.2. Instrumenation

Tecnai: Initial experimental data to identify graphene layers and some of the high resolution data of the metal-graphene system were obtained on an FEI Tecnai F30 microscope operated at 300 kV.

SuperSTEM: Most of the experimental data were acquired on an aberration corrected dedicated STEM, the Nion UltraSTEM It is equipped with a cold field emission gun with an energy spread of 0.3 eV and operated at 60 kV to prevent knock-on damage in graphene. The UltraSTEM microscope has a UHV design throughout, allowing pressures at the sample of below $5x10^{-9}$ torr for contamination-free observation. The estimated probe size (full width at half-maximum) is about 1.2 Å for the typical operating conditions for graphene investigations at 60 kV acceleration voltage and 30 mrad probe convergence semi-angle. Detectors with different geometries are used to collect electrons scattered to different angular ranges. For example, the BF detector used to obtain phase contrast images can collect electrons scattered up to 6 mrad. The so-called medium angle annular dark field (MAADF) detector collects electrons scattered from 45 to 190 mrad and the HAADF detector used to record the Z-contrast images has inner and outer radii of 70 mrad and 210 mrad, respectively. The instrument is equipped with a Gatan Enfina EEL spectrometer and its acceptance semi-angle was calibrated at 35 mrad for core-loss spectroscopy.

4. TEM characterisation of pristine graphene

TEM has been widely used in graphene research for applications as varied as: number of layers determination [33], elemental identification [22, 34], surface roughness (ripples) visualization [33, 35], detection of sheet edge types [36], observation of defects [37], and stacking faults [38, 39], impurities, such as contamination and individual ad-atoms [40] effects of radiation [41] and investigation of graphene based heterostructures [42].

Due to the substantial amount of hydrocarbon contamination, foreign ad-atoms and other residues from the chemicals used during fabrication and transfer, graphene layers were found not to be perfectly clean as also evidenced by low magnification STEM micrographs. Although the presence of contamination can be reduced with careful and cleaner fabrication and transfer procedures, 'unintentional' contamination (due to handling of the samples in air or to residual gases in the microscope column [21] albeit to lower amounts), was observed to exist on the surface at all times. An unprocessed HAADF image of a reasonably large area of a single layer graphene flake is shown figure 4a. It can be clearly seen that clean graphene regions (residue free areas) are limited in size (from few to few hundred nm^2) and surrounded by contamination, which was found to consist of H, C, O and Si in our EDX analysis. Very dark, grayish and bright regions in the image represent the graphene, hydrocarbon contamination and foreign ad-atoms, respectively. Contaminations are also clearly visible due to higher scattering in regions with heavier atoms and larger thickness. A close-up of figure 4a is presented in figure 4b which shows the atomically resolved graphene lattice.

TEM electron diffraction is a straightforward technique to distinguish single- from double- and few-layer graphene by comparing the intensities of the first and second ring of the diffraction spots [20, 33]. For single layer graphene the intensity of the outer hexagon spots is roughly the same as, or less than that of the inner one. In contrast, for double layer graphene the outer hexagon intensity is higher than the inner one. Figure 5 shows diffraction patterns

http://dx.doi.org/10.5772/56640

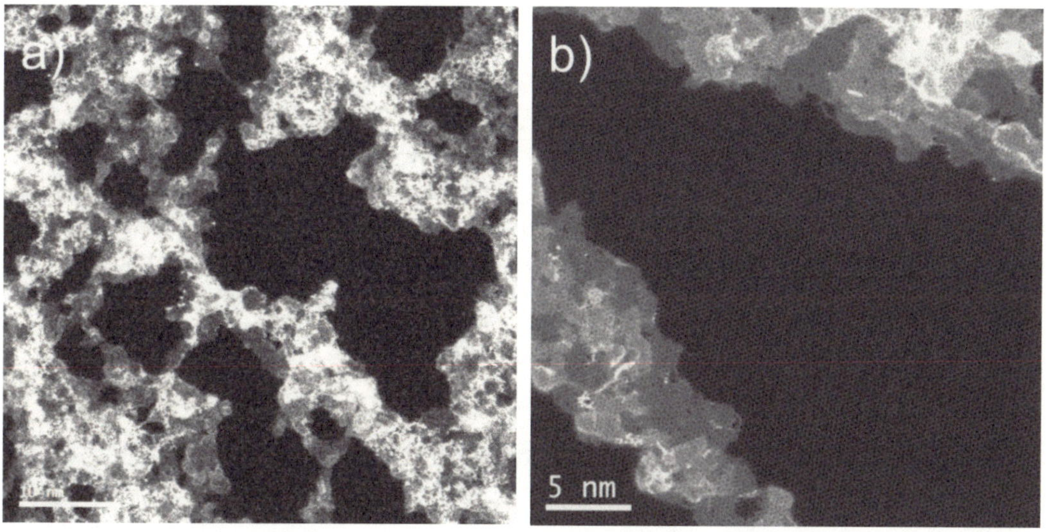

Figure 4. a) An overview unprocessed HAADF and b) High resolution HAADF image of pristine single layer graphene.

for single- and AB-stacked double- layer graphene and their intensity profiles. In addition, diffraction peak intensities vary only weakly with the tilt angle between the graphene membrane and the incident beam for single layer graphene, whereas for double layers, tilting of a few degrees leads to strong variations in the diffraction intensities [20].

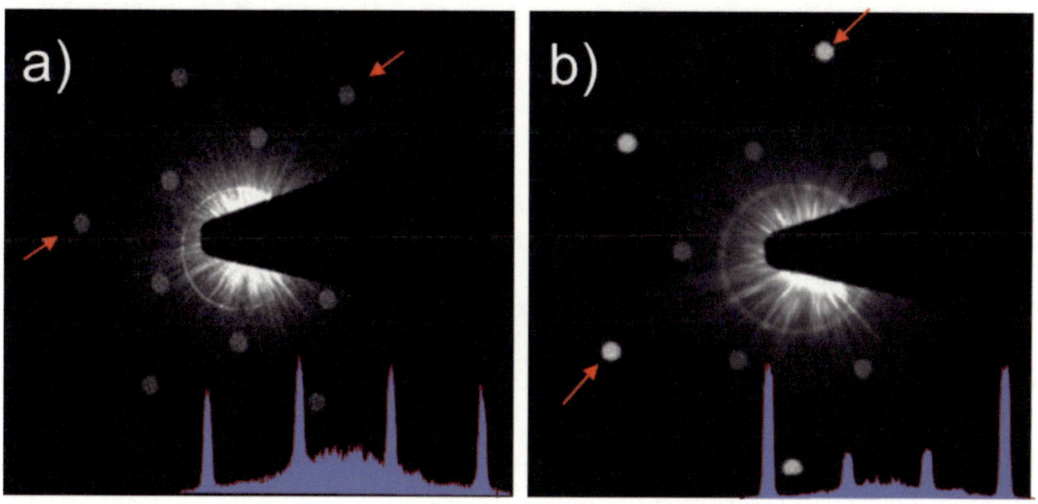

Figure 5. Diffraction patterns of a) single layer and b) double layer graphene. Intensity profile plots taken between the red arrows are shown in the inset. The diffraction patterns were acquired with a Tecnai F30.

After identifying single layer graphene, high resolution BF and HAADF micrographs were recorded. Although both BF and HAADF images can be used to visualize the graphene structure, HAADF images are more powerful to identify individual atoms in the lattice due to its approximate Z^2 sensitivity. Thus, atomic resolution STEM was used to visualize the hexagonal atomic structure of graphene as well as individual carbon atoms in the graphene lattice. Atomic resolution STEM- BF and HAADF images of single layer graphene are shown in figures 6a and d, respectively. Although the images are unprocessed and acquired at 60 kV, the hexagonal structure and individual carbon atoms, especially in the HAADF image, are visible. Carbon atoms appear bright and holes dark in the HAADF image and *vice versa* in the BF image. Simulated BF and HAADF images with the respective STEM parameters are also shown in figures 6b and c, respectively. The simulations were performed with the Kirkland TEMSIM program using multislice method to calculate the electron wave function exiting the specimen [43].

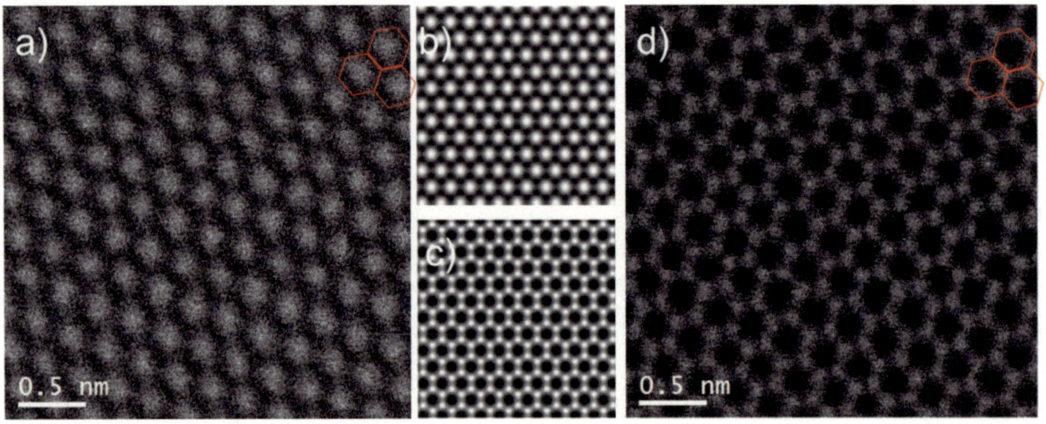

Figure 6. Unprocessed atomic resolution a) experimental raw BF, b) simulated BF, c) simulated HAADF and d) raw experimental HAADF images of pristine single layer graphene at 60kV.

HAADF images are found to be very useful for observing foreign ad-atoms on/in pristine graphene and most of them can be identified via intensity/contrast analysis by utilizing the sensitivity of HAADF images to the atomic number [22, 32]. However, HAADF imaging in combination with EELS dramatically increases the accuracy of chemical analysis. For example, Si, one of the most abundant and stable contaminants in graphene, was observed and identified by simultaneous HAADF imaging and EELS analysis shown in figure 7. The point spectrum was taken on the red circled bright atom, which was identified to be Si by its $L_{2,3}$ absorption edge.

Diffraction patterns are also utilized to identify and observe disordered or misoriented graphene membranes known as turbostratic graphene. This misalignment in between layers might be formed during exfoliation of graphene with different orientations on top of each other or during the growth process, in particular for graphene grown on nickel substrates. However, it is not perfectly clear whether they originate from the fabrication process or intrinsic stacking faults. The turbostratic graphene layers were easily identified by multiple diffraction spots in the diffraction pattern. Rather than having only 6 spots in each ring as in the case of single and

http://dx.doi.org/10.5772/56640

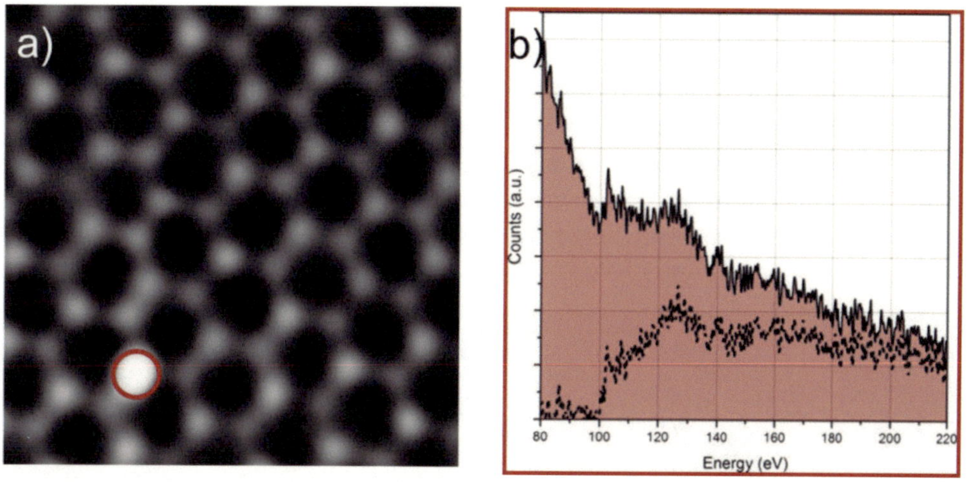

Figure 7. a) Noise filtered HAADF image of single layer graphene with bright atom at bottom left side, b) EEL spectra taken on the red circled bright atom in a) and background subtracted spectra shown as dotted line revealing that the bright atom is Si.

Bernal-stacked few layer graphene, several sets of 6 spots appear for misoriented layers (6, 12, 18 etc.) where each 6-spot pattern represents one layer [38]. The rotation angle between the layers can be determined by measuring the angle between neighboring diffraction spots in the same ring of the diffraction pattern and the angles between the layers have been found to vary from 1° to 59°. The angle between neighboring spots in the same ring of the diffraction pattern (hexagonal) of graphite (graphene layers) is 60°. Thus when the rotation angle between layers is 60° the spots from different layers overlay and rotation cannot be distinguished. When graphene layers are rotated with respect to each other it gives rise to Moiré patterns in which a larger periodicity is superimposed onto the single- or AB-stacked multi- layer graphene. It has been demonstrated both, theoretically and experimentally, that the graphene layers become electrically decoupled when the layers are misoriented [44]. The parabolic band behavior of AB-stacked double layer graphene changes to a linear dispersion when the layers are decoupled, meaning a change in the electronic structure upon layer rotation. Additionally, an external electric field does not open a band gap in the spectrum in case of turbostratic double layer graphene unlike in AB-stacked double layer graphene.

A diffraction pattern for turbostratic double layer graphene is shown in figure 8a. There are 12 spots in each ring rather than just 6 and each group of 6 spots arises from different layers. The rotation between layers is 12° as measured from neighboring spots. Although diffraction patterns can be used as a straightforward method to identify the presence of turbostratic graphene, it is important to witness the Moiré pattern at a reasonably high magnification because single, double (AB-stacked) and turbostratic graphene layers appear quite similar to each other when observed at low magnification. High magnification atomically resolved BF images of turbostratic double layer graphene are shown in figures 8b-d and rotation angles are measured as 5°, 8°, 15° for b, c, and d, respectively form their Fourier transforms shown as insets. As can be seen in figures 8b-d, the Moiré periodicity in the turbostratic graphene is larger than the lattice periodicity of single layer or AB-stacked few layer graphene [45]. Additionally, the periodicities are getting smaller as the rotation angle between the layers increases. Furthermore, individual graphene layers of turbostratic graphene can be reconstructed as shown in figure 8e and f by masking different sets of hexagonal spots as shown in the inset. The sum of the separate (reconstructed) images in figures 8e and f is then shown as inset to make sure the reconstruction of layers led to the correct image.

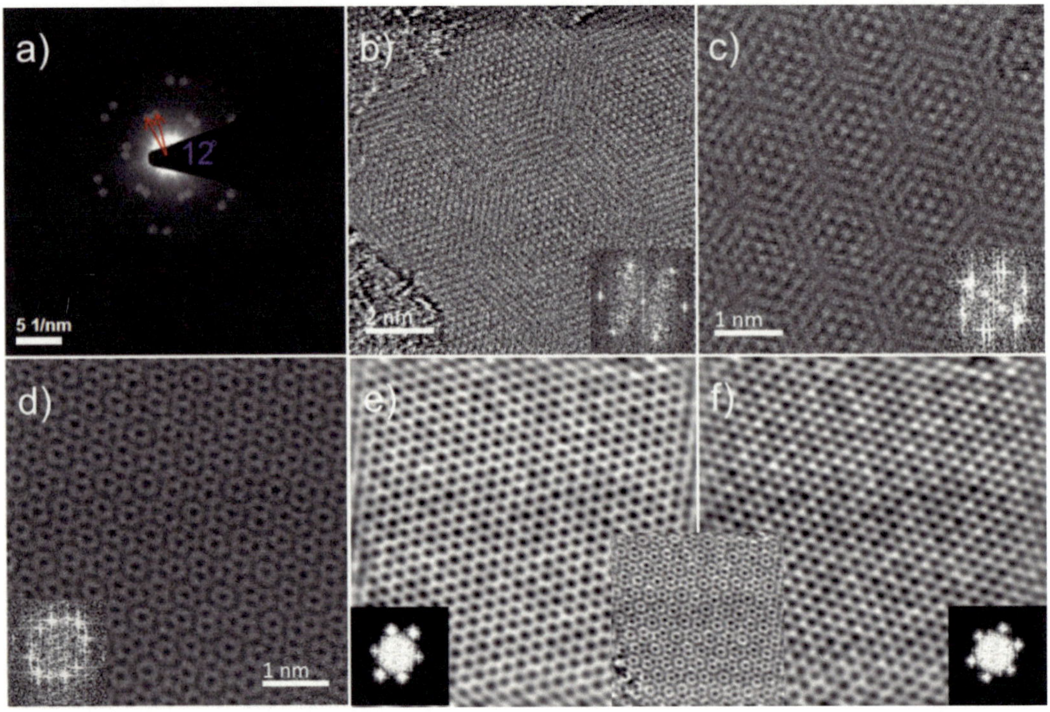

Figure 8. a) Diffraction patterns of turbostratic graphene, the angle between the spots is 12°; b). c) and d) turbostratic double layer graphene unprocessed BF images taken at 60kV with rotation angles between the layers of, respectively, 5°, 8° and 15° ; the rotation angles are measured on their respective FFTs, shown insets. e) reconstructed first layer by masking 6 spots in the FFT (inset), (f) reconstructed second layer by masking the other 6 spots (inset), and summed reconstructed layers (inset).

5. TEM characterisation of pristine h-BN

Single layer h-BN HAADF experimental (raw and filtered) images are presented in figure 9. B and N in both images are easily discernible as the atomic number of N is higher than B and thus N atoms appear more intense than B atoms; this gives triangular patterns. For a better demonstration of the difference between B and N, a part of figure 9a is filtered and magnified

http://dx.doi.org/10.5772/56640

as shown in figure 9b. The B-N sequence can also be revealed by the intensity profiles, e.g., like the one taken along the red line in the filtered image that shown in figure 9c.

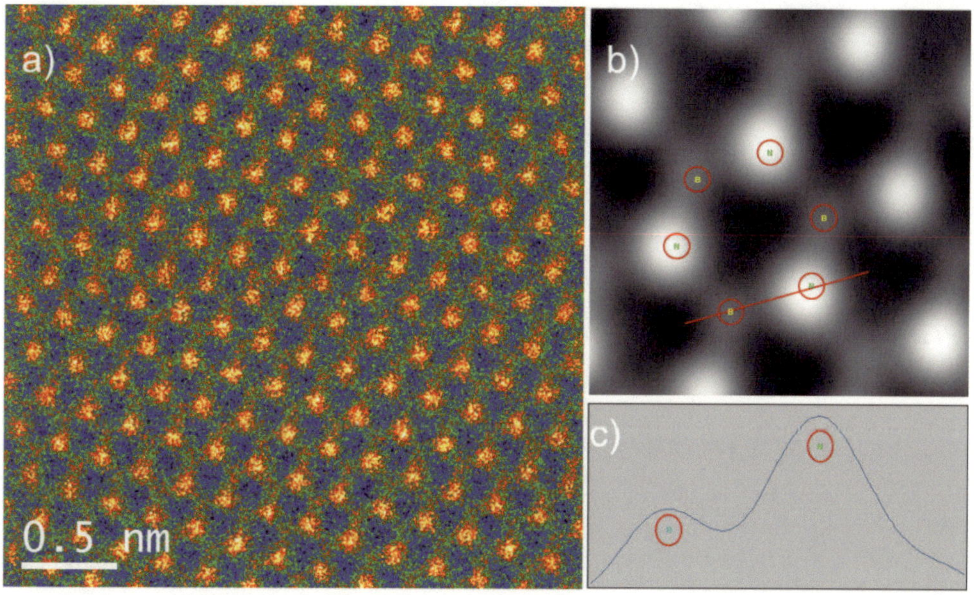

Figure 9. HAADF single layer h-BN images at 60 kV. (a) unprocessed (color inverted), (b) filtered (and magnified of a), (c) B and N can be distinguished as demonstrated by the intensity profile across the image (b).

AA′ stacking is the well-known stacking configuration for h-BN, with boron on top of nitrogen and vice versa. AA′ stacked double layer h-BN is shown in figure 10a. As can be seen the intensities in neighboring atomic positions are similar to each other due to the same number of atoms giving the same sum of Z^2 in the atomic columns. Although imaging was performed at and energy of 60 keV, which is lower than the calculated displacement threshold energy for the atoms in the h-BN structure, we were able to etch thick h-BN region layer-by-layer with the scanning electron beam. An example of this is shown in figure 10b, which was taken few minutes after figure 10a. The etching happened in triangular shapes as is always the case for h-BN because of the difference in the knock-on threshold energies of B and N [7, 46]. As a result of etching, a single layer region appeared revealing itself by the above mentioned triangular intensity pattern (see figure 10b).

Misorientations are also observed between h-BN layers similar to graphene, giving rise to Moiré patterns, which reveal themself by multiple spots in diffraction patterns or in Fourier transforms of the micrograph [39]. A noise filtered STEM-BF image of double layer turbostratic h-BN is shown in figure 11a. A magnified image of the red square in figure 11a is presented in figure 11b, which provides a closer look into turbostratic appearance of h-BN layers. The rotation angle between the layers was found to be 6° from measurements of the angle between neighboring spots in the Fourier transform as shown in figure 11c.

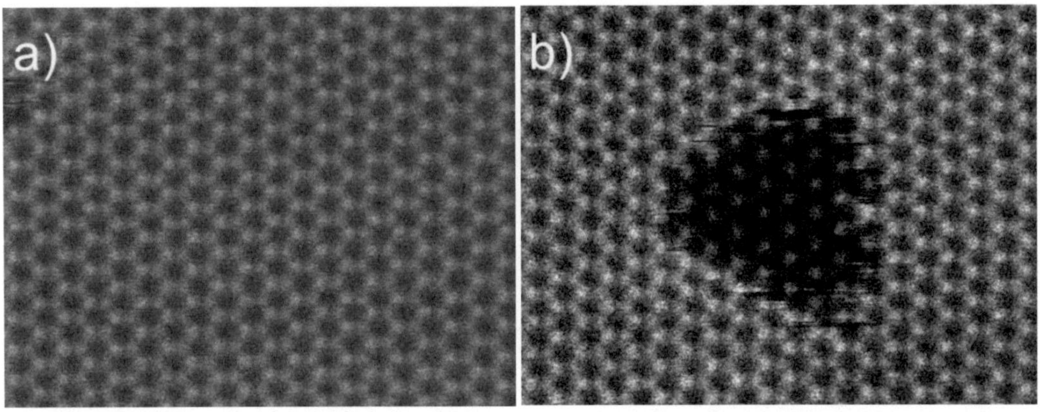

Figure 10. Unprocessed HAADF double layer h-BN images at 60 kV a) before and b) after electron beam induced etching, revealing a single layer region.

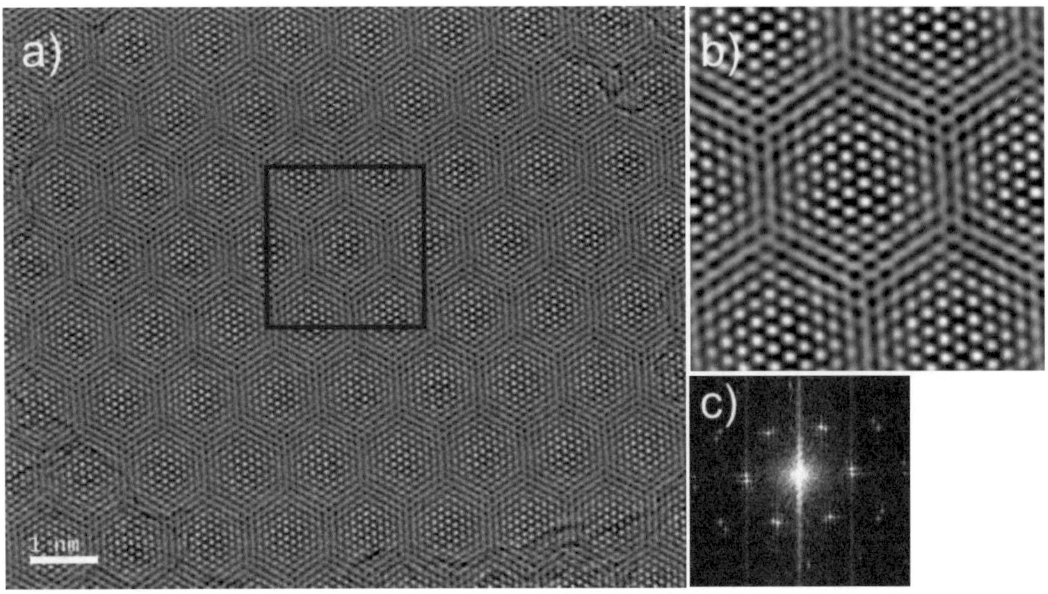

Figure 11. a) Unprocessed STEM BF image of turbostratic h-BN, b) magnified and c) FFT of (a).

6. Metal-graphene interactions

Any electronic device irrespective of whether macro or nano sized has metal contacts to connect it to external device components. Thus, metal behavior on material surfaces is an important subject of study. The interfacing of graphene with metal contacts is a recurring theme for graphene based devices. This area of graphene research is still ripening and the consequences of particular dopants on graphene's electronic properties are still being investigated. Understanding of metal-graphene interactions on a fundamental level is of significance to the improvement of graphene based devices. Thus, distribution, bonding, doping and stability of metals on the graphene surface needs to be understood. For this purposes we studied various metals on single and few layer graphene.

Figures 12a and b shows simultaneously taken BF and HAADF images of 2Å gold evaporated on single layer graphene. Gold atoms and clusters segregate to the hydrocarbon contamination, which indicates the extreme mobility of gold on the residue free pristine (clean) graphene surfaces (upper right corner of figures 12a and b).We never observed gold atoms on single or misoriented (turbostratic) graphene layers [47]. Single gold atoms can only be observed on few-layer graphene films, where the bonding contribution from sub-surface layers appears to suppress their diffusion and enhance the sticking probability to the graphene surface. The gold atoms, which have appeared on few-layer graphene (figures 12c and d; some of them are numbered), were dissociated from the bigger gold cluster (left, figure 12d) as a result of repeated scanning in the STEM [48]. The atoms stay for a few scans in their positions. By comparing locations of gold atoms in the HAADF image (figure 12d) with identical positions in the BF image, (figure 12c), the sites can be identified as on top of carbon atoms in the graphene honeycomb structure. The position of metal impurities with respect to the hexagon structure depends on the metal's electron affinity and the electron density on the hexagon structure; these factors determine whether gold atoms sit on top of carbon atoms, on the bond bridging two carbon atoms or centered at the middle of the hexagon.

Initial investigations of gold atoms on pristine graphene revealed that none of the metals studied reacted with the clean (residue free) graphene surface. Thus to be able to amend this behavior researchers proposed to attach hydrogen to carbon atoms in the graphene lattice. Hydrogen ad-atoms can trap small molecules to specific sites on the graphene surface and these molecules remain stable at room temperature [49]. This suggestion stimulated attempts of experimental proof of gold deposition on functionalized graphene with hydrogen [48]. Hydrogenation breaks graphene sp^2 bonds and leads to sp^3 bond formation, which leads to opening of a band gap. We have carried out gold deposition on hydrogenated graphene surfaces to find out whether the latter increase the sticking probability for gold atoms. To conduct a fair comparison, gold evaporation was performed for the same amount (2Å) and simultaneously on hydrogenated and pristine graphene surfaces. TEM-BF images of gold deposited pristine and hydrogenated graphene taken at the same magnification are presented in figures 13a and b, respectively. As can be seen in the figure the hydrogenated sample has a higher gold coverage, and gold cluster distributions and cluster sizes are of higher uniformity than in the pristine graphene sample. In the latter case less than half of the area is covered with gold and cluster sizes vary. However, similar to pristine graphene, gold is only retained in the contamination on the hydrogenated sample [48].

In the process of investigating the interaction between metal species and graphene in the (S)TEM, an intriguing etching phenomenon was observed. The etching occurred in the microscope column in UHV conditions at room temperature. This behavior was distinct from

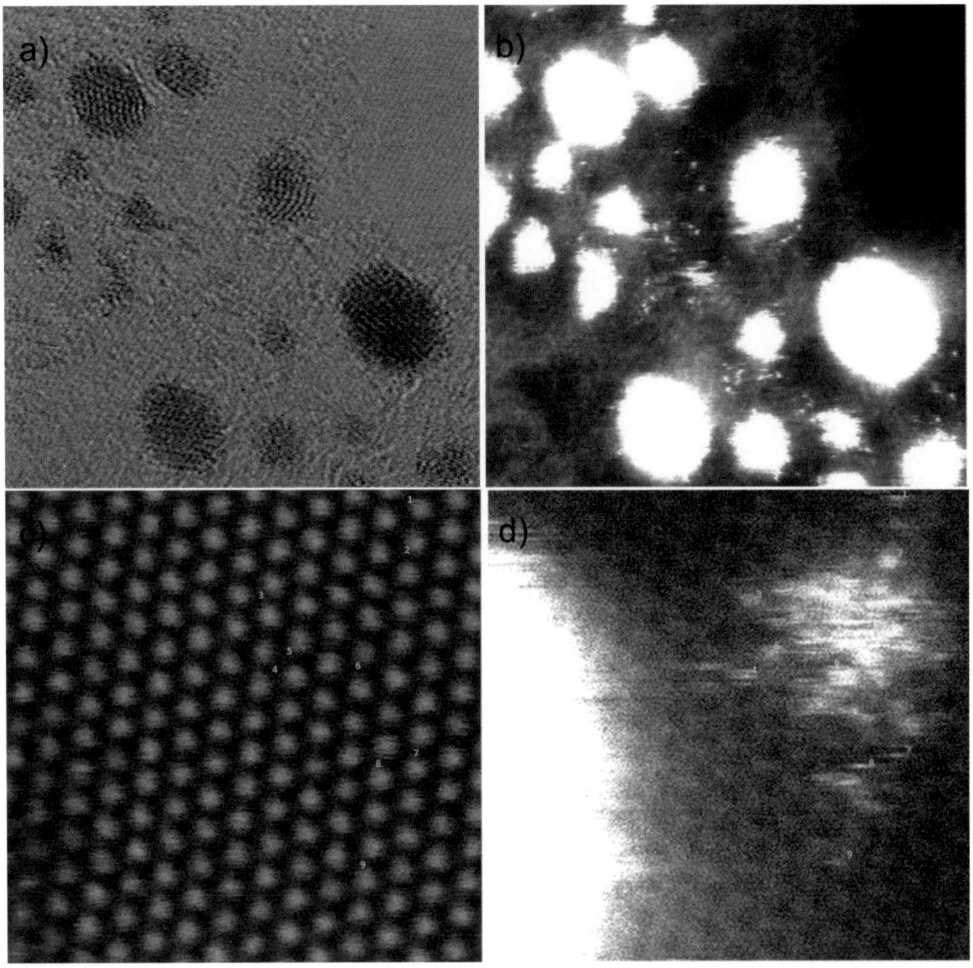

Figure 12. Unprocessed simultaneously taken a) BF and b) HAADF images of 2Å Au evaporated single layer graphene, simultaneously taken c) BF and d) HAADF images of 5Å Au evaporated few-layer graphene. All obtained at 60kV. Frame widths are 15 nm in a) and b) and 3nm in c) and d), respectively.

existing reports of graphene/graphite etching via metals at high temperature and under gas flow, a well-established method to cut graphitic layers [50, 51].

As seen in the images of figures 13 and 14a the metal clusters sit preferentially in the middle of the hydrocarbon contamination, however, after a few scans of the electron beam some of the clusters and/or individual atoms can be dragged by the beam to the edge of the contamination layer. Figure 14b shows an HAADF image of such a Ni cluster positioned in the middle of the contamination and figure 14c was taken after few scans that show the clusters were dragged to very edge of a contamination area. A magnified image of the red square in figure 14c is presented in figure 14d, which shows individual Ni atoms on a region of clean single layer graphene at the edge of the particle. While some carbon chains may still be present

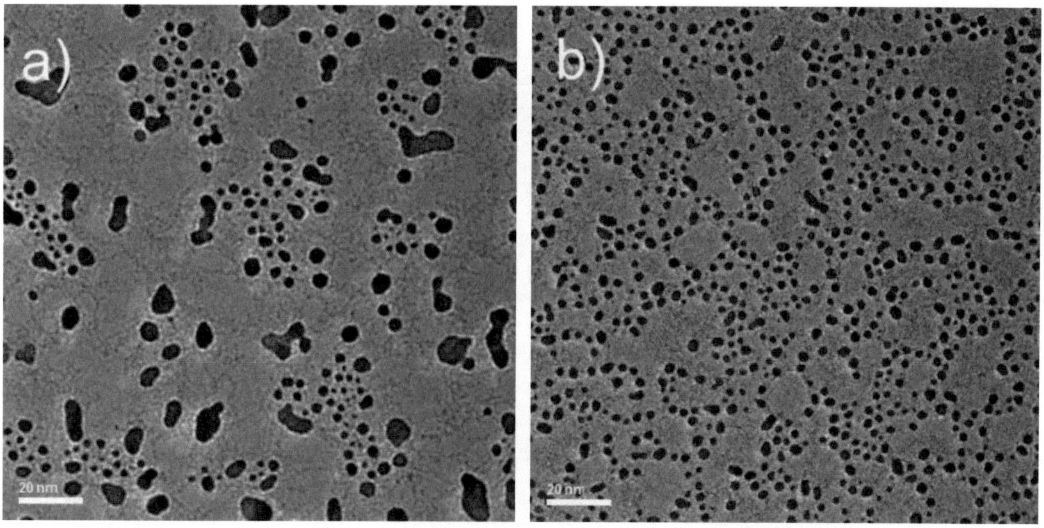

Figure 13. Unprocessed TEM images of 2Å gold evaporated onto a) pristine and b) hydrogenated graphene shown at the same magnification (the scale bar is 20 nm).

immediately under these Ni atoms, the here observed configuration offers a great proximity between pure graphene and metal atoms. After a few additional scans of the beam, a hole has formed (figure 14e), decorated by individual Ni atoms as evidenced by the clear Ni $M_{2,3}$ signature in the EEL spectrum (figure 14f), and acquired by placing the electron probe exactly on top of the bright atom marked in figure 14e. After the initial hole formation, individual Ni atoms are observed to jump onto the exposed edge with subsequent enlargement of the hole. The holes expand in proximity of the hydrocarbon contamination in which metal clusters reside. A strong indication that the drilling process is indeed metal-mediated arises from the observation that when no Ni atom is decorating the hole, the latter merely re-shapes dynamically but does not grow further in size. In other words, the drilling stops when the local reservoir of metal impurity atoms is exhausted, until more Ni atoms are drawn towards the energetically unstable edges of the hole *via* surface diffusion owing to the high mobility of single metal impurities on single layer graphene [52].

Further demonstration of the etching process which resulted in the hole evolution and expansion in the presence of metals (2Å Aluminum), is presented in figure 15. Figure 15a shows a clean, intact graphene patch (black) surrounded by hydrocarbons (gray) with Al clusters (white). Various stages of hole formation and enlargement are shown in subsequently taken images in figures 15b–d; figure 15e shows the hole after etching has more or less ceased. Once a hole is formed as result of interaction between Al atoms and graphene (figure 15b), the hole is decorated by newly arriving Al atoms, which are dragged by the electron beam, leading to further enlargement (figure 15c and d); no such atoms can be observed in figure 15e, indicating that the reservoir is exhausted. An overview at smaller magnification of an intermediate etching stage together with the aluminium distribution is shown in figure 15f. Figure 15 demonstrates clearly that the etching propagates from the border of the contamination into

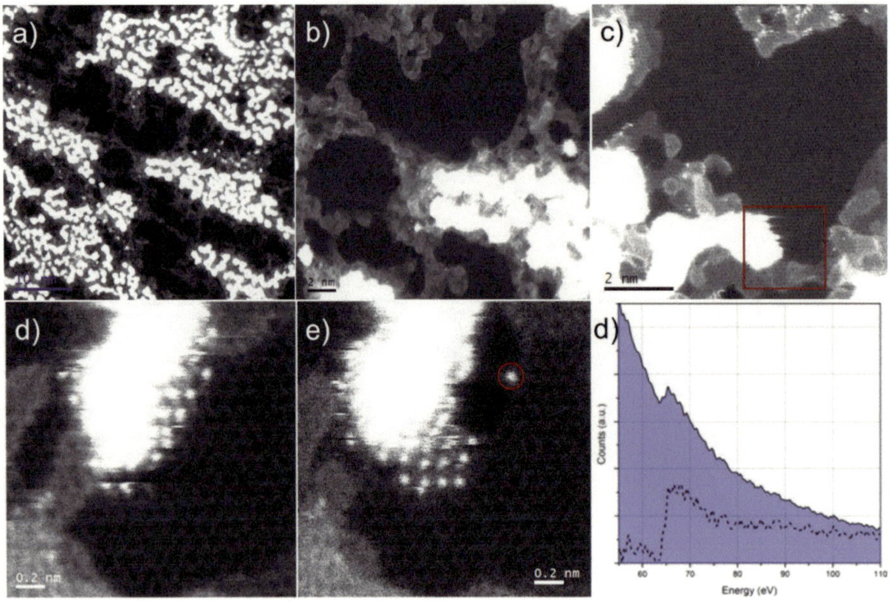

Figure 14. Unprocessed (a), (b) and (c) HAADF images of Ni clusters on graphene; (d) Ni-cluster sitting on the edge of the hydrocarbon contamination. Single Ni atoms have been dragged by the beam from the cluster and are in contact with the graphene single layer. (e) After a few more scans, a hole has formed, whose edges are decorated with single Ni atoms, identified by their Ni $M_{2,3}$ EELS signal, shown in f). EELS spectrum acquired by positioning the beam for 1s on the bright atom circled in (e)

clean graphene as long as metal atoms are supplied to the edge of the hole. In the absence of metal atoms at the hole, no progression of the etching is observed [53].

The migration of the metal atoms under the beam to the edge of the contamination layer and etching of graphene was observed for other metals on single layer graphene samples as well, e.g., for palladium (figure 16a) and titanium (figure 16b). In both cases, the nature of the atoms decorating the edges of the newly formed holes was confirmed by placing the electron probe directly on top of these atoms and recording an EELS spectrum, as shown on figure 16 below the corresponding HAADF images [52].

As mentioned earlier metals have been used as catalysts for patterning of graphene devices in hydrogen or oxygen flow at high temperatures [50, 54]. However, neither gas environment nor high temperatures were used in our study, which, to our knowledge, provides the first experimental evidence of electron beam- assisted drilling of graphene through direct interaction with metals. This destructive behavior was predicted specifically for Fe, Co, Ni and Al by calculations showing a drastic lowering of the formation energy for mono- and di-vacancies in single layer graphene when metal ad-atoms are present on the graphene surface [55]. In addition to that a constant trait of all those elements is their propensity to form oxides (except Au), which suggests that oxidation could be playing a major role in the effects we are observing. This hypothesis is further strengthened by the fact that by contrast no hole-forming was observed on Au-deposited samples, Au being of course not prone to oxidation except in very

http://dx.doi.org/10.5772/56640

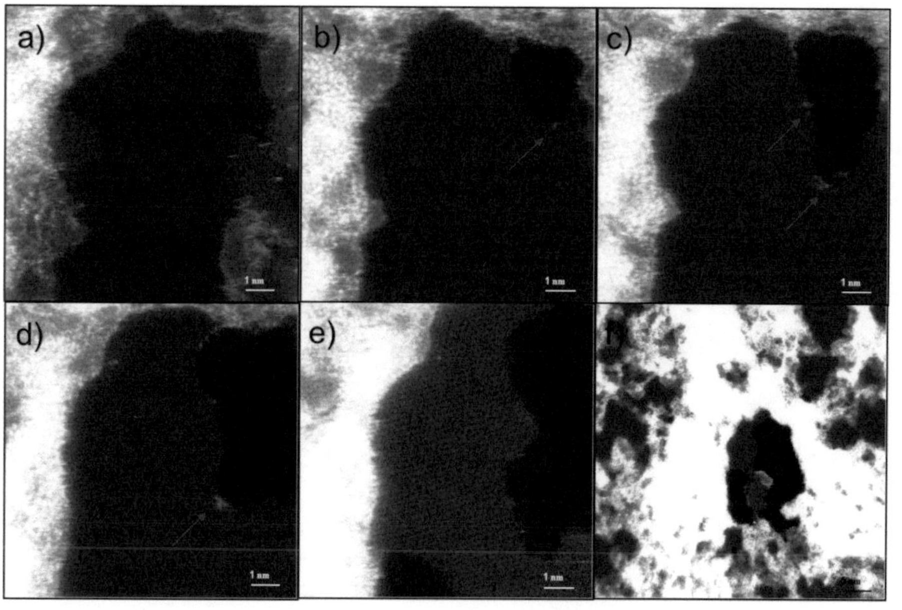

Figure 15. Unprocessed HAADF images of graphene etching in the presence of Al; (a) before etching, (b) after the start of the hole formation, (c) after hole enlargement in subsequent scans, (d) after continued etching as a result of a sustained supply of Al atoms to the hole's edge (some Al atoms are indicated by red arrows in (b–d), and (e) after the etching process has almost stopped because the Al atom supply has ceased; (f) lower magnification overview of the Al distribution and hole evolution. The scale bar is the same in (a–e), 1 nm.

specific circumstances. Nevertheless, even in the mono-vacancy formation model for the destruction of graphene proposed via DFT calculations Au is not expected to lead to a major loss in stability as its defect formation energy remains high so this observation cannot definitively point to a role of O in this etching process. The most likely sources of oxygen could be metal evaporation (relatively poor vacuum condition, ~10^{-7} torr), transport of the samples (exposing the sample to the air), hydrocarbon contamination (on the sample) and residual O_2 gases in the microscope column (with relatively high partial pressure of O). However, the latter can be ignored as the sample chamber (microscope column) was systematically at its base pressure of <$5x10^{-9}$torr.

7. Metal-h-BN interactions

Our metal-graphene interaction studies have inspired us to carry out micro structural studies of the metal-h-BN system to compare metal behaviors between these 2D systems. Similar to the graphene case Au, Ti and Ni metals have been investigated on the h-BN flakes.

Figure 17 shows simultaneously taken STEM- HAADF and BF images of 1Å gold evaporated on few-layer h-BN. Neither individual gold atoms nor clusters were observed on pristine parts of the h-BN sheets. However, ad-atoms and gold clusters can be seen in the hydrocarbon

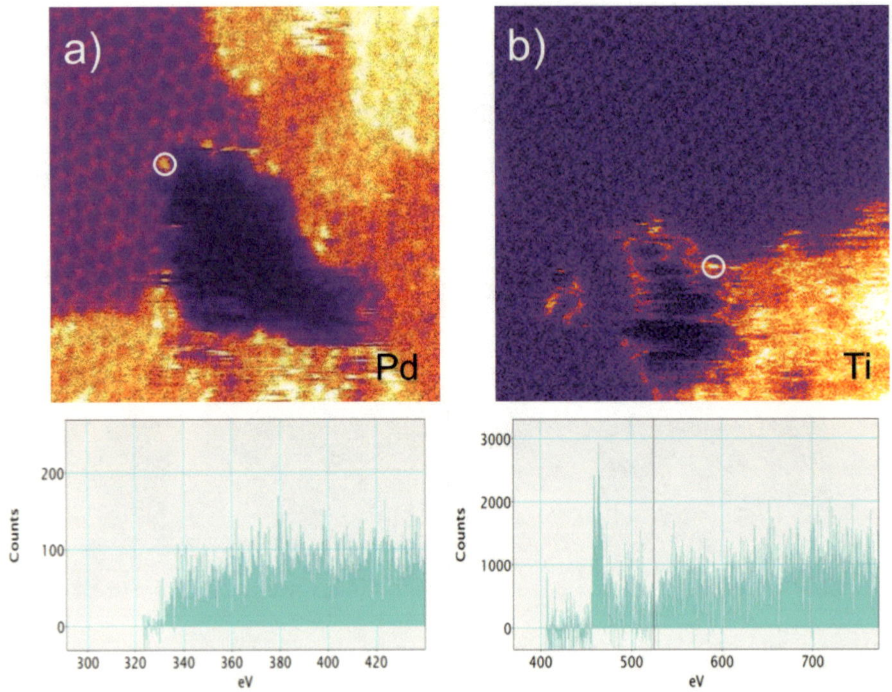

Figure 16. Unprocessed HAADF images (color inverted) of holes formed in single layer graphene through metal mediated etching with (a) Pd and (b) Ti. EEL spectra acquired by positioning the beam for 1s on the atoms encircled in the images are shown below the corresponding image.

contamination, which resembles the Au behavior on the graphene surfaces. This indicates the extreme mobility of gold on the clean h-BN surface. Single gold atoms have been observed in contaminated areas, as a result of dissociation from bigger gold cluster by the scanning beam in the STEM; these can be seen in figure 17a around the middle cluster, but no atoms can be seen on the clean h-BN surfaces. Another observation is the 'peeling' of BN layers under the continuous scanning electron beam. This has been shown earlier: the holes that occur are triangular in shape as indicated by the dashed white triangle in the HAADF image in figure 17a and in the corresponding BF image in figure 17b.

Another metal, which has been evaporated onto h-BN layers for comparison with graphene, is Ti as it has the highest coverage among the studied metals on the graphene surfaces for the same amount of metal. HAADF images of 1Å Ti evaporated double layer h-BN are shown in figure 18. A low magnification overview image in figure 18a shows the distribution of Ti on the h-BN surface. Ti is clustered and resides in the hydrocarbon chains similar to the graphene case, however, the distribution is less homogeneous compare to the Ti on graphene. A close-up image of figure 18a is shown in figure 18b; here the position of the Ti atoms and clusters can be clearly seen in the upper-right corner of the image. Although these clusters are dissociated and have moved towards edge of the hydrocarbon chains (bordering clean h-BN regions) during the scan, h-BN etching was not observed. This is in contrast to the Ti-graphene

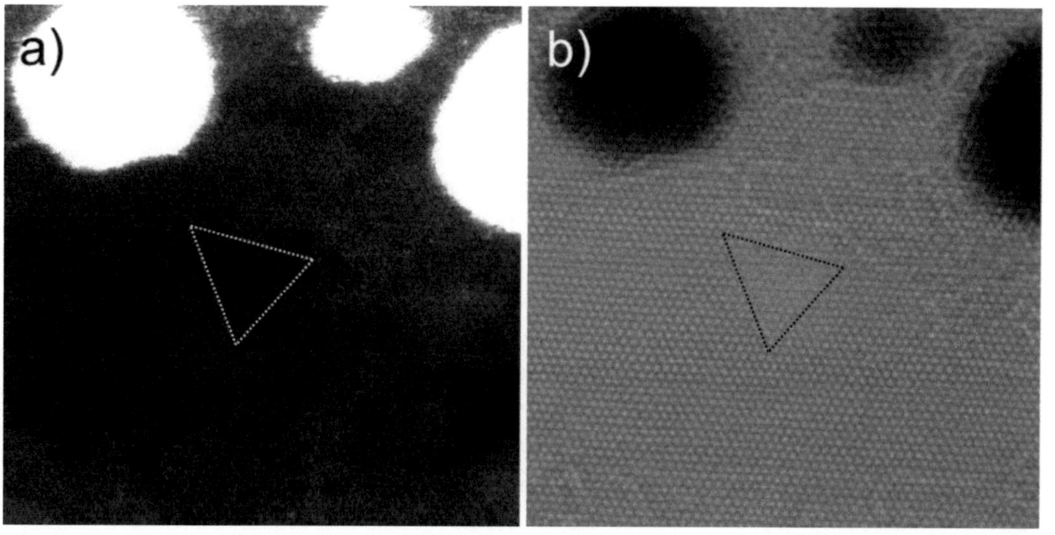

Figure 17. Unprocessed a) HAADF, b) BF image of 1Å Au doped h-BN, obtained at 100kV. Frame widths are 15nm.

interaction. Etching of h-BN as observed in both, figures 18b and c, is beam induced (knock-on damage), and is not metal mediated or related. Further evidenced for this is that it was not observed on the edge of the hydrocarbon chains, where the etching initiated in the case of metal-graphene interactions.

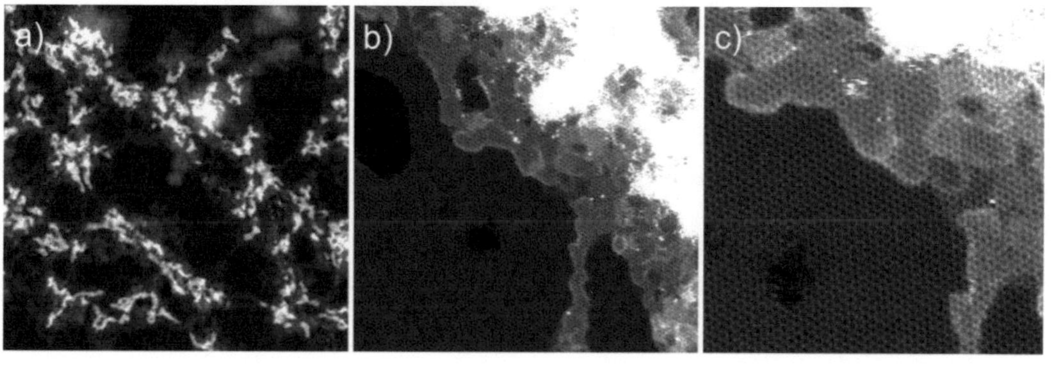

Figure 18. Unprocessed HAADF of 1Å Ti doped h-BN at different magnification, obtained at 60kV.

HAADF images of 1Å Ni doped h-BN and simultaneously taken corresponding BF images are shown in figure 19. Figures 19a, b and c are subsequently acquired HAADF images with increasing magnification at 60 kV acceleration voltage; magnified areas are indicated by blue squares in figures 19a and b. All HAADF images clearly show that there are no Ni atoms or cluster on clean h-BN surfaces. This is not obvious in the corresponding BF images, which reveal disturbances (Ni-islands) of the h-BN lattice; the impurities are in fact not sitting on the h-BN lattice itself, but in/on the contamination layer between Ni atoms and the h-BN surface. Exact Ni-atom positions can only be revealed in the HAADF images, which is the reason why HAADF imaging is more powerful than phase contrast (BF) imaging in terms of revealing constellations and positions of impurities and ad-atoms. In particular, individual Ni atoms can be observed in the HAADF image in figure 19c to sit on a very fine contamination layer, through which the h-BN structure is still clearly visible. Without the guidance of the HAADF images it would not be possible to identify the sites of the ad-atoms on the honeycomb structure.

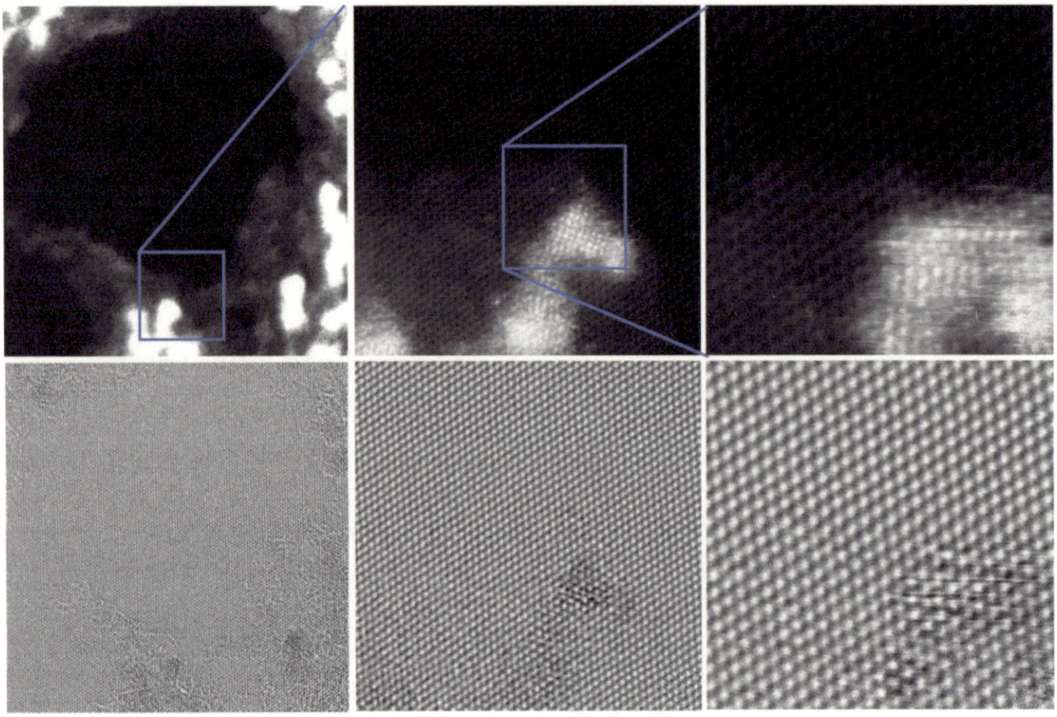

Figure 19. Unprocessed HAADF and BF images of 1Å Ni doped BN at different magnification, obtained at 60kV. a), b) and c) HAADF, d), e) and f) corresponding BF images. Frame widths are 25, 10, 5 nm in a) and b), c) and d), e) and f), respectively.

8. Conclusions

In conclusion, we used an aberration-corrected STEM, to investigate the structural properties of pristine graphene and h-BN layers on the atomic scale, and additionally the metal behaviors on these layers in terms of coverage, distribution and interaction.

Graphene and h-BN layer identification and atomic scale investigations were performed in TEM and STEM. We have shown that graphene and h-BN layers are not perfectly clean and the presence of hydrocarbon contamination appears to be an issue with graphene structures in that it might has an effect on the charge transfer between graphene and metals. Single, two and few layer graphene were differentiated via diffraction patterns as was turbostratic graphene and the rotation angles between the layers. Hexagonal periodicity and individual carbon, boron and nitrogen atoms in graphene and h-BN layers were observed in TEM and STEM operated at an acceleration voltage below the knock-on threshold of carbon. Cleanliness, surface contamination and foreign ad-atoms on the graphene layers were identified in HAADF and EELS in the STEM.

After studies of pristine graphene in the (S)TEM, the behavior of the metals Au; Cr; Ti; Ni; Al; Pd on suspended graphene and Au; Ti; Ni on h-BN surfaces was investigated. As a general observation, all studied metals did not interact with the clean graphene and h-BN surface as none of them were observed on clean parts of them. The metals were mainly clustered and found to be on the hydrocarbon chains. The size and distribution of the clusters were found to vary on the graphene for different metals, for example Au clusters were the biggest and Ti the smallest in size for the same amount of metal deposition, which indicates that Ti has a better distribution than Au. Unequal distributions and irregular sizes of Au clusters were amended upon chemical modification of the graphene via hydrogenation. However, Au ad-atoms or clusters were still not observed on clean part of the graphene similar to the pristine case. The weak interaction between the metal and graphene and the preponderance of metals to reside on the hydrocarbons might have effects on the electrical transport properties of graphene (charge transfer between graphene and metals), because the hydrocarbon contamination on the graphene surface is unavoidable unless annealed at high temperatures and kept in UHV.

On the other hand, interesting results have been witnessed for all studied metals (except Au) in that they etch the graphene layer during STEM investigations. Although the metals nucleated on the hydrocarbons they were dragged by the scanning electron beam to clean parts of the graphene surface. As soon as the metal interacted with carbon atoms of the graphene a vacancy appeared followed by creation of further vacancies which coalesce to form a hole. This was observed although the microscope column was kept in UHV and operated at 60 kV, i.e., below the damaging threshold of carbon. Although it is known that the defect formation energies in graphene are substantially lowered in the mere presence of metal atoms, the actual etching mechanism at play here is not fully resolved. An oxidation mechanism may provide the most likely explanation, whereby oxygen released from nearby contamination may lead to a C-C bond dissociation mediated by the oxidation of the metal atoms. Furthermore, the observed metal-mediated etching of graphene could provide an explanation for the degradation of device performance over time. This etching could also be exploited in controlled tailoring and self-assembly processes for future graphene based devices as the etched holes can be healed spontaneously.

HAADF images reveal that both Au, Ti and Ni metals do not stick to the clean (contamination-free) surface of h-BN, however, they form ad-atoms and clusters on the contamination. That indicates the interaction between metals and BN is much weaker than the metal-metal interaction. On the other hand, we haven't observed any metal-mediated h-BN etching at all in contrast to metal-graphene systems (even metals cannot be dragged towards clean bits under continuous scanning electron beam). That can be explained with the tendency of graphene to oxidation with yields of CO or CO_2 as the contaminations (even metal clusters) are rich from oxygen. It was also found that h-BN is a very good transparent support for the study of nanocrystals in the S/TEM.

Author details

Recep Zan[1,2*], Quentin M. Ramasse[3], Rashid Jalil[2] and Ursel Bangert[1]

*Address all correspondence to: recep.zan@manchester.ac.uk

1 School of Materials, The University of Manchester, Manchester, UK

2 School of Physics and Astronomy The University of Manchester, Manchester, UK

3 SuperSTEM Laboratory, STFC Daresbury Campus, Daresbury, UK

References

[1] Novoselov KS, Geim AK, Morozov SV, Jiang D, Zhang Y, Dubonos SV, et al. Electric Field Effect in Atomically Thin Carbon Films. Science. 2004 October 22, 2004;306(5696):666-9.

[2] Novoselov KS, Jiang D, Schedin F, Booth TJ, Khotkevich VV, Morozov SV, et al. Two-dimensional atomic crystals. Proceedings of the National Academy of Sciences of the United States of America. 2005;102(30):10451-3.

[3] Geim AK, Novoselov KS. The rise of graphene. Nat Mater. 2007;6(3):183-91.

[4] Xia F, Farmer DB, Lin Y-m, Avouris P. Graphene Field-Effect Transistors with High On/Off Current Ratio and Large Transport Band Gap at Room Temperature. Nano Letters. 2010;10(2):715-8.

[5] Dean CR, Young AF, Meric I, Lee C, Wang L, Sorgenfrei S, et al. Boron nitride substrates for high-quality graphene electronics. Nat Nano. 2010;5(10):722-6.

[6] Britnell L, Gorbachev RV, Jalil R, Belle BD, Schedin F, Mishchenko A, et al. Field-Effect Tunneling Transistor Based on Vertical Graphene Heterostructures. Science. 2012;335(6071):947-50.

[7] Jin C, Lin F, Suenaga K, Iijima S. Fabrication of a Freestanding Boron Nitride Single Layer and Its Defect Assignments. Physical Review Letters. 2009;102(19):195505.

[8] Novoselov KS, Falko VI, Colombo L, Gellert PR, Schwab MG, Kim K. A roadmap for graphene. Nature. 2012;490(7419):192-200.

[9] Bae S, Kim H, Lee Y, Xu X, Park J-S, Zheng Y, et al. Roll-to-roll production of 30-inch graphene films for transparent electrodes. Nat Nano. 2010;5(8):574-8.

[10] Stoller MD, Park S, Zhu Y, An J, Ruoff RS. Graphene-Based Ultracapacitors. Nano Letters. 2008;8(10):3498-502.

[11] Yoo E, Kim J, Hosono E, Zhou H-s, Kudo T, Honma I. Large Reversible Li Storage of Graphene Nanosheet Families for Use in Rechargeable Lithium Ion Batteries. Nano Letters. 2008;8(8):2277-82.

[12] Grigorenko AN, Polini M, Novoselov KS. Graphene plasmonics. Nat Photon. 2012;6(11):749-58.

[13] Stankovich S, Dikin DA, Dommett GHB, Kohlhaas KM, Zimney EJ, Stach EA, et al. Graphene-based composite materials. Nature. 2006;442(7100):282-6.

[14] Schedin F, Geim AK, Morozov SV, Hill EW, Blake P, Katsnelson MI, et al. Detection of individual gas molecules adsorbed on graphene. Nat Mater. 2007;6(9):652-5.

[15] Pantelic RS, Meyer JC, Kaiser U, Stahlberg H. The application of graphene as a sample support in transmission electron microscopy. Solid State Communications. 2012;152(15):1375-82.

[16] Pi K, McCreary KM, Bao W, Han W, Chiang YF, Li Y, et al. Electronic doping and scattering by transition metals on graphene. Physical Review B. 2009;80(7):075406.

[17] Bolotin KI, Sikes KJ, Jiang Z, Klima M, Fudenberg G, Hone J, et al. Ultrahigh electron mobility in suspended graphene. Solid State Communications. 2008;146:351-5.

[18] Zan R, Muryn C, Bangert U, Mattocks P, Wincott P, Vaughan D, et al. Scanning tunnelling microscopy of suspended graphene. Nanoscale. 2012;4(10):3065-8.

[19] Morozov SV, Novoselov KS, Katsnelson MI, Schedin F, Elias DC, Jaszczak JA, et al. Giant Intrinsic Carrier Mobilities in Graphene and Its Bilayer. Physical Review Letters. 2008;100(1):016602.

[20] Meyer JC, Geim AK, Katsnelson MI, Novoselov KS, Booth TJ, Roth S. The structure of suspended graphene sheets. Nature. 2007;446(7131):60-3.

[21] Gass MH, Bangert U, Bleloch AL, Wang P, Nair RR, Geim AK. Free-standing graphene at atomic resolution. Nat Nano. 2008;3(11):676-81.

[22] Krivanek OL, Dellby N, Murfitt MF, Chisholm MF, Pennycook TJ, Suenaga K, et al. Gentle STEM: ADF imaging and EELS at low primary energies. Ultramicroscopy. 2010;110(8):935-45.

[23] Gorbachev RV, Riaz I, Nair RR, Jalil R, Britnell L, Belle BD, et al. Hunting for Monolayer Boron Nitride: Optical and Raman Signatures. Small. 2010;7(4):465-8.

[24] Blake P, Hill EW, Neto AHC, Novoselov KS, Jiang D, Yang R, et al. Making graphene visible. Applied Physics Letters. 2007;91(6):063124-3.

[25] Li X, Cai W, An J, Kim S, Nah J, Yang D, et al. Large-Area Synthesis of High-Quality and Uniform Graphene Films on Copper Foils. Science. 2009 June 5, 2009;324(5932): 1312-4.

[26] Ago H, Ogawa Y, Tsuji M, Mizuno S, Hibino H. Catalytic Growth of Graphene: Toward Large-Area Single-Crystalline Graphene. The Journal of Physical Chemistry Letters. 2012;3(16):2228-36.

[27] Liang X, Sperling BA, Calizo I, Cheng G, Hacker CA, Zhang Q, et al. Toward Clean and Crackless Transfer of Graphene. ACS Nano. 2011 2012/11/21;5(11):9144-53.

[28] Suk JW, Kitt A, Magnuson CW, Hao Y, Ahmed S, An J, et al. Transfer of CVD-Grown Monolayer Graphene onto Arbitrary Substrates. ACS Nano. 2011 2012/11/21;5(9): 6916-24.

[29] Crewe AV. The physics of the high-resolution scanning microscope. Reports on Progress in Physics. 1980;43(5):621.

[30] Hartel P, Rose H, Dinges C. Conditions and reasons for incoherent imaging in STEM. Ultramicroscopy. 1996;63(2):93-114.

[31] Krivanek OL, Zhou W, Chisholm MF, Idrobo JC, Lovejoy TC, Ramasse QM, et al. Gentle STEM of Single Atoms: Low keV Imaging and Analysis at Ultimate Detection Limits. Low Voltage Electron Microscopy: John Wiley & Sons, Ltd; 2012. p. 119-61.

[32] Krivanek OL, Chisholm MF, Nicolosi V, Pennycook TJ, Corbin GJ, Dellby N, et al. Atom-by-atom structural and chemical analysis by annular dark-field electron microscopy. Nature. 2010;464(7288):571-4.

[33] Meyer JC, Geim AK, Katsnelson MI, Novoselov KS, Obergfell D, Roth S, et al. On the roughness of single- and bi-layer graphene membranes. Solid State Communications. 2007;143(1-2):101-9.

[34] Bangert U, Gass MH, Bleloch AL, Nair RR, Eccles J. Nanotopography of graphene. Physica Status Solidi (a). 2009;206(9):2115-9.

[35] Bangert U, Gass MH, Bleloch AL, Nair RR, Geim AK. Manifestation of ripples in free-standing graphene in lattice images obtained in an aberration-corrected scanning transmission electron microscope. Physica Status Solidi (a). 2009;206(6):1117-22.

[36] Girit CO, Meyer JC, Erni R, Rossell MD, Kisielowski C, Yang L, et al. Graphene at the Edge: Stability and Dynamics. Science. 2009 March 27, 2009;323(5922):1705-8.

[37] Banhart F, Kotakoski J, Krasheninnikov AV. Structural Defects in Graphene. ACS Nano. 2011;5(1):26-41.

[38] Zan R, Bangert U, Ramasse Q, Novoselov KS. Imaging of Bernal stacked and misoriented graphene and boron nitride: experiment and simulation. Journal of Microscopy. 2011;244(2):152-8.

[39] Warner JH, Rummeli MH, Gemming T, Buchner B, Briggs GAD. Direct Imaging of Rotational Stacking Faults in Few Layer Graphene. Nano Letters. 2008;9(1):102-6.

[40] Meyer JC, Kisielowski C, Erni R, Rossell MD, Crommie MF, Zettl A. Direct Imaging of Lattice Atoms and Topological Defects in Graphene Membranes. Nano Letters. 2008;8(11):3582-6.

[41] Chuvilin A, Meyer C. J, Algara-Siller G, Kaiser U. From graphene constrictions to single carbon chains. New Journal of Physics. 2009;11(8):083019.

[42] Haigh SJ, Gholinia A, Jalil R, Romani S, Britnell L, Elias DC, et al. Cross-sectional imaging of individual layers and buried interfaces of graphene-based heterostructures and superlattices. Nat Mater. 2012;11(9):764-7.

[43] Kirkland EJ. Advanced Computing in Electron Microscopy. 1. ed.: Plenum Press; 1998.

[44] Hass J, Varchon F, Millan-Otoya JE, Sprinkle M, Sharma N, de Heer WA, et al. Why Multilayer Graphene on 4H-SiC(0001) Behaves Like a Single Sheet of Graphene. Physical Review Letters. 2008;100(12):125504.

[45] Pong W-T, Durkan C. A review and outlook for an anomaly of scanning tunnelling microscopy (STM): superlattices on graphite. Journal of Physics D: Applied Physics. 2005;38(21):R329-55.

[46] Meyer JC, Chuvilin A, Algara-Siller G, Biskupek J, Kaiser U. Selective Sputtering and Atomic Resolution Imaging of Atomically Thin Boron Nitride Membranes. Nano Letters. 2009;9(7):2683-9.

[47] Zan R, Bangert U, Ramasse Q, Novoselov KS. Metal-Graphene Interaction Studied via Atomic Resolution Scanning Transmission Electron Microscopy. Nano Letters. 2011;11(3):1087-92.

[48] Zan R, Bangert U, Ramasse Q, Novoselov KS. Evolution of Gold Nanostructures on Graphene. Small. 2011;7(20):2868-72.

[49] Erni R, Rossell MD, Nguyen M-T, Blankenburg S, Passerone D, Hartel P, et al. Stability and dynamics of small molecules trapped on graphene. Physical Review B. 2010;82(16):165443.

[50] Campos LC, Manfrinato VR, Sanchez-Yamagishi JD, Kong J, Jarillo-Herrero P. Anisotropic Etching and Nanoribbon Formation in Single-Layer Graphene. Nano Letters. 2009;9(7):2600-4.

[51] Booth TJ, Pizzocchero F, Andersen H, Hansen TW, Wagner JB, Jinschek JR, et al. Discrete Dynamics of Nanoparticle Channelling in Suspended Graphene. Nano Letters. 2011;11(7):2689-92.

[52] Ramasse QM, Zan R, Bangert U, Boukhvalov DW, Son Y-W, Novoselov KS. Direct Experimental Evidence of Metal-Mediated Etching of Suspended Graphene. ACS Nano. 2012;6(5):4063-71.

[53] Zan R, Bangert U, Ramasse Q, Novoselov KS. Interaction of Metals with Suspended Graphene Observed by Transmission Electron Microscopy. The Journal of Physical Chemistry Letters. 2012 2012/03/22;3:953-8.

[54] Severin N, Kirstein S, Sokolov IM, Rabe JP. Rapid Trench Channeling of Graphenes with Catalytic Silver Nanoparticles. Nano Letters. 2008;9(1):457-61.

[55] Boukhvalov, D. W.; Katsnelson, M. I. Destruction of Graphene by Metal Adatoms. Appl. Phys. Lett. 2009, 95, 023109.

Section 2

Graphene Synthesis

Chapter 2

Graphene Formation on Metal Surfaces Investigated by *In-situ* STM

Guocai Dong, Dirk W. van Baarle and
Joost W. M. Frenken

Additional information is available at the end of the chapter

http://dx.doi.org/10.5772/56435

1. Introduction

Graphene, single-layer graphite, has drawn great attention recently, both because of its special properties [1-5], and because of its wide-range potential applications, for example in future-generation electronics [4, 6-8]. Chemical vapor deposition (CVD) of hydrocarbons on transition metals (TMs) provides a practical method for graphene production [9-17]. With this chapter we would like to summarize our investigations with in-situ scanning tunneling microscopy (STM), aimed at understanding the growth of graphene on metal surfaces. The 'fruit fly' of our work has been graphene formation on Rh(111). Observations on this surface are generalized into a simple physical picture that should be suitable to interpret the formation of graphene also on other metals.

The first experimentally produced and verified graphene was obtained in 2004 by mechanical exfoliation of graphite [18]. Soon in 2010 The Nobel Prize in Physics was awarded to Andre Geim and Konstantin Novoselov at the University of Manchester "for groundbreaking experiments regarding the two-dimensional material graphene". Scotch tape was used to repeatedly split graphite crystals into progressively thinner pieces. Many experiments on graphene have been performed using this method [5, 19]. It nicely provides experimentalists with the best quality in topographical flatness [19] and electron mobility [5]. The drawback of this method is, of course, reproducibility. From an application point of view, it is hard to imagine any production facility using microscopes to find tiny bits of graphene for electronic devices.

Since 2004, researchers have tried out many ways for a reproducible, bottom-up production of graphene. One method is to heat silicon carbide to high temperatures (>1100 °C), in order to make it sublime silicon and thereby reduce its surface to graphene [20, 21]. The choice of the face of the silicon carbide that is used for graphene creation, either silicon-terminated or carbon-terminated, strongly influences the thickness (i.e. the number of layers), mobility, and carrier density of the graphene. This process produces a graphene sample size that is dependent on the size of the SiC substrate. However, the size of the domains is still in the order of hundreds of nanometers [20, 21]. Although the quantum Hall effect (QHE) could be measured with the graphene samples created by this method [20], the electronic mobility in this graphene only reached 2700 $cm^2\ V^{-1}s^{-1}$, [22] which is much lower than for the graphene obtained by the scotch tape method.

A very promising alternative method to produce precisely mono-layer of graphene is chemical vapor deposition (CVD) on metal surfaces [9-17, 23-28]. Actually, graphitic layers on metals had been investigated much earlier [11, 29-33], before specific attention was paid to graphene. It is the method that we have concentrated on in our work. Metal surfaces as favorable substrates for graphene growth by CVD have catalysts for hydrocarbon decomposition and they strongly bind to the graphene that forms, while newly arriving hydrocarbon molecules do not stick to or decompose on the graphene that has already formed. Also as the solubility of carbon in a metal often depends strongly on temperature, graphene can also be made using segregated carbon [9, 12, 16, 27, 30, 34]. After samples have been prepared at high temperature and cooled down subsequently, it is not easy to distinguish which kind of carbon, segregated or deposited, has been responsible for the formation of the graphene layer.

As a consequence of its strong interaction with the substrate, graphene on top of a metal is very different from the idealized free-standing graphene. It does not have the interesting electronic properties mentioned before, but it does have its own special appeal, for example, being the template for the self assembly of additional overlayers [35], or the graphene-metal contact serving a spin filter [36, 37], etc. In order to produce free-standing graphene, several recipes have been introduced for transferring graphene from metals onto other substrates [15, 34, 38, 39]. With these recipes, CVD-grown graphite layers on metals have become a good candidate for producing high-quality graphene. A promising carrier mobility of 4,000 to 7,350 $cm^2\ V^{-1}s^{-1}$ has already been reported for graphene, obtained via this route [40-43]. The size of graphene made by CVD method is reported up to 30 inch [43].

No matter in what form the graphene is produced, free-standing or on a substrate, for many applications it is the structural quality of the graphene that limits its properties and therefore the quality of the final products in which it can be applied. The typical types of imperfections introduced by the CVD method on metals are impurities and structural defects. The problem of impurities can be solved by using a sufficiently clean production setup and sufficiently clean materials (metal substrate and precursor gasses). The structural defects include other carbon phases (e.g. metal carbide), domain boundaries between different graphene patches, point defects [13] (e.g. pentagons and heptagons [44]) and thickness variations (voids or multilayers). It has been demonstrated that such defects can influence the electronic properties of graphene [44]. To reduce and eventually completely avoid the occurrence of structural defects, full control of the formation mechanism of graphene is necessary. This can only come through a thorough, atomic-scale understanding of the process. Once we have acquired a full understanding of the basic mechanisms of graphene formation and possible competing processes, we can use that to subsequently develop an optimal growth recipe.

This chapter will start with a brief introduction of the in-situ STM technique and compare it with other in-situ techniques, e.g. low-energy electron microscopy (LEEM) and scanning electron microscopy (SEM). Ethylene deposition on Rh(111) is taken as an example to demonstrated the formation of graphene on metal surfaces. This system is ideal to demonstrate the interaction between carbon and metals, as all effects are present 'at the same time': in addition to graphene formation, we observe carbide formation and dissolution of carbon into the substrate, which later re-surfaces and deteriorates the quality of the graphene. A universal physical picture for graphene formation and competing phenomena on metal surfaces will be provided. This picture is based on very general surface-science knowledge, which makes it easy to be adapted to other CVD procedures and to graphene formation on other metals. Solid experimental evidences will be shown to confirm the picture. Guided by this picture, we will demonstrate several paths to improve the quality of graphene formed on a metal surface. Each path shows its own advantages and disadvantages. We hope these demonstrations will help the readers understand the universal picture better and use it to improve graphene quality for different applications.

2. STM for in-situ measurements

Direct real-space microscopic observation during the formation of graphene is obviously important for understanding the kinetic processes and thermal dynamics as well as for improving the quality of graphene. However, due to the high growth temperature, to realize such observation one has to face many technical challenges. Firstly, any temperature changes in the macro structures of the microscope will cause dramatic thermal drift on the atomic scale in the imaging process. Secondly, even though one can image at a constant, high temperature, it is more difficult to stabilize the whole system at higher temperatures than for example room temperature or lower.

Many common microscopy techniques, such as STM, LEEM, and SEM, etc. have been performed *in situ* to investigate the graphene formation at high temperatures on metals [9, 16, 38, 45-47]. In table 1, the working conditions and the performances of these techniques are listed for comparison. With some special designs, all of these methods satisfy the required conditions for graphene growth, i.e., sample temperature and gas pressure. Although there are arguments about the definition of the resolution, high-temperature STM is superior in spatial resolution. In fact, it is the only technique which can provide images in which the moiré pattern between graphene and metal substrate (about 3 nm) can be clearly distinguished even at the growth condition of graphene [46]. The higher resolution certainly provides more information for understanding the growth kinetics of graphene. On the other hand, LEEM and SEM can image much larger areas, which gives a macroscopic knowledge of the growth. In STM, the measurement itself has less influence the formation, because the energy of the tunneling electrons is typically not more a few eV instead of keV which is typical for a SEM. Considering the advantages and disadvantages, we chose a special designed STM for our investigation of graphene growth.

	Spatial resolution at high temperature *	Maximum field of view	Sample temperature	Maximum vacuum pressure
STM	1 nm	3 μm	~1200 K	1 x 10^{-3} mbar
LEEM	2 nm at room temperature	75 μm	~1200 K	1 x 10^{-6} mbar
SEM	2 nm at room temperature	4 mm	~1200 K	1 x 10^{-4} mbar

* There are some debates in the definition of resolution. Only general numbers are list here.

Table 1. Comparison of microcopies that can be used at the conditions for graphene growth on metals. The specifications of each microscopy are very dependent on the specific design. E.g. differential pumping stages might increase the maximum vacuum pressure.

The STM we used had been optimized for (fast) scanning at high sample temperatures and also during substantial temperature variations [48, 49]. Like most other STMs, our Variable-Temperature STM (VT-STM) is using a piezoelectric tube, in our case with a length of 12 mm, to scan regions up to typically 3 x 3 μm^2, with sub-nanometer resolution. The macroscopic dimensions of the piezo element, the sample and other components in the mechanical path between the tip and the sample make that even modest temperature variations can cause a significant drift in an STM image. More dramatically, when the component of this drift in the z-direction, along the tip axis, were to exceed the control range of the piezo element, this would lead to a situation where either the surface drifts out of range or a crash of the tip into the sample surface occurs. For these reasons, most STMs are used either at room temperature or at a constant, low or (modestly) elevated temperature, at which the sample and microscope are first allowed to equilibrate for a long time, typically several hours, prior to the coarse tip-sample approach and imaging. The STM we were using could compensate for thermal drifts by its special design [48]. With this design, neither the expansion of the piezo element and other components of the microscope nor that of the sample and sample holder can give rise to a significant tip-sample distance variation. In practice, experiments have been performed without the need for any mechanical adjustments in the tip height, while the sample temperature was being changed from 300 K to approximately 1200 K.

In addition to the drift-minimizing mechanical design, the STM electronics have been optimized to scan fast. It has been demonstrated that the STM can image surfaces at video rate (24 frames per second) [49] (A commercial version of this instrument is marketed by Leiden Probe Microscopy BV, www.leidenprobemicroscopy.com). Because of the critical experimental conditions, in particular the high temperatures during deposition, video-rate scanning is very difficult. Therefore, the high-speed capabilities of the VT-STM have not been exploited directly in the work described in this chapter. On the other hand, the low noise level that the high-

http://dx.doi.org/10.5772/56435

speed electronics has at low frequencies and the modular design of the electronics provided stable scanning conditions plus possibilities for customization of the scans, which played an important role in this work.

3. Experimental detail

Our special-purpose, VT-STM [48, 49], enables us to follow the reaction and growth of graphene *in situ*. Ethylene decomposition on Rh(111) is taken as a generic example of carbon deposition on metals. The interaction between carbon and Rh can lead to carbide formation [50], carbon dissolution [51], and graphene formation [29]. These are also the typical 'products' that can be formed on other metals. Actually, carbon is the main contaminant in Rh crystals.

The complexity of the carbon-on-Rh(111) system makes it a good example for demonstrating, in general, the interaction between carbon and metals. The 3-fold (111) orientation was chosen, to act as a template for graphene crystallization. In addition, graphene forms an easily recognizable moiré pattern on Rh(111). A simple calculation shows that the moiré pattern 'magnifies' the misfit defects between the lattice of the graphene layer and that of the Rh substrate. For example, 1 degree of actual graphene rotation with respect to the Rh lattice results in a 10 degree apparent rotation in the moiré pattern. In this way, atomic information about graphene can be achieved even without actually achieving atomic resolution in STM.

All measurements in this chapter were performed with a STM-setup which consisted of a single ultrahigh vacuum (UHV) chamber, equipped with a scanning tunneling microscope (STM), a low-energy electron diffraction (LEED) apparatus, an Auger electron spectroscopy (AES) instrument, and a quadrupole mass spectrometer (QMS), for analysis purposes. The base pressure of the vacuum system was 1.5×10^{-11} mbar. During measurements, the vacuum level was kept below 2×10^{-10} mbar. To achieve and maintain this vacuum, a single 170 liter/second magnetically levitated turbo-molecular pump, a 410 liter/second ion getter pump, and a titanium sublimation pump, integrated into a cold trap, were connected to the UHV chamber. Also, the entire vacuum system was extensively degassed by a bake-out procedure up to 450 K and the sample holder and STM components were further degassed by prolonged annealing of the Rh sample to 800K.

The Rh sample was cleaned by cycles of Ar^+ ion sputtering, followed by flash annealing to 1300 K and then exposure of the sample to 2 to 3 $\times 10^{-7}$ mbar of O_2 for 1 to 2 hours at temperatures of 700 to 800 K. This latter step proved necessary in order to remove C surface contamination emanating from the bulk. Residual O was removed from the Rh surface by flashing the sample to 1000 K. After this cycle had been repeated at least 10 times, no C contamination could be detected by Auger Electron Spectroscopy, and the density of visible impurities in the STM images were typically below 1/μm^2. The dosage of ethylene gas, from ALDRICH™, with a purity of 99.99 %, was through a precision leak valve.

4. A universal description of graphene formation on metals

In this section, we introduce a new physical picture (Figure 1), which descripts the general principles of graphene and carbide formation on metal surfaces. It is based on the combination of our direct STM observations and the considerations published by McCarty *et al.* [12]. In Figure 1, the black line represents the density of carbon adatoms on the Rh surface, dictated by equilibrium with carbon dissolved in the bulk. The shape and position of this line is determined by the carbon concentration in the substrate, and by the dissolution energy of carbon [12]. The solid red line represents the carbon adatom density derived from equilibrium with the graphene islands. The point where the black line crosses the solid red line gives the dissolution temperature of graphene. We now add the solid blue line, which represents the carbon adatom density which would establish equilibrium with the carbide islands. The positions of the solid red and blue lines are determined by the edge energies of graphene and carbide. The lower position of the line means the edge energy is lower and the island is more stable. The dashed red and blue lines are the supersaturation necessary for observable nucleation of graphene and carbide. The dashed lines are not only determined by the edge energies of graphene and carbide, but also by the critical nucleation size of each product. So it is not necessary that the dashed blue line is higher than the dashed red line, just because the solid blue line is higher than the solid red line. In fact, our experiment shows the dashed red line is higher than the dashed blue one, while the solid blue line is higher than the solid red line. In the following sections, we will experimentally determine the relative position of each line in this figure. Guided by this figure, we show several paths for improving the quality of graphene.

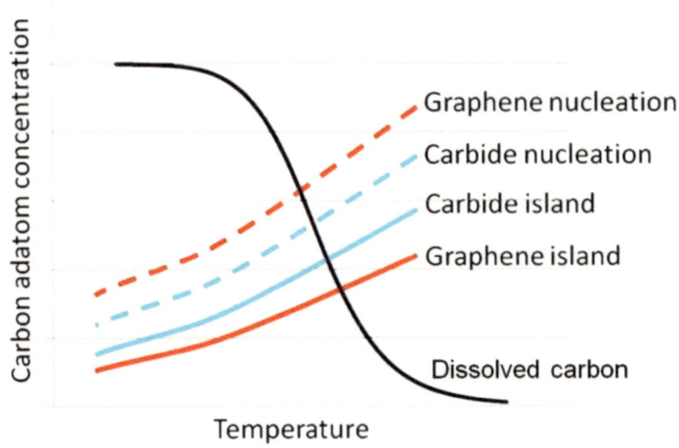

Figure 1. Schematic diagram, showing the temperature dependence of various equilibrium adatom concentrations of carbon on the metal surface. The black line is the carbon adatom concentration, which is in equilibrium with carbon in the bulk. The solid red line is the adatom concentration for equilibrium with graphene islands (after [12]); the solid blue line is the adatom concentration for equilibrium with carbide islands; the locations where the solid lines cross the black line are the dissolution temperatures for graphene and carbide; the dashed red and blue lines indicate the supersaturation levels required for nucleation of graphene and carbide, respectively, on a clean metal surface.

http://dx.doi.org/10.5772/56435

5. The temperature dependence of graphene and carbide formation

In order to obtain a first impression of the temperature dependence of graphene formation, we exposed the Rh surface to a high dose of ethylene at room temperature and slowly heated up the sample, while the surface was monitored continuously with STM. The initial, saturated layer of ethylene was obtained by exposing the freshly cleaned Rh(111) surface [51] in the UHV chamber to 3 x 10^{-5} mbar s of ethylene gas at room temperature. The result of this exposure is a pronounced, atomic-scale roughness, decorating the entire surface (Figure 2A), that we associate with the disordered overlayer of ethylene molecules. During the initial stages of the temperature ramp, clusters formed without any preference for specific edge orientations (Figure 2B). These clusters grew in size when the temperature was increased. Even though we have directly observed the motion and coalescence of the clusters, the drifting during temperature changes makes is difficult to conclude that this was the sole ripening mechanism. Previous research showed ethylene decomposition on Rh to proceed in various stages at different temperatures up to ~800 K [29, 52]. Based on this, we interpret the irregular structures in Figure 2B to be clusters of carbon or CH_x. At around 870 K, some islands were observed with a hexagonal shape, and the characteristic moiré pattern inside, indicating that, at this temperature, graphene had already been formed, and that the domains were large enough to appear as moiré patterns. The restructuring of the overlayer into graphene necessarily should start with small domains, smaller than one unit of the moiré pattern (2 x 12 x 12 carbon atoms). Ripening processes make some domains grow at the expense of others, to become larger than this size. Because any ripening process requires the relocation of many carbon atoms, this ripening should be relatively slow [53]. The combination of this slow ripening with the fast temperature ramp up to 870 K of 0.2 K/s, implies that the moiré pattern observed at 870 K is a sign that graphene formation had started already at a temperature below 870 K. Indeed, Figure 2C shows our lowest-temperature observation of a moiré pattern, at 808 K, (temperature ramp rate was 0.05 K/s) close to the temperature range of 700-800 K, necessary for complete decomposition of ethylene [29, 52]. Up to 969 K, the ripening of the graphene islands continued, making the islands larger, more compact and more hexagonal (Figure 2D). Also, the orientations of different domains became similar. Nevertheless, we still observed more than a single moiré pattern at this high temperature, which indicates that several overlayer orientations may have similarly low energies.

In a separate experiment we found that the graphene overlayer dissolved into the Rh substrate, starting at a temperature of ~1050 K. The information obtained from these heating experiments can be cast in the form of the upper temperature bar in Figure 2E, displaying that, from 808 K to 1053 K, graphene can be formed on a Rh surface. This 250 K temperature range for stable graphene on Rh is much wider than the 50 K window that has been reported previously [24]. This process, combining adsorption of ethylene at room temperature with a subsequent temperature ramp, also yields a lower graphene formation temperature than the temperature of ~1100 K, reported in case ethylene was dosed directly at high temperature [54].

From this experiment, we can conclude that graphene is more stable than carbide and that in Figure 1 the solid red line should be lower than the blue one. In this experiment, clustering of

CH_x must make the surface carbon concentration locally very high, beyond that shown by the dashed blue and red lines. Under such conditions, both carbide and graphene nuclei can be formed. With time, the carbide islands disappear and the graphene islands become larger, since they are energetically more favourable. This is also why the carbide may form first on metal surfaces, but after long periods of annealing (e.g., 400 minutes) at high temperature, it can transform to graphene [55].

Another experiment was designed to determine the relative position of dashed red and blue lines in Figure 1. In the experiment above, the carbon adatom density in CH_x is high enough for the nucleation of both graphene and carbide; the two nucleation process cannot be distinguished. To discriminate, we fixed the temperature of a clean Rh sample at 830 K, which is suitable for the formation of graphene and carbide, then we started adding the ethylene starting from zero. On the clean Rh surface, the carbon adatom density was increasing from almost zero, while the pressure of ethylene was being increased. Because there was no nucleus of graphene or carbide on the surface, carbon would form the product (graphene or carbide) which requires the lower nucleation carbon adatom density, during the increasing of the density of carbon adatoms. With STM it was seen that an overlayer formed, starting from the Rh step edges on the surface, and proceeding until it covered the entire surface. As can be seen in Figure 3A, this overlayer had a structure that was different from that of graphene. After the entire surface had been covered by this layer, we heated the sample to higher temperatures, while monitoring the structure with the STM. It was found that the overlayer did not transform into the graphene structure even up to the temperature of 1016 K, at which it disappeared. The stability of the structure indicates that it did not consist of CHx clusters, which would otherwise have transformed into graphene at and above 808 K. We interpret the disappearance again as the dissolution of the deposited material into the Rh substrate. The fact that the dissolution temperature differs significantly from the value of 1050 K, mentioned above for graphene, indicates that in this case the overlayer was not graphene. AES (Figure 3B) on this layer shows that the KLL of carbon peak had shifted from 272 eV to 275 eV and the MNN peak of Rh had also undergone a change in shape and position. Both shifts indicate the formation of a new compound [56]. We suggest that this is a rhodium carbide overlayer. Comparing this AES spectrum with a reference spectrum taken on a one-monolayer graphene-covered Rh surface (Figure 3B), we see that the C-to-Rh peak ratios for these two cases were similar. Assuming that the Rh peak in the AES spectrum of graphene-covered Rh corresponds to effectively one atomic layer of Rh, we find that the C-to-Rh ratio in the carbide must have been in the order of 2:1. In Figure 2E, in analogy to the temperature bar for a graphene overlayer, the lower bar shows the growth and stability regime of a carbide layer on Rh(111). The fact that carbide was formed first in this experiment means that the require carbon adatom density for nucleation of graphene is higher than that for carbide, which makes the position of the dashed red line higher than dashed blue line in Figure 1. From these two experiments, we can determine the relative position of the lines in Figure 1. It also naturally explains why a lower dissolving temperature was measured for carbide. That is simple because that the solid blue line crosses the black line at a lower temperature than the red one.

http://dx.doi.org/10.5772/56435

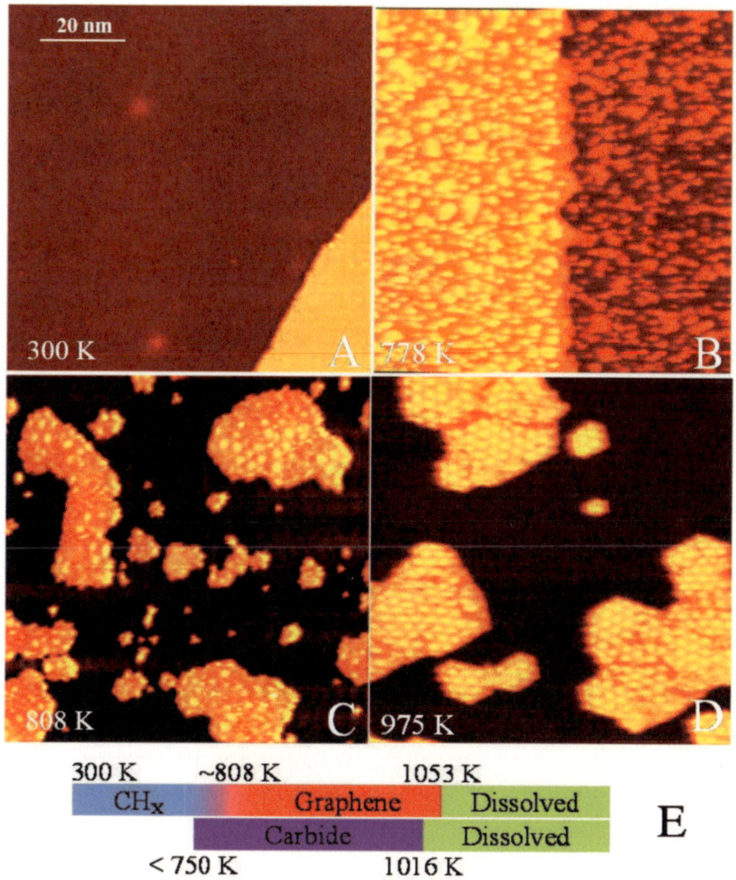

Figure 2. STM images measured during continuous heating from 300 K to ~1000 K, after room temperature ethylene deposition. Panels A, B and D are from one series of experiments. (A) The Rh(111) surface directly after exposure to 3 x 10^{-5} mbar s of ethylene at room temperature. A mono-atomic step on the Rh surface crosses the image. The saturated ethylene adsorbed layer causes the rough appearance of the upper and lower terraces. (B) At 778 K, the overlayer organized into irregular clusters, but no moiré pattern was found at this temperature. (C) Starting from 4 x 10^{-7} mbar s ethylene exposure at room temperature, this image shows the lowest temperature where the moiré pattern of graphene was found. (D) At 975 K, the graphene had ripened into larger islands, with similar orientations. (E) Temperature ranges for the observed stability of graphene and carbide on Rh(111). The STM images all have a size of 85 nm × 85 nm. They have been taken at sample voltages of V_b = 0.05 V, 1.16 V, 1.43 V, and -1.84 V for panels A, B, C and D, respectively, and at a tunnel current of I_t = 0.05 nA.

6. Towards the optimal growth recipe

6.1. Direct carbon deposition at high temperature

To reduce the nucleation density of graphene therefore reduce the defect lines between different nuclei, dosing carbon at a high sample temperature is an obvious solution, since the high temperature largely enhances the diffusion of carbon adatoms. However, the experiment

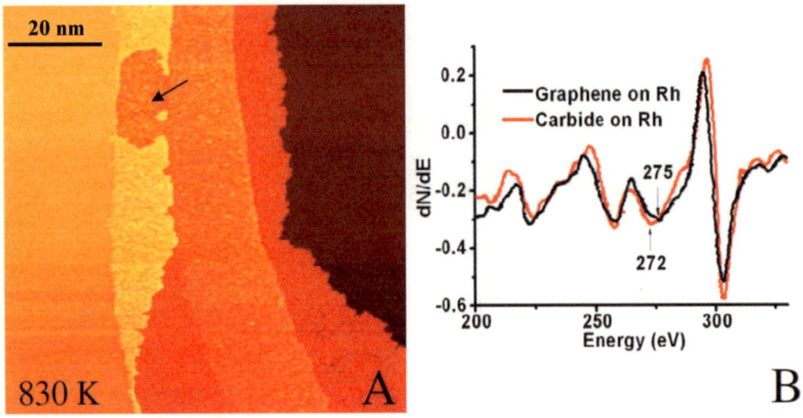

Figure 3. Carbide formation on Rh surface by direct ethylene dosing on a clean Rh. (A) An STM image measured at 830K, during deposition of ethylene. We interpret the rough layer that started to grow from the Rh steps as a carbide film (see text). Before nucleation of the film, the Rh steps were modified, as is indicated by the arrows. Nucleation started when the ethylene pressure reached 4.4 x 10^{-8} mbar. (B) Comparison between two Auger electron spectra of Rh(111); red: with a single-monolayer carbide covered Rh, formed by ethylene deposition at 750K, and black: with single-layer graphene. The size of the STM image is 85 nm × 85 nm. It has been taken at sample voltages of V_b = 2.79 V, and at a tunnel current of I_t = 0.05 nA.

above shows that carbide will be formed first on the metal surface before graphene is formed. Interestingly, there is a narrow temperature window, between 1016 K and 1053 K, in which a carbide layer is no longer stable, but a graphene monolayer is. In order to explore this window, we held the Rh at 1035 K, while dosing with ethylene. Indeed, a graphene layer was observed to form at this temperature. It covered the entire surface without a trace of carbide (Figure 4B). However, the moiré pattern of the graphene overlayer became distorted during the slow cooling down, after the deposition, as shown in Figure 4C. The low-energy electron diffraction (LEED) pattern that was taken afterwards at room temperature showed a weak Rh(111) pattern on a strong, diffuse background (inset Figure 4D). We attribute the deterioration of the overlayer during the cooling down to segregation of carbon. It is known that carbon, which dissolves into the near-surface region of the Rh at the growth temperature, can segregate back to the surface when the temperature is decreased [51]. This segregation could lead to the nucleation of islands of a second graphene or carbide layer, between the original graphene layer and the Rh substrate, which will locally push up the graphene overlayer and distort it.

One may augur that the distortion of graphene moiré pattern might be caused by the different thermal expansion coefficient between graphene and the Rh substrate (in this temperature regime graphene has a negative expansion coefficient [57, 58]). This effect can be excluded, since other experiments, in which the role of segregated carbon could be minimized (Figure 5), showed that the quality of the graphene was not affected solely by large changes in temperature.

Now we turn to the role of dissolved carbon atoms. When we exposed a clean Rh(111) substrate directly to ethylene at 1035 K, several dark spots appeared in the STM images, before the graphene was formed (Figure 4A). Graphene formation only started when the ethylene pressure reached 3.5 x 10^{-7} mbar, which we associate with the dashed red line in Figure 1. In the framework of Figure 1, it is natural to ascribe the dark spots in Figure 4A to dissolved carbon atoms, because the supersaturation required for graphene nucleation is high enough for noticeable carbon dissolution, just before nucleation. This dissolution leads to a relatively high concentration of carbon atoms in the substrate, close to the surface. Since the solubility of carbon in the Rh is heavily dependent on temperature, there is a strong driving force for the dissolved carbon atoms to segregate during cooling and accumulate between the Rh substrate and the graphene overlayer. The resulting accumulation layer distorted the regular graphene moiré patterns, as shown in Figure 4C. Other STM measurements of ours also showed that it was possible to re-dissolve these extra carbon atoms by heating the sample to the growth temperature. We speculate that this is only possible as long as the density of carbon in the accumulation layer is not sufficient for the formation of a second layer of graphene. In the latter case, we should have observed the top graphene layer dissolve together with the accumulated (graphene) layer. Note, that the high concentration of dissolved carbon atoms led to a higher dissolution temperature (1118 K) of graphene for this sample, in agreement with our expectation from Figure 1.

Figure 1 also shows that in order to have less carbon dissolved during the direct deposition; one should use a higher pressure. The argument for this counterintuitive approach is that by increasing the pressure the nucleation rate increases super-linearly, whereas the rate of dissolution will simply be proportional to the pressure. Following this idea, we have been able to improve the quality of the graphene overlayer on Rh as is demonstrated in Figure 4D. The temperature of the clean Rh sample was first raised to 1028 K, after which the ethylene pressure was quickly increased to a high value of 3.9 x 10^{-6} mbar. The graphene layer that formed under these conditions covered the whole surface with a much lower defect density than the graphene structure shown in Figure 4B. However, in spite of the high ethylene pressure used here, the moiré pattern still distorted due to segregated carbon when the sample was cooled to room temperature. The combination of the high solubility of carbon in Rh and the macroscopic thickness of our substrate (1 mm) made it practically impossible to completely avoid carbon dissolution at high temperatures and segregation of dissolved carbon during the cool-down.

One of the recipes reported for a single monolayer of graphene on Rh(111) is to expose the clean rhodium surface for 3 minutes to 2 x 10^{-7} mbar ethylene at an even higher temperature of 1100 K [54]. We have monitored this recipe with our STM at this high temperature, but graphene was not formed under these conditions and also not with longer exploring times up to 10 minutes. Instead, our images show that in this case a graphene layer was formed only after the exposure, during cooling of the sample, when the temperature had reduced to ~1000 K, showing that, in this recipe, the graphene film is assembled completely from segregated carbon. The STM images at room temperature did not show distortions, in accordance with the images in Wang et al. [54]. We attribute this absence of distortions to a nearly complete consumption of the dissolved carbon in the near-surface region by the graphene monolayer. The difficulty in this procedure is to precisely control the coverage, i.e., the total amount of segregated carbon. It is relatively easy to obtain large areas covered by graphene suitable for STM imaging or for a local spectroscopic measurement. But it is very difficult to reach full

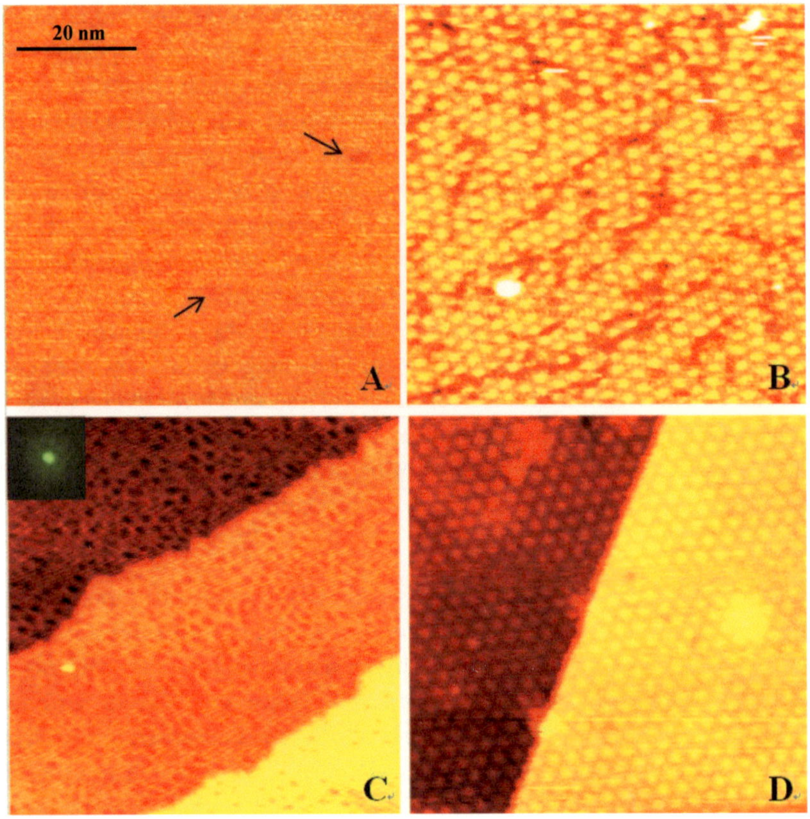

Figure 4. Graphene formation by direct ethylene dosing at high temperature. (A) STM image before the whole surface was covered by graphene. Some dark points (indicated by the arrows), which can be caused by dissolved carbon, are visible. The temperature of the sample was 1035 K and the pressure of ethylene was 3.5 x 10^{-7} mbar. (B) A STM image of graphene-covered Rh, which was achieved after panel A. The image was taken at the growth temperature. The 'defect' lines in the image were formed already before the graphene overlayer nucleated. We associate these lines with dissolved carbon atoms. (C) STM image of the sample of (B), after it was cooled to 340 K. The moiré pattern of the graphene was deformed. The inset in (C) shows one integer-order Rh LEED spot (same view as in inset of Figure 5B) and the near absence of superstructure spots around it, recorded at room temperature. (D) STM image of graphene-covered Rh, which was achieved by direct ethylene deposition and imaging at 1028 K. The pressure of ethylene was quickly raised to 3.9 x 10^{-6} mbar. The STM images all have a size of 65 nm × 65 nm. They have been taken at sample voltages of V_b = 2.5 V, 2.5 V, 3.7 V, and 4.8 V for panels A to D, respectively, and at a tunnel current of I_t = 0.05 nA.

coverage of the surface with graphene of precisely one layer thickness, i.e., without any excess carbon.

6.2. Graphene formation on a seeded surface

The above experiments demonstrated that the temperature window for producing graphene by direct deposition is quite limited, and graphene formed in this temperature window distorts when the substrate is subsequently cooled to room temperature. On the other hand, the first experiment, which started with deposition at room temperature, showed a wider temperature window for graphene formation (upper bar in Figure 2E), starting below and ending above the carbide window (lower bar in Figure 2E). The differences between the two experiments in Figure 2 and Figure 3 suggest that the choice which the growing overlayer had to make between carbide and graphene, in the temperature interval from 800 K to 1016 K, is fully determined by the structure of the initial overlayer nuclei; in other words, graphene patches will continue to grow as graphene, whereas carbide patches will continue to grow as carbide. This interpretation of the experiments also coincides with the theory described in Figure 1. If the carbon adatom density and the temperature is set between the solid lines and dashed lines and the surface is already partially covered by graphene/carbide, carbon adatoms can only contribute the growth of the existing graphene/carbide islands without forming new nuclei. In turn, this suggests a refinement of our approach, namely to separate the stage of nucleation from that of further growth. For example, one could start with the graphene-seeded Rh surface from the first experiment, and expose that to further ethylene deposition under different conditions, in order to obtain full coverage by graphene. One may expect nucleation and growth of the graphene overlayer to follow the well-known rule, in which the nucleation density is determined by F/D, where F is the flux of arriving atoms and D is their surface diffusion coefficient [59]. If the deposition is carried out very slowly, so that the value determined by F/D is smaller than the existing seed density on the surface, the newly arriving carbon atoms will all be incorporated onto the edges of growing graphene islands, which prevents them from forming new nuclei [60]. The presence of the graphene seeds on the surface should have the additional effect that newly arriving carbon atoms have a strongly reduced probability to dissolve into the Rh, again because they will be incorporated in the graphene overlayer on a much shorter timescale, due to the difference in the rates of diffusion of the carbon atoms into the bulk and over the surface, which should be expected to be significant. On a non-seeded surface, dissolution has to compete only with the rather rare processes of nucleation of either graphene or of carbide. How 'difficult' graphene nucleation is on a hot Rh surface is evidenced by our findings in a separate experiment, on a clean Rh surface: at 1035 K, an ethylene pressure of to 3.5×10^{-7} mbar was required to observe the first graphene nuclei in our STM images.

The suggested approach, of ethylene deposition at high temperatures onto a Rh surface that has been pre-seeded with graphene by low-temperature deposition, was successful. Figure 5 shows the result obtained at a growth temperature of 975 K; even more informative is the corresponding STM movie, which can be found online. Starting with the end situation of the first experiment (Figure 2D), ethylene was deposited, at a pressure of 3×10^{-9} mbar, which was increased at the end of the procedure to 1×10^{-8} mbar, simply to accelerate the process. The newly arrived carbon atoms continued with the structure and orientation of the graphene that was already present, until the entire surface was covered with graphene. When a new kink formed at a concave corner between two differently oriented domains, it showed a preference for following the graphene domain which had the same, or a similar orientation, as the Rh crystal. In this way, the metal substrate guided the orientation of the graphene layer, and the density of domain boundaries became lower than that expected from the initial configuration of the graphene seeds. After this sample had been cooled down to room temperature, neither STM nor LEED (insert of Figure 5B) showed a deformation of the moiré pattern. The LEED pattern showed that the graphene layer had a preferred orientation; otherwise, a ring of

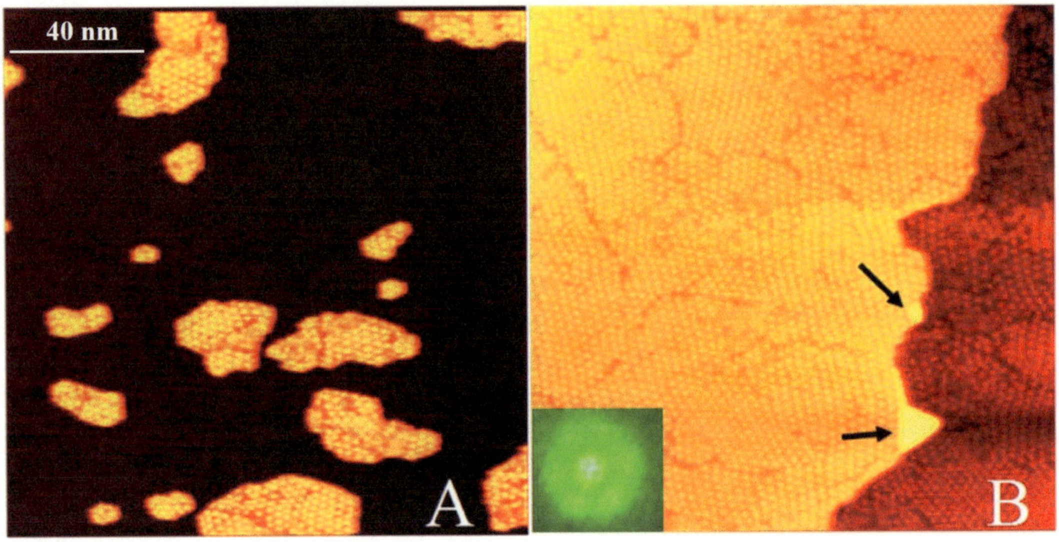

Figure 5. STM images of graphene formation, starting with a seeded surface. (A) The graphene-seeded Rh surface achieved by annealing a pre-deposited sample from room temperature to 975 K. Most of the graphene islands had same orientation. But superstructure domain boundaries can be found within individual islands. (B) Graphene-covered Rh, after ethylene deposition at 975K on the seeded surface, at pressures ranging from 3×10^{-9} to 1×10^{-8} mbar over a period of 76 minutes. During this procedure, The domain boundary density became ~30% lower than that in the starting situation. (A). Two Rh double-layer defects are indicated by the arrows. The inset in (B) shows the superstructure spots around one Rh LEED spot. All images are 160 nm × 160 nm and have been taken at a sample voltage of V_b = -1.84 V and a tunneling current of I_t = 0.05 nA.

intensity should have been observed around the integer-order maxima from the Rh, rather than the six superstructure spots that are clearly present in the LEED pattern. Although the superstructure spots are not very sharp, the orientation difference between graphene domains is seen to be minor, because the moiré pattern is an amplifier of all variations in position and orientation. The bright regions, indicated by black arrows in Figure 5B, were formed in the final stage of the deposition. They will be explained later.

6.3. Using segregated carbon to form graphene

In the previous section, it was shown that the deposition of ethylene can be controlled to form exclusively graphene rather than carbide, and to bring the process to a lower temperature. In this section we show that as an alternative to direct deposition, also the segregation of dissolved carbon atoms could be used in a controlled way, in order to form graphene. By cooling down the sample very slowly, the effective flux of segregated carbon is kept low. These carbon atoms also tend to follow the structure and orientation of the pre-existing seeds; if these have the graphene structure, this procedure results in the growth of graphene. After the segregation has come to an end, the coverage of the graphene overlayer can be brought to completion by a final ethylene exposure. This lower process temperature will be associated with a lower density of carbon dissolved in the substrate and, if it is of any importance, also with a lower

http://dx.doi.org/10.5772/56435

strain due to the differential thermal expansion between the graphene overlayer and the Rh substrate [57, 58].

This alternative procedure using segregated carbon to form graphene was also successful, as is shown in Figure 6A, where a part of the surface was followed from 862 to 829 K. This is very promising for achieving 2D single crystalline graphene, using dissolved carbon. One difference, relative to deposited carbon, is that the graphene layer also chose the growth direction (Figure 6E) where the Rh step has to retract, to allow further graphene growth. To emphasize that this proceeds against the direction of regular step-flow growth, we refer to this as 'anti-step flow'; in some papers it is called 'growth by etching step edges' [61]. In this temperature regime, Rh atoms should be expected to diffuse much faster over the surface than the carbon atoms can in graphene. When sites become temporarily available, as a result of step fluctuations of the Rh, newly arriving carbon atoms may fill up the vacant sites, the strong cohesion in the graphene preventing the Rh step from fluctuating completely back to its original shape and location. This anti-step-flow mechanism resulted in a large number of Rh islands enclosed by graphene. Figure 6C shows one of Rh islands enclosed by graphene, to which more ethylene had been added. The carbon introduced by the ethylene deposition was found to also be accommodated via the anti-step-flow mechanism, instead of forming graphene on top of the enclosed Rh. In this stage, Rh atoms were pushed up and nucleated a higher atomic layer; the graphene stopped growing when the complete enclosed Rh island had a double-layer height (Figure 6A). The reason that, in Figure 6A, the Rh did not become a double-layer structure, is that, in that case, there was a narrow passage in the graphene layer, through which the Rh could diffuse away. In separate experiments, in which the Rh was exposed to ethylene at a low temperature of 864 K, the growth of graphene, both at the free edges of graphene islands ('step-flow' growth) and growth against the Rh steps ('anti-step-flow') was observed. At this temperature, the growth rates for these two mechanisms were comparable. At the higher temperatures of the experiments discussed before, e.g. at 975 K, the anti-step-flow growth rate was much lower than the step-flow growth rate. That the mechanism was still active is illustrated by the appearance of double-layer defects in the final stages of the graphene growth, as is indicated by the two arrows in Figure 5B. The temperature dependence of the ratio between the two growth rates can be explained as follows. Assume that decomposition of ethylene or the growth of graphene has a lower energy barrier at the step edges of Rh than on the terraces. At 864 K, the resulting higher decomposition or growth rate of ethylene at the Rh step makes the amount of carbon atoms available for anti-step-flow higher than that for step-flow growth. At 975 K, all ethylene on the Rh terraces rapidly decomposes, which means that the carbon production rate is not limited, at that temperature, by the decomposition, but by the deposition of ethylene. Since the steps represent a much smaller effective area than the terraces, and the anti-step-flow requires an additional process of moving Rh atoms, the ratio between anti-step-flow and step-flow should become low at high temperatures. Nevertheless, anti-step-flow will still occur, and it will lead to Rh double-layer defects. The only way to reduce their density is by further elevating the temperature or by making the average distance between the graphene nuclei larger than the average distance between the steps on the Rh substrate.

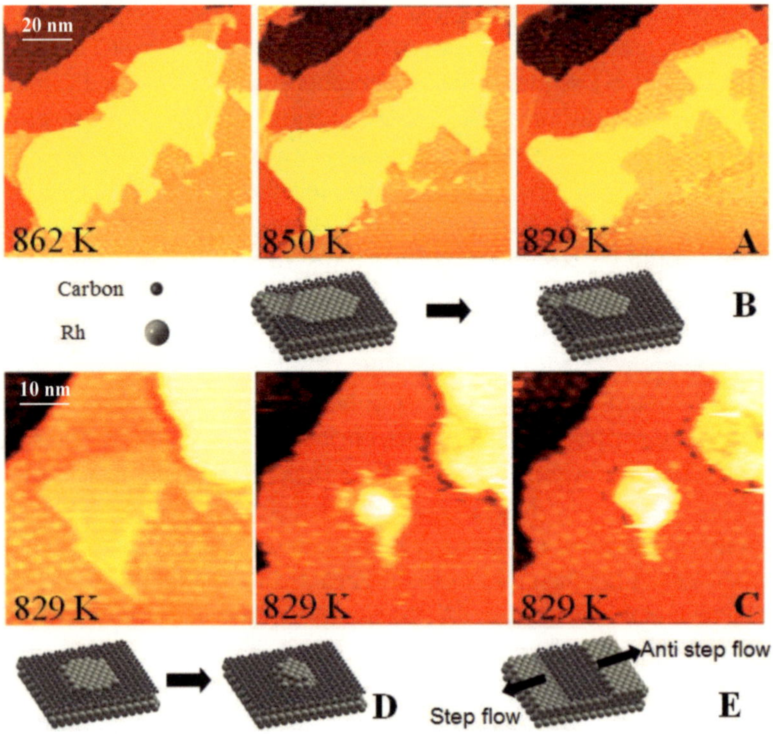

Figure 6. Graphene formation by dissolved carbon and Rh double layer defect formation. (A) STM images of the same area during the cooling down of a Rh sample, partly covered by graphene, after adsorption of ethylene at room temperature and annealing to 977K. The Rh island indicated by the red arrow is in the same atomic level as the graphene indicated by black arrow. They seem to be in different levels because of the sample voltage setting. The segregated carbon followed the orientation of the existing graphene. Image size: 100 × 100 nm^2. (B) A simple atomic model of the graphene formation by segregated carbon. The Rh island became smaller, because Rh diffused away via the small opening in the graphene layer. (C) Graphene growth, by adding more ethylene while the sample was held at 829K. These images show how a Rh island, enclosed by graphene, was pushed into a double layer island with half of the original area. Growth of the graphene stopped after the whole island had become a double layer. Image size: 50 × 50 nm^2. (D) A sketch of the formation of the Rh double layer island. (E) The concept of step flow and anti step flow. In the latter case, Rh atoms have to move to the upper atomic level, in order for the surface to accommodate the additional carbon.

7. Conclusions

In this chapter we presented information with both scientific and technical aspects. Technically, the performance of the VT-STM was demonstrated, showing what kinds of extra information can be achieved with this kind of setup. Scientifically, we introduced the schematic diagram in Figure 1 which can explain the reported experimental results. We have used in-situ measurements to determine the correctness of the picture, and use it to guide the further improvement of graphene quality. This picture can easily be adapted to other CVD procedures on metals. The key elements that may vary from system to system are the stable phases, their formation energies and step energies, the dissolution energy, and the critical nucleation size.

Guided the picture presented, we show three different paths to improve the quality of graphene, namely direct ethylene dosing on clean Rh, first nucleation and growth, and using segregated carbon to form graphene. All of these methods can effectively avoid the formation of carbide. For the method of direct ethylene dosing, the temperature window to avoid carbide formation is quite narrow. The size of the produced graphene domain is quite large. Although the dissolution of carbon cannot be avoided and the segregation of carbon between graphene-metal interface which distorted the moiré pattern of graphene on Rh, the top graphene layer stayed unchanged. This was confirmed by the observation that the moiré pattern became visible again when the sample was heated to the growth temperature. If one wants to produce graphene with larger domains and one does not care about the carbon between the graphene and the metal, it is an easy recipe to follow. In the case of seeding followed by growth, the system is kinetically trapped in a 'graphene-forming' mode. Graphene islands are already present and only a low carbon adatom density is needed to continue the growth. Because of the low ethylene pressure and the rapid incorporation of the deposited carbon into the graphene matrix, the density of adatoms cannot become high enough for nucleation of a carbide (or new graphene islands). Additionally, the dissolved carbon problem is also partially avoided. So all this taken together, the final result is a surface fully covered by graphene. The disadvantage using the seeding method is that the domain density of the graphene is much higher than the one made direct ethylene dosing. The quality of the graphene overlayer is limited by defects in the form of two different classes of domain boundaries. One class results from the merger of neighboring graphene islands with different orientations. On the atomic scale, this type of domain boundary must involve defects with respect to the hexagonal structure of perfect graphene, for example arrays of pentagons and/or heptagons [13] instead of the normal hexagons. The other class of domain boundaries originates from the merger of islands with precisely the same orientation. Due to the mismatch in lattice spacing between the overlayer and the substrate, there are still 288 translationally inequivalent possibilities to position the graphene on the Rh(111). This makes the probability for a perfect fit lower than 0.4%, so that most of these mergers will be accompanied by a phase defect line in the moiré pattern. Although such lines may look 'dramatic', we speculate that this phase defect is not leading to dangling bonds, i.e., the defect is present only in the moiré period of the super-structure but not in the topology of the graphene network [13]. An alternative path to reduce the domain boundary density to a value which is even lower than the one made by direct ethylene dosing is to reduce the initial nucleation. Other groups also have noticed the importance of initial nucleation [44, 60] and used the two step growth method. However, the nucleation conditions (i.e. precursor pressure and temperature) were normally used for the second step of growth after seeding. In this way, the nucleation density in the final product cannot be reduced lower than the intrinsic nucleation density ruled by the growth condition. We now propose that the growth condition should be controlled between the solid and dashed lines. In this case, the carbon adatom density is only high enough to continue the growth of graphene, the intrinsic nucleation density by this adatom density can be very low. If the initial nucleation is also set to be very low, e.g. one on entire metal surface, the nucleation density in the final product can be much lower than the graphene made by direct ethylene dosing. With carful control, one single graphene domain over a macroscopic area is very well possible. Our attempts to use segregated carbon plus further ethylene deposition at low temperature to form a complete, undistorted graphene overlayer were successful: the added graphene follows the graphene phase and orientation of the initially nuclei. This procedure forms a very promising route for achieving 2D single-crystalline graphene. However, in the case of Rh substrate, the low growth temperature results in a high growth rate of graphene in the anti-step-flow direction. This results in many Rh double-layer defects, which form holes in the final graphene overlayer.

How to control the thickness of the graphene, in particular how to avoid the nucleation of the second layer below the first layer, now seems evident: after the (first) graphene layer has been formed on the metal surface, the nucleation of a second layer can be avoided by cooling down the sample very rapidly. Also by choosing a metal with a low solubility for carbon, for example, Cu, nucleation of the second layer can be further suppressed.

The physical picture we have shown here explains also why the reported recipes are sometimes not very reproducible. For example, the carbon adatom density is a function of the bulk carbon concentration. In a non-equilibrium case, it also depends on the time during which the sample was held at elevated temperature, and the rate of increase or decrease of the temperature. These parameters vary from case to case and are often not specified.

The central message of present work is that by following an appropriate reaction path through the diagram in Figure1 one can control the formation of the three carbon phases: graphene, carbide, and dissolved carbon. Recipes for producing high-quality graphene should not only consider temperature and pressure, but also the initial situation, the rate of temperature increase/decrease, and the bulk carbon concentration.

Acknowledgements

This work is part of the research program of the Foundation for Fundamental Research on Matter (FOM), which is part of the Netherlands Organization for Scientific Research (NWO).

Author details

Guocai Dong[1,2], Dirk W. van Baarle[1] and Joost W. M. Frenken[1]

1 Kamerlingh Onnes Laboratory, Leiden University, Leiden, The Netherlands

2 JiangNan Graphene Research Institute, Changzhou, China

References

[1] Wallace, P. R. *The Band Theory of Graphite.* Physical Review, (1947)., 622.

[2] Geim, A. K, & Novoselov, K. S. *The rise of graphene.* Nature Materials, (2007)., 183-191.

[3] Novoselov, K. S, et al. *Room-Temperature Quantum Hall Effect in Graphene,* (2007)., 1379.

[4] Castro NetoA.H., et al., *The electronic properties of graphene.* Reviews of Modern Physics, (2009)., 109-162.

[5] Du, X, et al. *Fractional quantum Hall effect and insulating phase of Dirac electrons in graphene.* Nature, (2009)., 192-195.

[6] Westervelt, R. M. *Applied physics- Graphene nanoelectronics.* Science, (2008)., 324-325.

[7] Britnell, L, et al. *Field-Effect Tunneling Transistor Based on Vertical Graphene Heterostructures.* Science, (2012)., 947-950.

[8] Lin, Y, et al. *GHz Transistors from Wafer-Scale Epitaxial Graphene.* Science, (2010)., 662.

[9] Loginova, E, et al. *Factors influencing graphene growth on metal surfaces.* New Journal of Physics, (2009)., 063046.

[10] Martoccia, D, et al. *Graphene on Ru(0001): A 25x25 supercell.* Physical Review Letters, (2008)., 126102.

[11] Oshima, C, et al. *A heteroepitaxial multi-atomic-layer system of graphene and h-BN.* Surface Review and Letters, (2000)., 521-525.

[12] Mccarty, K. F, et al. *Kinetics and thermodynamics of carbon segregation and graphene growth on Ru(0001).* Carbon, (2009)., 1806-1813.

[13] Coraux, J, et al. *Structural coherency of graphene on Ir(111).* Nano Letters, (2008)., 565-570.

[14] Oshima, C, & Nagashima, A. *Ultra-thin epitaxial films of graphite and hexagonal boron nitride on solid surfaces.* Journal of Physics-Condensed Matter, (1997)., 1-20.

[15] Kim, K. S, et al. *Large-scale pattern growth of graphene films for stretchable transparent electrodes.* Nature, (2009)., 706-710.

[16] Sutter, P. W, Flege, J. I, & Sutter, E. A. *Epitaxial graphene on ruthenium.* Nature Materials, (2008)., 406-411.

[17] Wintterlin, J, & Bocquet, M. L. *Graphene on metal surfaces.* Surface Science, (2009)., 1841-1852.

[18] Novoselov, K. S, et al. *Electric Field Effect in Atomically Thin Carbon Films.* Science, (2004). , 666-669.

[19] Lui, C. H, et al. *Ultraflat graphene.* Nature, (2009). , 339-341.

[20] Emtsev, K. V, et al. *Towards wafer-size graphene layers by atmospheric pressure graphitization of silicon carbide.* Nature Materials, (2009). , 203-207.

[21] Hass, J, et al. *Highly ordered graphene for two dimensional electronics*, in *Appl. Phys. Lett*(2006). AIP. , 143106.

[22] Berger, C, et al. *Electronic Confinement and Coherence in Patterned Epitaxial Graphene.* Science, (2006). , 1191-1196.

[23] Rut'kovE.V. and N.R. Gall, *Role of edge atoms of graphene islands on metals in nucleation, growth, alkali metal intercalation.* Physics of the Solid State, (2009). , 1738-1743.

[24] Rut'kovE.V., A.V. Kuz'michev, and N.R. Gall, *Carbon interaction with rhodium surface: Adsorption, dissolution, segregation, growth of graphene layers.* Physics of the Solid State, (2011). , 1092-1098.

[25] Dai, B, et al. *Rational design of a binary metal alloy for chemical vapour deposition growth of uniform single-layer graphene.* Nat Commun, (2011). , 522.

[26] Gao, L, et al. *Repeated growth and bubbling transfer of graphene with millimetre-size single-crystal grains using platinum.* Nat Commun, (2012). , 699.

[27] Liu, N, et al. *Universal Segregation Growth Approach to Wafer-Size Graphene from Non-Noble Metals.* Nano Letters, (2010). , 297-303.

[28] Pan, Y, et al. *Highly Ordered, Millimeter-Scale, Continuous, Single-Crystalline Graphene Monolayer Formed on Ru (0001) (21pg 2777, 2009).* Advanced Materials, (2009). , 2739-2739.

[29] Castner, D. G, Sexton, B. A, & Somorjai, G. A. *Leed and Thermal Desorption Studies of Small Molecules (H2,O2,Co,Co2,C2h4C2h2 and C) Chemisorbed on Rhodium (111) and (100) Surfaces.* Surface Science, (1978). , 519-540.

[30] Isett, L. C, & Blakely, J. M. *Segregation Isosteres for Carbon at (100) Surface of Nickel.* Surface Science, (1976). , 397-414.

[31] Shelton, J. C, Patil, H. R, & Blakely, J. M. *Equilibrium Segregation of Carbon to a Nickel (111) Surface- Surface Phase-Transition.* Surface Science, (1974). , 493-520.

[32] Oshima, C, et al. *A hetero-epitaxial-double-atomic-layer system of monolayer graphene/monolayer h-BN on Ni(111).* Solid State Communications, (2000). , 37-40.

[33] Oshima, C, et al. *Hetero-epitaxial double-atomic-layer system of monolayer graphene monolayer h-BN on Ni(111) studied by HREELS.* Microbeam Analysis 2000, Proceedings, (2000). , 313-314.

[34] Yu, Q. K, et al. *Graphene segregated on Ni surfaces and transferred to insulators.* Applied Physics Letters, (2008)., 113103.

[35] Goriachko, A, & Over, H. *Modern Nanotemplates Based on Graphene and Single Layer h-BN.* Zeitschrift Fur Physikalische Chemie-International Journal of Research in Physical Chemistry & Chemical Physics, (2009)., 157-168.

[36] Bertoni, G, et al. *First-principles calculation of the electronic structure and EELS spectra at the graphene/Ni(111) interface.* Physical Review B, (2005)., 075402.

[37] Karpan, V. M, et al. *Theoretical prediction of perfect spin filtering at interfaces between close-packed surfaces of Ni or Co and graphite or graphene.* Physical Review B, (2008)., 195419.

[38] Sutter, P. *EPITAXIAL GRAPHENE How silicon leaves the scene.* Nature Materials, (2009)., 171-172.

[39] Reina, A, et al. *Large Area, Few-Layer Graphene Films on Arbitrary Substrates by Chemical Vapor Deposition.* Nano Letters, (2009)., 30-35.

[40] Li, X, et al. *Large-Area Graphene Single Crystals Grown by Low-Pressure Chemical Vapor Deposition of Methane on Copper.* Journal of the American Chemical Society, (2011)., 2816-2819.

[41] Liu, W, et al. *Synthesis of high-quality monolayer and bilayer graphene on copper using chemical vapor deposition.* Carbon, (2011)., 4122-4130.

[42] Li, X. S, et al. *Large-Area Synthesis of High-Quality and Uniform Graphene Films on Copper Foils.* Science, (2009)., 1312-1314.

[43] Bae, S, et al. *Roll-to-roll production of 30-inch graphene films for transparent electrodes.* Nat Nano, (2010)., 574-578.

[44] Yu, Q, et al. *Control and characterization of individual grains and grain boundaries in graphene grown by chemical vapour deposition.* Nature Materials, (2011)., 443-449.

[45] Gunther, S, et al. *Single Terrace Growth of Graphene on a Metal Surface.* Nano Letters, (2011)., 1895-1900.

[46] Dong, G. C, et al. *Graphene formation on metal surfaces investigated by in-situ scanning tunneling microscopy.* New Journal of Physics, (2012)., 053033.

[47] Loginova, E, et al. *Evidence for graphene growth by C cluster attachment.* New Journal of Physics, (2008)., 093026.

[48] Hoogeman, M. S, et al. *Design and performance of a programmable-temperature scanning tunneling microscope.* Review of Scientific Instruments, (1998)., 2072-2080.

[49] Rost, M. J, et al. *Scanning probe microscopes go video rate and beyond.* Review of Scientific Instruments, (2005)., 053710.

[50] Chen, J. G, et al. *Controlling surface reactivities of transition metals by carbide formation.* Journal of Molecular Catalysis a-Chemical, (1998)., 285-299.

[51] Delouise, L. A, & Winograd, N. *Carbon-Monoxide Adsorption and Desorption on Rh(111) and Rh(331) Surfaces.* Surface Science, (1984)., 417-431.

[52] Dubois, L. H, Castner, D. G, & Somorjai, G. A. *The Chemisorption of Acetylene and Ethylene on Rh(111)- a Low-Energy Electron-Diffraction (Leed), High-Resolution Electron-Energy Loss (Els), and Thermal-Desorption Mass-Spectrometry (Tds) Study.* Journal of Chemical Physics, (1980)., 5234-5240.

[53] Ibach, H. *Physics of Surfaces and Interfaces.* (2006). Jülich: Springer-Verlag Berlin Heidelberg. 525., 10

[54] Wang, B, et al. *Coupling Epitaxy, Chemical Bonding, and Work Function at the Local Scale in Transition Metal-Supported Graphene.* ACS Nano, (2010)., 5773-5782.

[55] Lahiri, J, et al. *Graphene Growth on Ni(111) by Transformation of a Surface Carbide.* Nano Letters, (2011)., 518-522.

[56] Sinharoy, S, & Levenson, L. L. *The formation and decomposition of nickel carbide in evaporated nickel films on graphite.* Thin Solid Films, (1978)., 31-36.

[57] Röhrl, J, et al. *Raman spectra of epitaxial graphene on SiC(0001).* Applied Physics Letters, (2008)., 201918.

[58] Zakharchenko, K. V, Katsnelson, M. I, & Fasolino, A. *Finite Temperature Lattice Properties of Graphene beyond the Quasiharmonic Approximation.* Physical Review Letters, (2009)., 046808.

[59] Venables, J. A. *Introduction to Surface and Thin Film Processes.* (2001). Cambridge: Cambridge University press.

[60] Van Gastel, R, et al. *Selecting a single orientation for millimeter sized graphene sheets.* Applied Physics Letters, (2009)., 121901.

[61] Starodub, E, et al. *Graphene growth by metal etching on Ru(0001).* Physical Review B, (2009)., 235422.

Chapter 3

Synthesis and Biomedical Applications of Graphene: Present and Future Trends

Ajay Kumar and Chee Huei Lee

Additional information is available at the end of the chapter

http://dx.doi.org/10.5772/55728

1. Introduction

In 1959 great physicist Richard P. Feynman through his statement: "There is a plenty of room at the bottom", predicted a new era of technology known as nanotechnology, where manipulation of the properties of a device is possible at atomic, molecular and macromolecular scales [1, 2]. Nanotechnology is basically a technology based on structures of sizes varying between 1 nm to 100 nm. Figure 1 illustrates the scale of nanomaterials with comparison with some biological systems. At nanoscale, materials properties are changed drastically due to quantum mechanical phenomena, which come into play at these sizes [3-5]. Nanomaterials exhibit higher surface to volume ratios which is inversely proportional to the size of the nanomaterials. There are also numerous properties which obey the same scaling law, e.g. melting and other phase transition temperatures. Atoms in these structures lie near to the surface, which are also known as higher energy sites. The behaviour of atoms at these higher energy sites has a significant influence on the properties of the material. Edge and corner atoms in the nanomaterials have lower coordination and bind foreign atoms and molecules more tightly [6].

Among other nanomaterials, graphene has become a rising star in condensed matter physics and material research. Graphene is defined as an atomic thick planar sheet of sp^2-hybridized carbon atoms that pack into a two-dimensional (2D) honeycomb lattice made out of hexagons, as presented in Figure 2a. Because of its novel and unique properties, tremendous fundamental and technological studies have been stimulated. Graphene has a unique band structure in which the conduction band and valence band just touch each other, forming an exactly zero-band gap semiconductor (also known as semi-metal).The charge carriers behave as Dirac fermions, which has a zero effective mass at the K and K′ point (named as Dirac points) at the corners of its Brillouin zone [8]. This gives rise to many extraordinary phenomenon, for examples ultrahigh carrier mobility (theoretical prediction up to 200,000$cm^2V^{-1}s^{-1}$) [9-11], half-

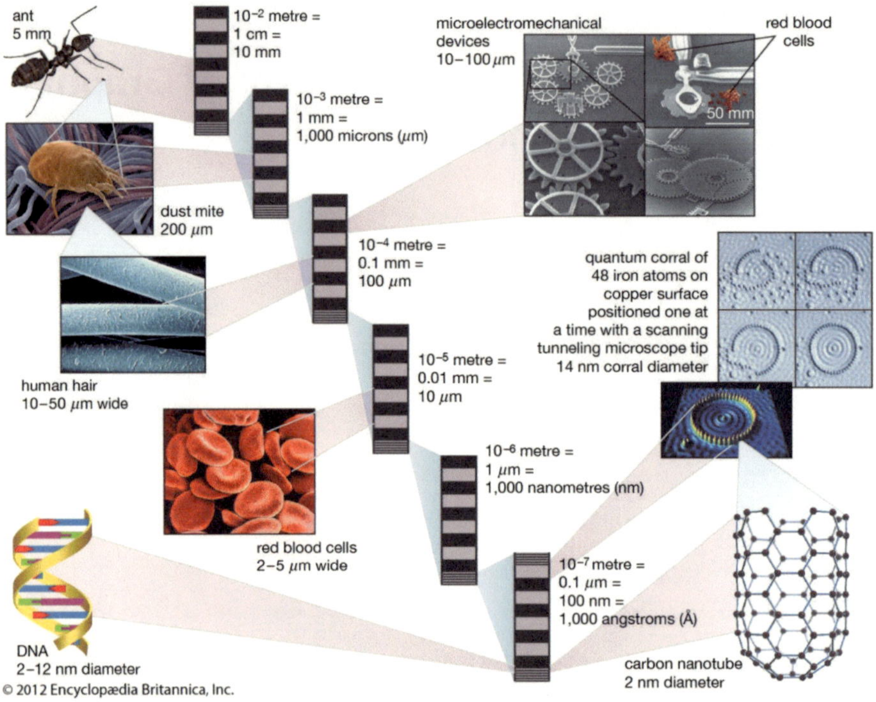

Figure 1. Scale of nanomaterials [7].

integer quantum Hall effect [12, 13], absorption of 2.3% of visible light [14], a high thermal conductivity [15], the correspondent Young's modulus and intrinsic strength of 1.0 TPa and 130 GPa respectively (the breaking strength is about 200 times larger than that of steel) [16]. Furthermore, graphene can be seen on a "magic substrate" (~300nm SiO_2/Si) with a standard optical microscope [17]. These interesting properties have been shown to have huge potential applications in many areas, such as graphene electronic transistors [18, 19], integrated circuits [20], transparent and flexible electronics [21, 22] as well as supercapacitors [23]. In a recent study, detection of the adsorption of a single gas molecule was done using graphene sensor, as graphene is highly sensitive to any change in the electrical resistance attributed to the local changes in carrier concentration. The importance of these structures is not only due to their future technological applications, but also because they may define a stage in the formation of other nanostructures, such as nanotubes [24]. Some of the very promising biomedical applications of these materials can be such as joint replacement prostheses, vascular grafts, and intraocular lenses/other ophthalmological devices.

2. Overview of graphene synthesis

Atomic layers of graphene were first obtained in 2004 [25]. Andre K. Geim and Konstantin S. Novoselov used a regular Scotch tape to extract thin layers of graphite from highly oriented

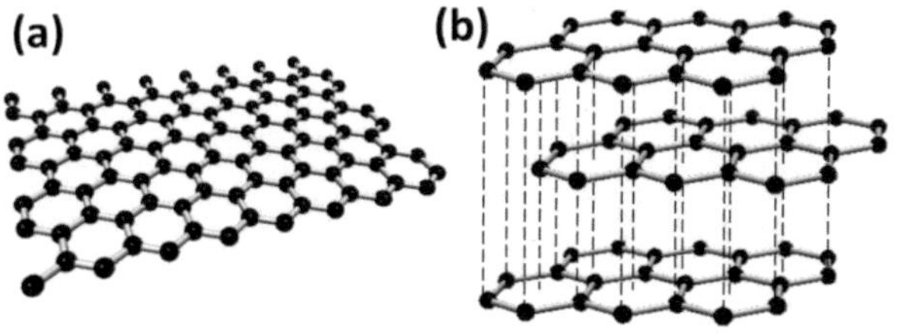

Figure 2. a) Graphene is a single layer honeycomb lattice of carbon atoms. b) Graphite viewed as a stack of graphene layers.

pyrolitic graphite (HOPG) and then transferred these layers to a silicon substrate. Since then, this technique is referred as "mechanical exfoliation", which so far provides the best quality graphene in terms of structural integrity. Many fascinating physics have been observed based on this type of sample preparation [12, 26, 27]. Both scientists have been awarded a Nobel Prize in Physics 2010, for producing, isolating, identifying and characterizing graphene [28]. However, this technique is only limited for scientific interest, as the size, thickness and location are mainly uncontrollable, in which large scale production and applications are hindered.

Another graphene synthesis method involves solution based exfoliation of graphene oxide (GO) [29-31]. Particularly for large scale applications, such as supercapacitors, composite materials, gas sensors [32, 33] and flexible electronic materials, this solution based method becomes very promising. In general, simple graphite powders are used as a starting material. They are oxidized by chemical modification, the so-called Hummers' method to produce water dispersible GO. GO can be easily exfoliated in water by the addition of mechanical energy (such as sonication) because of the interaction of water with the oxygen-containing (epoxide and hydroxyl) functional groups introduced during the oxidation process. GO can then be substantially reduced or restored to graphene network by thermal annealing or chemical reducing agents treatment. This "chemically derived graphene" is usually referred as reduced graphene oxide (rGO). The synthesis process can easily be scaled-up to produce gram quantities and disperse rGO in solution. However, the graphite materials undergo serious alteration during the process of oxidation and reduction. It has been recognized that solution based production of graphene contributes significant structural defects and uncompleted reduction process, leaving oxygen functional groups on graphene flakes. Nevertheless, rGO can be still useful for various applications as mentioned above.

Another relatively "straight forward" way to produce graphene is a conversion of SiC to graphene via sublimation of silicon atoms at high temperature (usually at ~1300°C in ultrahigh vacuum condition) [34, 35]. It has been shown that graphene via this method exhibits high mobility and remarkable 2D electron gas (2DEG) behavior. This method has the potential for large scale integration of nanoelectronic devices. However, the requirement of ultrahigh vacuum and high temperature may limit the accessibility of this method, due to a higher cost. Further review on epitaxial graphene growth using SiC substrates can be found in [36].

In the last few years, a lot of research interest has also been paid to the multilayer graphene nanoflakes films (MGNFs), also known as carbon nanowalls, nanoflakes, nanosheets or petals.

These are few nm thick graphene nanosheets, terminated with vertically aligned ultrathin graphene edges having lateral dimension in the micron range. They can be grown without the need of a catalyst. These structures exhibit a large amount of open graphitic edge planes responsible for their higher surface activity. Due to large surface area, sharp edges and vertical orientation, MGNFs hold promise as templates for various nanostructure materials e.g. magnetic, metal and oxides nanoparticles having applications as batteries, data storage media, and field emission devices. The freestanding (non-surface bound) nature and absence of catalyst made MGNFs very attractive as they also possess the combined advantages of large surface area, highly electrochemical activity and stable mechanical strength. These nanostructures have demonstrated excellent electron transfer properties, highly electrocatalytic activity and good selectivity for a number of biomolecules [37-40].

At a glance, the current graphene synthesis methods are summarized in Table 1, including the CVD method, which will be discussed in Section 3.

Synthesis Method	Brief Description/Remarks
Mechanical Exfoliation	• Using a regular Scotch tape to peel off graphene from HOPG. • Atomic layer of graphene can be seen on ~300 nm SiO_2 substrates under an optical microscope. • Pristine graphene with highest quality of electrical properties. • The size, thickness and location are uncontrollable, with limited practical applications.
Solution based exfoliation of graphene oxide (GO)	• Graphite powders are initially oxidized by chemical modification (Hummers' method) to be dispersed in solution. • GO are subsequently reduced to graphene by thermal annealing or chemical reducing agents. • Large scale production for bulk applications, such as supercapacitors, composite materials, etc. • Significant structural defects and leaving oxygen functional groups on the product.
Epitaxial growth using SiC substrates	• A conversion of SiC substrate to graphene via sublimation of silicon atoms on the surface. • Done at high temperature (~1300°C) and ultrahigh vacuum condition. • Limited accessibility due to high-end equipment.
CVD growth Graphene	• Most promising, inexpensive and feasible method for single layer or multi-layers graphene production. • Using transition metal (Ni, Cu, etc.) substrates or thin films as catalyst. • Flowing carbon source (CH_4) and reactant gases (H_2) at high temperature (~1000°C) for the nucleation of graphene. • Single layer graphene can usually be obtained on Cu. • Can be scaled up for large area graphene production for practical applications, such as transparent electrode applications.

Table 1. A Brief Summary of Graphene Synthesis Methods

3. Chemical Vapor Deposition (CVD) synthesis

Among these approaches, chemical vapor deposition (CVD) using transition metal substrates has been considered the most promising, inexpensive and feasible method to produce single layer or multi-layers graphene. Graphene grown on Ni [21], Pd [41], Ru [42], Ir [43] and Cu [44] have been demonstrated in the past few years. CVD using Cu is one of the fastest developing processes to produce single layer graphene due to the low solubility of carbon in Cu which leads to a self-limited process [44]. It has been shown that large area growth and excellent device properties can be achieved by this method [22]. Figure 3 demonstrates a schematic of an experimental setup of CVD which is commonly employed to produce single layer graphene by Cu or Ni catalysts. It basically consists of a tube furnace for high temperature heating, a quartz vacuum chamber, a vacuum and pressure control system for the growth condition adjustment, and several mass flow controllers (MFC) to provide carbon source and reactant gases with a necessary flow rate.

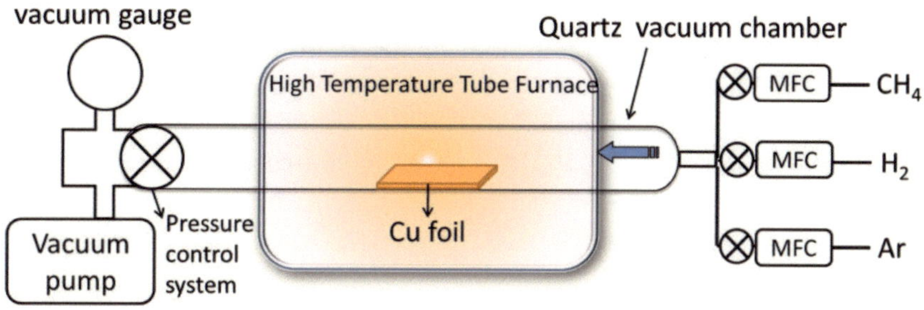

Figure 3. Schematic of a common setup for chemical vapor deposition of graphene.

3.1. Nickel as a catalyst

Learning from huge amount of studies on carbon nanotubes, different research groups have demonstrated using Ni layer as catalysts for large scale graphene synthesis [21, 45-47]. At a glance, ultrathin graphene films (1 to ~10 layers) grown on evaporated polycrystalline Ni surface have been performed in an atmospheric pressure or low pressure CVD. In brief, Ni films with 100-500 nm is first deposited by electron beam (e-beam) evaporation or sputtering on a SiO_2/Si. It was then annealed to form polycrystalline Ni grains with atomically flat surfaces and sizes of 1-20 μm. The CVD growth is carried out at 900 °C or 1000 °C under a highly diluted flow of methane (CH_4) in H_2 under ambient pressure for 5 to 10 minutes. The graphene on Ni can then be transferred to different substrates, as will be discussed in the transfer mechanism section. The geometry of graphene film can be patterned by the standard lithographic process after the transfer process. Alternatively, the Ni catalytic layer can be first pre-patterned to produce the same graphene geometry at desired locations.

It is critical to control the synthetic graphene by its parameters, for examples temperature, gas composition, gas flow rate, deposition time as well as heating and cooling rate. It has been found that using diluted CH_4 is important to produce single and few-layer graphene. Using concentrated methane will lead to multilayers of graphene (> 5 layers) [47]. Cooling rate also significantly affects the thickness and the amount of defects of graphene on Ni. It has been reported that a fast-cooling process can suppress the amount of precipitated C, leading to single of few layers graphene. However, as more freedom participates in the deposition process, the control of deposition is of more difficulties. It is suggested that a moderate cooling rate of 10 °C/s is the optimum condition for thin layer graphene growth. Generally, CVD by Ni catalysts still yields films with a wide range of graphene layer thicknesses, from one to a few tens of layers and with defects associated with fast cooling.

3.2. Copper as a catalyst

CVD graphene using Cu as a transition metal substrate for single layer graphene synthesis was first demonstrated by Rouff's group in 2009 [44]. In fact, it was a relatively new technique compared to mechanical exfoliation and the desorption of Si from single crystal SiC. Since then, there are a lot of research activities using Cu to grow single layer graphene and transfer graphene onto an insulating substrate for device fabrication and testing. In principle, graphene on Cu is grown by the decomposition of methane gas in dilute hydrogen environment over the surface at 1000°C. The thickness of the Cu substrate is usually 25-50 μm. For a typical growth process, the Cu foil is first annealed at 1000 °C in dilute H_2 environment for 30 minutes. Then, a small flow rate of methane (CH_4) as the carbon source gas is introduced for about 30 minutes.

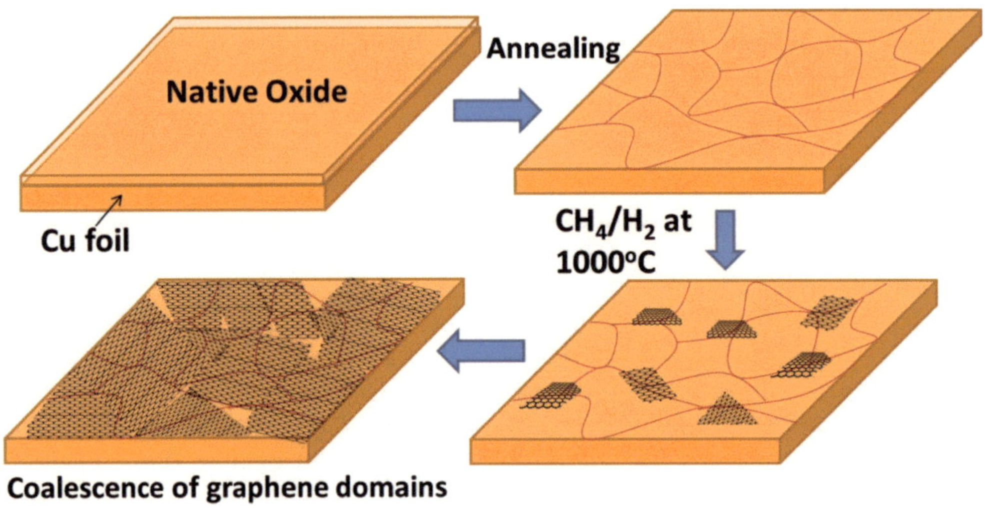

Figure 4. Schematic illustrating the proposed growth mechanism of graphene on Cu substrates by CVD: (a) copper foil with native oxide; (b) Native Cu oxide is reduced while Cu develops grains on the surface after annealing at high temperature in H_2 environment; c) The exposure of the Cu foil to CH_4/H_2 atmosphere at 1000 °C leading to the nucleation of graphene islands; (d) enlargement of the graphene flakes and coalescence of graphene domains with different lattice orientation [48].

After exposure to CH_4, the furnace is cooled to room temperature. Because of it self-limited surface process [44], the graphene growth by Cu is robust for different growth conditions, compared to the growth by Ni. The cooling rate does not result in discernible differences in the number of layers growth.

Figure 4 illustrates the proposed growth mechanism of graphene on Cu. The annealing at high temperature in H_2 environment is to remove the native oxide layer on the Cu surface, while Cu grains will also develop. With the exposure of Cu foil in CH_4/H_2 environment, nucleation of graphene islands start taking place randomly but preferentially at the grain boundary of Cu surface. As the exposure to CH_4 continues, the graphene domains grow in size to cover the whole area of Cu substrates and eventually aggregate into a continuous graphene film.

It is noted that these initial graphene domains could have different lattice orientations depending on the crystallographic orientation of the underlying Cu grains. Therefore, it is not surprising that the large area graphene is indeed formed by multiple domains of tens of μm^2 [49]. Graphene may grow across metal steps and grain boundaries [43, 45]. This may also explain why graphene can grow laterally even on a rough Cu foil. The domains usually do not align in perfect lattice structure, therefore give rise to the inter-domain defects [50]. Despite these defects, large area graphene in industrial scale has been demonstrate for transparent and flexible electrodes to fabricate a touch screen display [22]. Many efforts have been carried out to produce CVD-graphene which has comparable quality with those produced by mechanical exfoliation. It is possible to increase the domain size by understanding the growth parameters and modifying the process control [50]. The effect of different facets of Cu has been studied extensively [51]. Generally, there is a consensus that Cu(111) facet facilitates monolayer and uniform graphene growth, due to the close lattice matching between hexagonal graphene (lattice constant of graphene, a = 2.46Å) while hexagonal Cu(111)'s, a = 2.56Å) [51, 52]. Furthermore, it is also important to understand the growth mechanism of graphene using metal catalysts under different growth conditions. It has been shown that Cu-graphene can be grown under atmospheric pressure [53] or low pressure [44, 54]. However, the kinetics of graphene growth is very much different under different conditions, giving rise to the uniformity issue of large area graphene.

3.3. Transfer mechanism

Graphene grown on transition metals must be transferred onto insulating substrates for device fabrication and electronic characterization [55-57]. Different methods have been demonstrated to transfer the as-grown graphene on metal substrates onto desired insulating substrates, such as polymeric foils (polyethylene terephthalate (PET)), glass and SiO_2/Si. Up to this date, the relatively straightforward way is to chemically etch the metal away to obtain a free standing graphene membrane. This membrane can then be scooped on a desired substrate. A general transfer technique of CVD-graphene onto a target substrate is illustrated in Figure 5. Typically, the transfer is first conducted by spin-coating a thin polymeric layer, such as poly(methylmethacrylate) (PMMA) or Polydimethylsiloxane (PDMS) on top of the as-grown graphene. This polymer provides a supportive framework for graphene before the transfer. The underneath Cu substrate is then etch away by iron chloride ($FeCl_3$) solution. Other etching agents that can be employed include HCl, HNO_3, $Fe(NO_3)_3$, $(NH_4)_2SO_8$, and $CuCl_2$. After the Cu is completely dissolved, the floating membrane can be scooped and placed on a desired substrate. After drying, the polymeric film is dissolved with acetone or chloroform. Another successful transfer of graphene has also been performed by the roll-to-roll process using a thermal release tape as the supportive layer [22]. In brief, a thermal release tape is first attached to the graphene film grown on copper, followed by the Cu etching and rising with deionized water. This assembly together with a target substrate is inserted into a roller and exposed to mild heat to release the graphene from the tape to the target substrate. A large area touch screen is further demonstrated with the transferred graphene on a flexible PET substrate using this transfer technique [22].

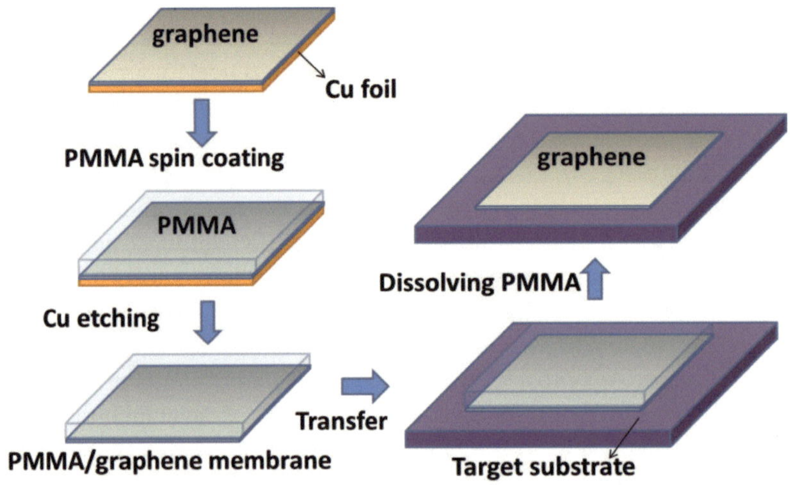

Figure 5. A cartoon showing the transfer process of graphene grown on Cu foil onto a target substrate.

Apparently, all these transfer methods of graphene from metal substrates to insulating substrates introduce certain degrees of structural defects, such as crack, wrinkles and ripples [21]. In addition, it is hard to completely remove the polymer layer, leaving certain amount of impurities on graphene. In this aspect, it is of great importance to perverse the quality of graphene during the transfer process for device performance, yield and uniformity. Additional processes such as PMMA heating, substrate treatment, high vacuum annealing, chloroform cleaning and so on, have also been introduced in order to achieved crackles and clean graphene with its intrinsic properties [58-61].

4. Biomedical applications of graphene

As mentioned in the Introduction section, graphene rises up as a promising material for many potential applications, such as graphene electronic transistors [18, 19], integrated circuits [20], transparent and flexible electronics [21, 22], composite materials [29], supercapacitors [23], as well as gas sensors [32, 33].

In addition, Dai et al was the first to study the biomedical applications of graphene by using graphene oxide as drug carrier to kill the cancer cells selectively [62]. After this study, a lot of research interest has been created to use graphene in many different applications such as the exploration of graphene-based nanomaterials for tissue engineering, molecular imaging, and drug/gene delivery applications. There are many biological interactions such as cellular uptake and cytotoxicity largely dependent on the size of nanomaterials. In addition to this, shape and surface of these nanomaterials also contribute towards their specific biomedical applications. In this section, we will discuss first the biosensing applications of graphene and in the next part graphene scaffolds will be discussed.

4.1. Biosensing

A biosensor is an analytical device that helps to gain understanding of the bio-composition, structure and function of target analytes by biological reactions. As shown in Figure 6, a biosensor mainly consists of a rocognition layer, a transducer as well as electronic components. The recognition layer determines the biological response which is further converted into an electrical signal with the help of the transducer. This electrical signal is then amplified and processed by the external electronic system. Biosensors are very useful tools in many applications such as medical care, environmental field monitoring etc.

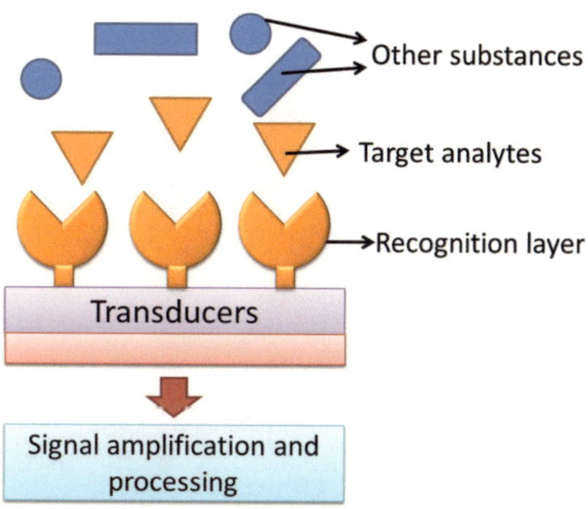

Figure 6. Schematic diagram of a biosensor

Electrochemical activity of nucleic acids was first reported by Palećek [63]. Since the discovery of electroactive behaviour of nucleic acids, a lot of research has been devoted to their quantification via electrochemistry. The aim is to use the nucleic acid recognition layers immobilized over a signal transducer (electrochemical, optical or piezoelectric), to fabricate a DNA biosensor for detecting target analytes.

Due to their interesting properties such as small size and excellent electrical conductivity, carbon based nanomaterials are very useful to manufacture the transducer electrode for electrochemical sensing applications. Britto et al. first reported the use of nanotubes for dopamine oxidation [64]. DNA electrochemical biosensing has become an attractive method due to its ease of miniaturization, low cost and direct readout of the electronic signals. Deoxyribonucleic acid (DNA) has contributed to the design of therapeutics and to many other applications via genetic engineering. Major studies on DNA have focused on the sequence-specific recognition and mutation of the single stranded DNA (ss-DNA) by various techniques [65-68]. However, DNA is rarely present in the single-stranded form in nature. The detection of double stranded DNA (ds-DNA) is important since it will allow the direct visualization of the genomic information in living cells and the development of cell based technology [69]. The electrochemical DNA biosensors can be distinguished into label free (based on intrinsic electroactivity of nucleobases) and labelled (where redox active species is used with ds-DNA) ones [70].

In case of the electrochemical label-free DNA sensors, sensing depends on the intrinsic electrochemical properties of the nucleic acid target. As guanine is the best redox active base, its oxidation signal at about +1 V (vs Ag/AgCl) is usually used to detect the hybridization process. Using this approach a detection limit below 100nM was achieved for a 19 base sequence containing 4 guanine residues [70]. Electrochemical impedance spectroscopy is another aspect where label free approaches has been used by recording any change in the faradaic impedance of a ss-DNA modified electrode to its hybridised version. Ferricyanide was used as a redox species.

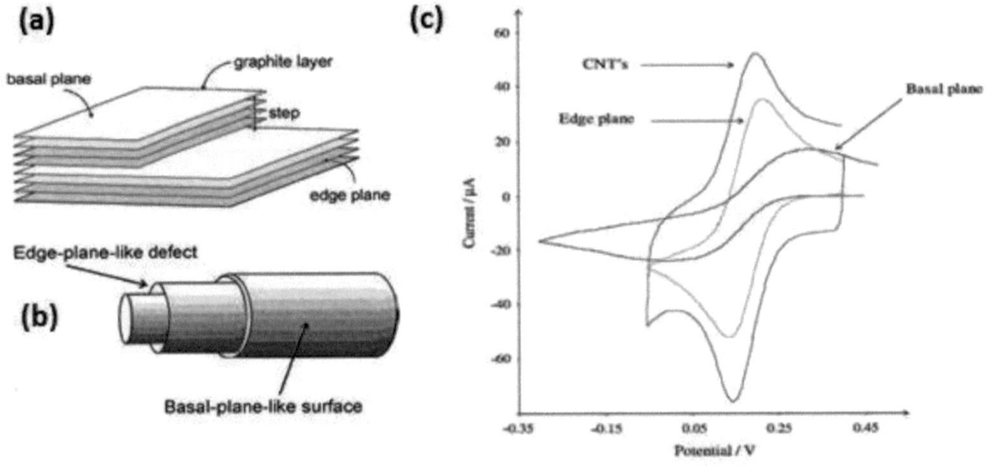

Figure 7. Schematic representation of (a) highly ordered pyrolytic graphite crystal, (b) multiwalled carbon nanotubes and (c) A comparison of electrochemical behaviour of nanotubes-modified electrode, edge- and basal plane pyrolytic graphite electrode. Reprinted with permission from Ref. [73]. Copyright 2006, John Wiley & Sons, Inc.

The higher charge transfer resistance was attributed to a repulsive force on ferricynide by the DNA duplex due to its negative charge. Thus label-free DNA sensing approach is a promising way to avoid any complications for the device fabrication. It is also a cost effective and more accurate method based on intrinsic electroactivity of nucleobases [71, 72].

http://dx.doi.org/10.5772/55728

Labelled approached DNA sensors are based on the use of redox indicators to make a distinction between the signal and ds-DNA on the electrode [71]. Millan et al. used Co(phen)3^{3+} as a redox indicator to probe the ssDNA hybridization with the glassy carbon electrode and observed a much better response in comparison to the ss-DNA on glassy carbon electrode. In another study, using the same indicator with carbon paste, Mycobacterium tuberculosis bacteria with detection limit as low as 3.4 nM was reported [72, 74]. Ferrocenyl-napthalene diimide, nanoparticles and methylene blue are other indicators being used for the labelled approaches to achieve lower detection limits [75, 76]. An enzyme indicator with magnetic separator was also used to avoid the selectivity problems [77]. This approach is more complicated and costly in comparison to the labelled approach.

Current challenges to the rapid development of low-cost, label free highly sensitive ds-DNA based electrochemical sensor technology are primarily high electrochemical oxidation potentials and low electron transfer rates associated with the deep entrapment of DNA bases in the stable rigid double helix structure, preventing immediate electrical contact between the nucleic base and the electrode surface. Adenine and guanine residues of DNA can electrochemically be detected with a poor signal and a low sensitivity at various electrodes such as gold, glassy carbon (GC), carbon nanotubes (CNTs), and polymer modified graphite/GC [78-83]. From a fundamental point of view, the main problem on evaluating the use of CNTs for electrochemical applications such as sensors and energy storage devices, is the presence of metallic impurities, as the synthesis of nanotubes is usually achieved with the use of metal catalyst. First issue with these impurities is that these are electrochemically active and can dominate the electrochemistry of CNTs. The second is the toxicological hazards as they can participate in the redox activities of biomolecules. The complete removal of these metallic impurities from the nanotubes is not achievable even after purification method such as acid treatment. Acid treatment is a time consuming process which needs to be optimized first otherwise nanotubes can be completely destroyed and also CNTs after acid treatment are full of defects which deteriorate their electrical performance. This drawback can be very critical to the observed electrochemical reaction. Thus, there is a need of a material possessing similar/better electrochemical response without any metallic impurities.

Currently, due to the different electrochemical activity and assays employed, the mechanism of the electrooxidation of ds-DNA based on different electrodes is not clear and the interaction of DNA with electrodes has a non consistent behaviour. For examples, Bollo et al [80], Pedano et al. [84] and Nowicka et al. [85] found that ds-DNA exhibited weak responses at polished GC electrodes, while two other research groups reported no electrochemical response of ds-DNA at bare GC electrodes [80, 86]. Obviously, this points to the fact that the oxidation dynamics of ds-DNA on such electrodes could relate with unknown factors such as the surface roughness, surface functional species, surface area, preferential facets, grain size, electroactive media, time of exposure to air etc, many of which are difficult to be controlled by simple mechanical polishing or electrochemical activation procedures. As for electrodes modified with substances capable of catalyzing the reaction of nucleic bases, their reproducibility is highly variable, due to the difficulty in controlling the quantity, thickness and active surface area of catalytic medias by a simple method such as the drop casting or dip coating etc.

4.2. Scaffolds for bioengineered organs

Tissue engineering is a very challenging area of research and highly desirable to improve the well-being of mankind. Many three-dimensional scaffolds for bioengineered organs have been studied for their suitability for tissue growth for example carbon nanomaterials (nanotubes, nano-diamond and fullerene) [87]. Biocompatibility is a very important issue for a material to be used in biomedical applications as it initiates the cell cultivation. In the commercially used metal implants, their elastic moduli are many times higher than tissues; this causes the bone to be insufficiently loaded and eventually can hamper the growth [88, 89].

Collagen sponge honeycomb have also been reported for cell culture applications but effective cell adhesion on their surface was not good due to bigger pore sizes [90]. To overcome this issue, Multiwall CNTs were coated on these, which lead to better cell proliferation [91]. CNTs are also found to be useful for improvement of the cell adhesion and differentiation [92]. Selective bone tumor therapy was also proposed by using fullerene as a drug carrier agent by inducing the photodynamic damage on biological systems [93].

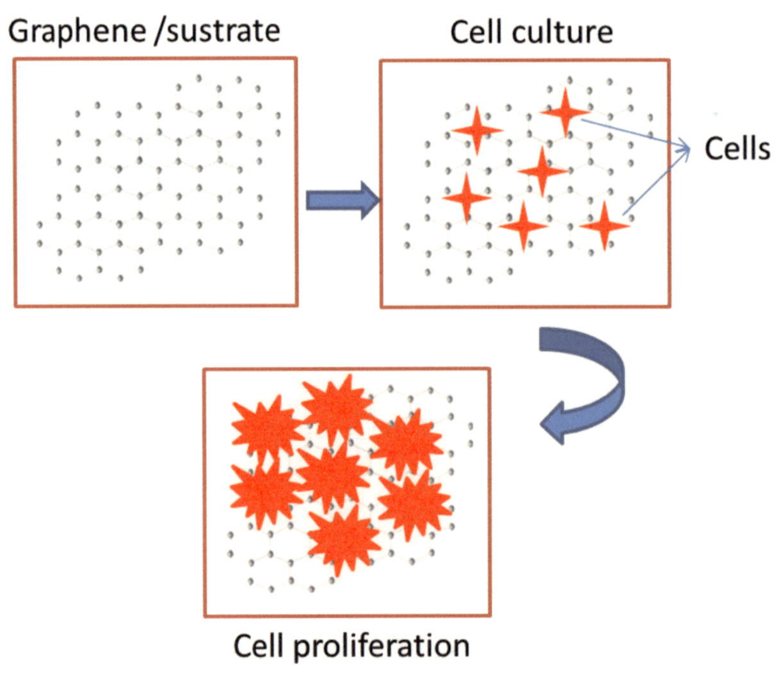

Figure 8. Schematic representation of Cell culture and proliferation on graphene platform.

Nanodiamonds are currently being used in many biomedical applications (such as drug delivery and surgical implants) due to its lack of cytotoxicity [94]. Surface area of a monolayer graphene is higher approximately by an order of magnitude in comparison to other biomaterials making it a very interesting biomaterial.

Graphene flat surface nature in addition to the many other mechanical properties can also be exploited for applications such as structural reinforcement of biocompatible films and scaffolds for bioengineered organs. Regarding their cell culture studies, an improvement of the osteogenic potential has been achieved using graphene-coated surfaces for the differentiation of human mesenchymal stem cells (hMSCs) and preosteoblasts into osteoblasts [95, 96]. Cell culture studies of graphene oxide sheets also show promise as the suitable implant material [97]. Although there have been studies on graphene for tissue engineering, there is a lack of studies for health impact issues as well as actual interaction mechanism of these materials with different cells. To address these, the studies should be more focused and different biomedical applications must be optimized.

5. Future trend and outlook

Chemical vapor deposition (CVD) using Cu as the catalyst substrate is undoubtedly a promising way to produce single layer graphene over a large area. By further understanding the growth mechaniscm and optimization of the growth condition, it can be foreseen that high quality of graphene can be routinely reproduced by this CVD technique. Note that this type of CVD process can be scalable easily and only limited by the size of the apparatus. Mass production is not an issue while over 30 inches diagonal size of graphene had been demonstrated. However, there are other obstacles that need to be overcome to produce as high quality as the pristine graphene from mechanical exfoliation. For examples, further improvement of the transfer process is still highly desired to minimize structural defects and impurities on graphene. On the other hand, direct deposition of graphene onto insulating substrates is still a challenge to the scientific community. There is defenitely an advantage of avoiding the transfer process that can be problematic as mentioned. Besides, the transfer process can be time consuming and not an environmental friendly process.

Up to this point, direct growth of graphene with definite number of layers is of difficulties. It will be a breakthrough if one can achieve specific layers of graphene growth for particular applications. Single layer graphene has no bandgap while double layers graphene can have a small bandgap. In addition, functionalization of graphene by physical and chemical method also provide the possibilities and challenges of tailoring the electronic and sensing properties of graphene for many applications. This manipulation includes sizes, geometries, band gaps, doping levels, functionalized chemical groups and so on [98]. There are also studies implementing plasma technology for producing and manipulating graphene [99-105]. Despite of current and future challenges, graphene research provides huge potential for material and functional applications and it is still progressively active around the world.

As discussed earlier, biomedical applications of nanomaterials can be further extended to as fluorescent markers in biodetection assays, chemotherapy and transplant materials. Immobilization, electro-oxidation and quantitative analysis of a monolayer of native DNA self-assembled on the vertically aligned graphene nanoflake films have been demonstrated by Pagona's group recently [39]. Graphene nanoflakes edge plans exhibits unique electronic structure and edge state properties. These features attribute to the exceptional electrocatalytic activity of these nanomaterials for ds-DNA detection. It can be envisioned that the use of MGNFs as nanoconnectors, which establish direct electrical communication between the graphene edge plane and the active site of DNA or other biomolecules, will create a new generation of graphene-based enabling biotechnology leading to the production of label-free DNA biodetection, biofuel cells and electrocatalytic devices. For the tissue enginering applications of graphene, a systematic optimization approach keeping in mind the envirnmental safety for different cells response is needed.

Acknowledgements

The authors acknowledge Prof. Mohan Sankaran from Chemical Engineering Department of CWRU and Prof. Xuan P. A. Gao from Physics Department of CWRU for their academic advices and financial support.

Author details

Ajay Kumar[1*] and Chee Huei Lee[2]

*Address all correspondence to: ajayepph@gmail.com, cheehuei80@gmail.com

1 Department of Chemical Engineering, Case Western Reserve University, Cleveland, Ohio, USA

2 Singapore University of Technology and Design, Singapore

References

[1] Feynman RP. There's Plenty of Room at the Bottom. Caltech's Engineering and Science Magazine 1960;23:5: 22.

[2] Drexler KE. Nanotechnology: From Feynman to Funding. Bulletin of Science, Technology & Society 2004;24(1): 21-7.

[3] Hodes G. When Small Is Different: Some Recent Advances in Concepts and Applications of Nanoscale Phenomena. Advanced Materials 2007;19(5): 639-55.

[4] Gleiter H. Nanoscience and Nanotechnology: The Key to New Studies in Areas of Science Outside of Nanoscience and Nanotechnology. MRS Bulletin 2009;34: 456-64.

[5] Holleitner AW. Nanotechnology: An Introduction to Technological and Scientific Aspects. TTN Forum 2006;5: 17-32.

[6] Roduner E. Size matters: why nanomaterials are different. Chemical Society Reviews 2006;35(7): 583-92.

[7] Carbon nanotube. [Art]: Encyclopædia Britannica Online; Available from: http://www.britannica.com/EBchecked/media/73082/Examples-from-biological-and-mechanical-realms-illustrate-various-orders-of (accessed 29 November 2012).

[8] Castro Neto AH, Guinea F, Peres NMR, Novoselov KS, Geim AK. The electronic properties of graphene. Reviews of Modern Physics 2009;81(1): 109-62.

[9] Chen J-H, Jang C, Xiao S, Ishigami M, Fuhrer MS. Intrinsic and extrinsic performance limits of graphene devices on SiO2. Nature Nanotechnology 2008;3(4): 206-9.

[10] Morozov SV, Novoselov KS, Katsnelson MI, Schedin F, Elias DC, Jaszczak JA, et al. Giant Intrinsic Carrier Mobilities in Graphene and Its Bilayer. Physical Review Letters 2008;100(1): 016602.

[11] Geim AK, Novoselov KS. The rise of graphene. Nature Materials 2007;6(3): 183-91.

[12] Zhang Y, Tan Y-W, Stormer HL, Kim P. Experimental observation of the quantum Hall effect and Berry's phase in graphene. Nature 2005;438(7065): 201-4.

[13] Du X, Skachko I, Duerr F, Luican A, Andrei EY. Fractional quantum Hall effect and insulating phase of Dirac electrons in graphene. Nature 2009;462(7270): 192-5.

[14] Nair RR, Blake P, Grigorenko AN, Novoselov KS, Booth TJ, Stauber T, et al. Fine Structure Constant Defines Visual Transparency of Graphene. Science 2008;320(5881): 1308.

[15] Balandin AA, Ghosh S, Bao W, Calizo I, Teweldebrhan D, Miao F, et al. Superior Thermal Conductivity of Single-Layer Graphene. Nano Letters 2008;8(3): 902-7.

[16] Lee C, Wei X, Kysar JW, Hone J. Measurement of the Elastic Properties and Intrinsic Strength of Monolayer Graphene. Science 2008;321(5887): 385-8.

[17] Blake P, Hill EW, Neto AHC, Novoselov KS, Jiang D, Yang R, et al. Making graphene visible. 2007;91(6): 063124.

[18] Kedzierski J, Pei-Lan H, Healey P, Wyatt PW, Keast CL, Sprinkle M, et al. Epitaxial Graphene Transistors on SiC Substrates. Electron Devices, IEEE Transactions on 2008;55(8): 2078-85.

[19] Schwierz F. Graphene transistors. Nature Nanotechnology 2010;5(7): 487-96.

[20] Park J-U, Nam S, Lee M-S, Lieber CM. Synthesis of monolithic graphene–graphite integrated electronics. Nature Materials 2012;11(2): 120-5.

[21] Kim KS, Zhao Y, Jang H, Lee SY, Kim JM, Kim KS, et al. Large-scale pattern growth of graphene films for stretchable transparent electrodes. Nature 2009;457(7230): 706-10.

[22] Bae S, Kim H, Lee Y, Xu X, Park J-S, Zheng Y, et al. Roll-to-roll production of 30-inch graphene films for transparent electrodes. Nature Nanotechnology 2010;5(8): 574-8.

[23] Stoller MD, Park S, Zhu Y, An J, Ruoff RS. Graphene-Based Ultracapacitors. Nano Letters 2008;8(10): 3498-502.

[24] Caetano EWS, Freire VN, dos Santos SG, Albuquerque EL, Galvão DS, Sato F. Defects in Graphene-Based Twisted Nanoribbons: Structural, Electronic, and Optical Properties. Langmuir 2009;25(8): 4751-9.

[25] Novoselov KS, Geim AK, Morozov SV, Jiang D, Zhang Y, Dubonos SV, et al. Electric Field Effect in Atomically Thin Carbon Films. Science 2004;306(5696): 666-9.

[26] Bolotin KI, Sikes KJ, Jiang Z, Klima M, Fudenberg G, Hone J, et al. Ultrahigh electron mobility in suspended graphene. Solid State Communications 2008;146(9-10): 351-5.

[27] Taychatanapat T, Watanabe K, Taniguchi T, Jarillo-Herrero P. Quantum Hall effect and Landau-level crossing of Dirac fermions in trilayer graphene. Nature Physics 2011;7(8): 621-5.

[28] "The Nobel Prize in Physics 2010 - Advanced Information". Nobelprize.org; Available from: http://www.nobelprize.org/nobel_prizes/physics/laureates/2010/advanced.html (accessed 12 November 2012).

[29] Stankovich S, Dikin DA, Dommett GHB, Kohlhaas KM, Zimney EJ, Stach EA, et al. Graphene-based composite materials. Nature 2006;442(7100): 282-6.

[30] Eda G, Fanchini G, Chhowalla M. Large-area ultrathin films of reduced graphene oxide as a transparent and flexible electronic material. Nature Nanotechnology 2008;3(5): 270-4.

[31] Hernandez Y, Nicolosi V, Lotya M, Blighe FM, Sun Z, De S, et al. High-yield production of graphene by liquid-phase exfoliation of graphite. Nature Nanotechnology 2008;3(9): 563-8.

[32] Fowler JD, Allen MJ, Tung VC, Yang Y, Kaner RB, Weiller BH. Practical Chemical Sensors from Chemically Derived Graphene. ACS Nano 2009;3(2): 301-6.

[33] Robinson JT, Perkins FK, Snow ES, Wei Z, Sheehan PE. Reduced Graphene Oxide Molecular Sensors. Nano Letters 2008;8(10): 3137-40.

[34] de Heer WA, Berger C, Wu X, First PN, Conrad EH, Li X, et al. Epitaxial graphene. Solid State Communications 2007;143(1–2): 92-100.

[35] Berger C, Song Z, Li X, Wu X, Brown N, Naud C, et al. Electronic Confinement and Coherence in Patterned Epitaxial Graphene. Science 2006;312(5777): 1191-6.

[36] Walt AdH, Claire B, Xiaosong W, Mike S, Yike H, Ming R, et al. Epitaxial graphene electronic structure and transport. Journal of Physics D: Applied Physics 2010;43(37): 374007.

[37] Ajay K, Abhijit G, Pagona P. Thermal stability study of nitrogen functionalities in a graphene network. Journal of Physics: Condensed Matter 2012;24(23): 235503.

[38] Chiou JW, Ray SC, Peng SI, Chuang CH, Wang BY, Tsai HM, et al. Nitrogen-Functionalized Graphene Nanoflakes (GNFs:N): Tunable Photoluminescence and Electronic Structures. The Journal of Physical Chemistry C 2012;116(30): 16251-8.

[39] Shang N, Kumar A, Sun N, Sharma S, Papakonstantinou P, Li M, et al. Vertical graphene nanoflakes for the immobilization, electrocatalytic oxidation and quantitative detection of DNA. Electrochemistry Communications 2012;25(0): 140-3.

[40] Ney A, Papakonstantinou P, Kumar A, Shang N-G, Peng N. Irradiation enhanced paramagnetism on graphene nanoflakes. Applied Physics Letters 2011;99(10): 102504.

[41] Kwon S-Y, Ciobanu CV, Petrova V, Shenoy VB, Bareño J, Gambin V, et al. Growth of Semiconducting Graphene on Palladium. Nano Letters 2009;9(12): 3985-90.

[42] Sutter PW, Flege J-I, Sutter EA. Epitaxial graphene on ruthenium. Nature Materials 2008;7(5): 406-11.

[43] Coraux J, N`Diaye AT, Busse C, Michely T. Structural Coherency of Graphene on Ir(111). Nano Letters 2008;8(2): 565-70.

[44] Li X, Cai W, An J, Kim S, Nah J, Yang D, et al. Large-Area Synthesis of High-Quality and Uniform Graphene Films on Copper Foils. Science 2009;324(5932): 1312-4.

[45] Reina A, Jia X, Ho J, Nezich D, Son H, Bulovic V, et al. Large Area, Few-Layer Graphene Films on Arbitrary Substrates by Chemical Vapor Deposition. Nano Letters 2008;9(1): 30-5.

[46] Yu Q, Lian J, Siriponglert S, Li H, Chen YP, Pei S-S. Graphene segregated on Ni surfaces and transferred to insulators. Applied Physics Letters 2008;93(11): 113103.

[47] De Arco LG, Yi Z, Kumar A, Chongwu Z. Synthesis, Transfer, and Devices of Single- and Few-Layer Graphene by Chemical Vapor Deposition. Nanotechnology, IEEE Transactions on 2009;8(2): 135-8.

[48] Mattevi C, Kim H, Chhowalla M. A review of chemical vapour deposition of graphene on copper. Journal of Materials Chemistry 2011;21(10): 3324-34.

[49] Li X, Cai W, Colombo L, Ruoff RS. Evolution of Graphene Growth on Ni and Cu by Carbon Isotope Labeling. Nano Letters 2009;9(12): 4268-72.

[50] Li X, Magnuson CW, Venugopal A, An J, Suk JW, Han B, et al. Graphene Films with Large Domain Size by a Two-Step Chemical Vapor Deposition Process. Nano Letters 2010;10(11): 4328-34.

[51] Wood JD, Schmucker SW, Lyons AS, Pop E, Lyding JW. Effects of Polycrystalline Cu Substrate on Graphene Growth by Chemical Vapor Deposition. Nano Letters 2011;11(11): 4547-54.

[52] Tao L, Lee J, Chou H, Holt M, Ruoff RS, Akinwande D. Synthesis of High Quality Monolayer Graphene at Reduced Temperature on Hydrogen-Enriched Evaporated Copper (111) Films. ACS Nano 2012.

[53] Gao L, Ren W, Zhao J, Ma L-P, Chen Z, Cheng H-M. Efficient growth of high-quality graphene films on Cu foils by ambient pressure chemical vapor deposition. Applied Physics Letters 2010;97(18): 183109.

[54] Bhaviripudi S, Jia X, Dresselhaus MS, Kong J. Role of Kinetic Factors in Chemical Vapor Deposition Synthesis of Uniform Large Area Graphene Using Copper Catalyst. Nano Letters 2010;10(10): 4128-33.

[55] Lee Y, Bae S, Jang H, Jang S, Zhu S-E, Sim SH, et al. Wafer-Scale Synthesis and Transfer of Graphene Films. Nano Letters 2010;10(2): 490-3.

[56] Chen JH, Ishigami M, Jang C, Hines DR, Fuhrer MS, Williams ED. Printed Graphene Circuits. Advanced Materials 2007;19(21): 3623-7.

[57] Li X, Zhu Y, Cai W, Borysiak M, Han B, Chen D, et al. Transfer of Large-Area Graphene Films for High-Performance Transparent Conductive Electrodes. Nano Letters 2009;9(12): 4359-63.

[58] Cheng Z, Zhou Q, Wang C, Li Q, Wang C, Fang Y. Toward Intrinsic Graphene Surfaces: A Systematic Study on Thermal Annealing and Wet-Chemical Treatment of SiO2-Supported Graphene Devices. Nano Letters 2011;11(2): 767-71.

[59] Liang X, Sperling BA, Calizo I, Cheng G, Hacker CA, Zhang Q, et al. Toward Clean and Crackless Transfer of Graphene. ACS Nano 2011;5(11): 9144-53.

[60] Suk JW, Kitt A, Magnuson CW, Hao Y, Ahmed S, An J, et al. Transfer of CVD-Grown Monolayer Graphene onto Arbitrary Substrates. ACS Nano 2011;5(9): 6916-24.

[61] Chan J, Venugopal A, Pirkle A, McDonnell S, Hinojos D, Magnuson CW, et al. Reducing Extrinsic Performance-Limiting Factors in Graphene Grown by Chemical Vapor Deposition. ACS Nano 2012.

[62] Liu Z, Robinson JT, Sun X, Dai H. PEGylated Nanographene Oxide for Delivery of Water-Insoluble Cancer Drugs. Journal of the American Chemical Society 2008;130(33): 10876-7.

[63] Paleček E. Past, present and future of nucleic acids electrochemistry. Talanta 2002;56(5): 809-19.

[64] Brett AMO, Serrano SHP. The Electrochemical Oxidation of DNA. Journal of the Brazilian Chemical Society 1995;6: 97-100.

[65] Palecek E, Fojta M. Electrochemical DNA Sensors. In: Willner I, Katz E, editors. Bioelectronics: From Theory to Applications: Wiley-VCH Verlag GmbH & Co. KGaA; 2005. p. 127-92.

[66] Odenthal KJ, Gooding JJ. An introduction to electrochemical DNA biosensors. Analyst 2007;132(7): 603-10.

[67] Wang J. Nanoparticle-based electrochemical DNA detection. Analytica Chimica Acta 2003;500(1–2): 247-57.

[68] Drummond TG, Hill MG, Barton JK. Electrochemical DNA sensors. Nature Biotechnology 2003;21(10): 1192-9.

[69] Ghosh I, Stains CI, Ooi AT, Segal DJ. Direct detection of double-stranded DNA: molecular methods and applications for DNA diagnostics. Molecular BioSystems 2006;2(11): 551-60.

[70] Gooding JJ. Electrochemical DNA Hybridization Biosensors. Electroanalysis 2002;14(17): 1149-56.

[71] Wang J, Kawde A-N, Musameh M. Carbon-nanotube-modified glassy carbon electrodes for amplified label-free electrochemical detection of DNA hybridization. Analyst 2003;128(7): 912-6.

[72] Millan KM, Mikkelsen SR. Sequence-selective biosensor for DNA based on electroactive hybridization indicators. Analytical Chemistry 1993;65(17): 2317-23.

[73] Banks CE, Crossley A, Salter C, Wilkins SJ, Compton RG. Carbon Nanotubes Contain Metal Impurities Which Are Responsible for the "Electrocatalysis" Seen at Some Nanotube-Modified Electrodes. Angewandte Chemie International Edition 2006;45(16): 2533-7.

[74] Millan KM, Saraullo A, Mikkelsen SR. Voltammetric DNA Biosensor for Cystic Fibrosis Based on a Modified Carbon Paste Electrode. Analytical Chemistry 1994;66(18): 2943 8.

[75] Erdem A, Kerman K, Meric B, Akarca US, Ozsoz M. DNA Electrochemical Biosensor for the Detection of Short DNA Sequences Related to the Hepatitis B Virus. Electroanalysis 1999;11(8): 586-7.

[76] Wang J, Rivas G, Cai X, Dontha N, Shiraishi H, Luo D, et al. Sequence-specific electrochemical biosensing of M. tuberculosis DNA. Analytica Chimica Acta 1997;337(1): 41-8.

[77] Wang J, Xu D, Erdem A, Polsky R, Salazar MA. Genomagnetic electrochemical assays of DNA hybridization. Talanta 2002;56(5): 931-8.

[78] Ferapontova EE, Domínguez E. Direct Electrochemical Oxidation of DNA on Polycrystalline Gold Electrodes. Electroanalysis 2003;15(7): 629-34.

[79] Pang D-W, Qi Y-P, Wang Z-L, Cheng J-K, Wang J-W. Electrochemical oxidation of DNA at a gold microelectrode. Electroanalysis 1995;7(8): 774-7.

[80] Bollo S, Ferreyra NF, Rivas GA. Electrooxidation of DNA at Glassy Carbon Electrodes Modified with Multiwall Carbon Nanotubes Dispersed in Chitosan. Electroanalysis 2007;19(7-8): 833-40.

[81] Wang J, Li M, Shi Z, Li N, Gu Z. Electrochemistry of DNA at Single-Wall Carbon Nanotubes. Electroanalysis 2004;16(1-2): 140-4.

[82] Wu K, Fei J, Bai W, Hu S. Direct electrochemistry of DNA, guanine and adenine at a nanostructured film-modified electrode. Analytical and Bioanalytical Chemistry 2003;376(2): 205-9.

[83] Özcan A, Şahin Y, Özsöz M, Turan S. Electrochemical Oxidation of ds-DNA on Polypyrrole Nanofiber Modified Pencil Graphite Electrode. Electroanalysis 2007;19(21): 2208-16.

[84] Pedano ML, Rivas GA. Immobilization of DNA on glassy carbon electrodes for the development of affinity biosensors. Biosensors and Bioelectronics 2003;18(2–3): 269-77.

[85] Nowicka AM, Zabost E, Donten M, Mazerska Z, Stojek Z. Electrooxidation of dissolved dsDNA backed by in situ UV–Vis spectroscopy. Bioelectrochemistry 2007;70(2): 440-5.

[86] Wang H-S, Ju H-X, Chen H-Y. Voltammetric Behavior and Detection of DNA at Electrochemically Pretreated Glassy Carbon Electrode. Electroanalysis 2001;13(13): 1105-9.

[87] Bacáková L, Grausová L, Vacík J, Lavrentiev V, Blazewicz S, Fraczek A, et al. Adhesion and Growth of Human Osteoblast-Like Cell in Cultures on Nanocomposite Carbon-Based Materials. Nanoscience and Nanotechnology Letters 2011;3(1): 99-109.

[88] Stoppie N, Van Oosterwyck H, Jansen J, Wolke J, Wevers M, Naert I. The influence of Young's modulus of loaded implants on bone remodeling: An experimental and numerical study in the goat knee. Journal of Biomedical Materials Research Part A 2009;90A(3): 792-803.

[89] Whiteside LA. The effect of stem fit on bone hypertrophy and pain relief in cementless total hip arthroplasty. Clinical Orthopaedics and Related Research 1989;247: 138-47.

[90] McKegney M, Taggart I, Grant MH. The influence of crosslinking agents and diamines on the pore size, morphology and the biological stability of collagen sponges and their effect on cell penetration through the sponge matrix. Journal of Materials Science: Materials in Medicine 2001;12(9): 833-44.

[91] Hirata E, Uo M, Nodasaka Y, Takita H, Ushijima N, Akasaka T, et al. 3D collagen scaffolds coated with multiwalled carbon nanotubes: Initial cell attachment to inter-

nal surface. Journal of Biomedical Materials Research Part B: Applied Biomaterials 2010;93B(2): 544-50.

[92] Zanello LP, Zhao B, Hu H, Haddon RC. Bone Cell Proliferation on Carbon Nanotubes. Nano Letters 2006;6(3): 562-7.

[93] Gonzalez KA, Wilson LJ, Wu W, Nancollas GH. Synthesis and In vitro characterization of a tissue-Selective fullerene: vectoring C60(OH)16AMBP to mineralized bone. Bioorganic & Medicinal Chemistry 2002;10(6): 1991-7.

[94] Mochalin VN, Shenderova O, Ho D, Gogotsi Y. The properties and applications of nanodiamonds. Nature Nanotechnology 2012;7(1): 11-23.

[95] Nayak TR, Andersen H, Makam VS, Khaw C, Bae S, Xu X, et al. Graphene for Controlled and Accelerated Osteogenic Differentiation of Human Mesenchymal Stem Cells. ACS Nano 2011;5(6): 4670-8.

[96] Kalbacova M, Broz A, Kong J, Kalbac M. Graphene substrates promote adherence of human osteoblasts and mesenchymal stromal cells. Carbon 2010;48(15): 4323-9.

[97] Ryoo S-R, Kim Y-K, Kim M-H, Min D-H. Behaviors of NIH-3T3 Fibroblasts on Graphene/Carbon Nanotubes: Proliferation, Focal Adhesion, and Gene Transfection Studies. ACS Nano 2010;4(11): 6587-98.

[98] Sun Z, James DK, Tour JM. Graphene Chemistry: Synthesis and Manipulation. The Journal of Physical Chemistry Letters 2011;2(19): 2425-32.

[99] Nandamuri G, Roumimov S, Solanki R. Remote plasma assisted growth of graphene films. Applied Physics Letters 2010;96(15): 154101.

[100] Yuan GD, Zhang WJ, Yang Y, Tang YB, Li YQ, Wang JX, et al. Graphene sheets via microwave chemical vapor deposition. Chemical Physics Letters 2009;467(4–6): 361-4.

[101] Peltekis N, Kumar S, McEvoy N, Lee K, Weidlich A, Duesberg GS. The effect of downstream plasma treatments on graphene surfaces. Carbon 2012;50(2): 395-403.

[102] Burgess JS, Matis BR, Robinson JT, Bulat FA, Keith Perkins F, Houston BH, et al. Tuning the electronic properties of graphene by hydrogenation in a plasma enhanced chemical vapor deposition reactor. Carbon 2011;49(13): 4420-6.

[103] Hopkins PE, Baraket M, Barnat EV, Beechem TE, Kearney SP, Duda JC, et al. Manipulating Thermal Conductance at Metal–Graphene Contacts via Chemical Functionalization. Nano Letters 2012;12(2): 590-5.

[104] Baraket M, Walton SG, Lock EH, Robinson JT, Perkins FK. The functionalization of graphene using electron-beam generated plasmas. Applied Physics Letters 2010;96(23): 231501.

[105] Matis BR, Burgess JS, Bulat FA, Friedman AL, Houston BH, Baldwin JW. Surface Doping and Band Gap Tunability in Hydrogenated Graphene. ACS Nano 2011;6(1): 17-22.

Chapter 4

Laser Based Fabrication of Graphene

Kai Wang

Additional information is available at the end of the chapter

http://dx.doi.org/10.5772/55821

1. Introduction

Graphene, two dimensional (2D) carbon-carbon atoms arranged in a hexagonally honeycomb lattice structure, is the latest carbon (C) allotrope to be discovered. It has gained tremendous scientific research interests due to some unique properties, such as quantum hall effect and nearly ballistic electronic transportation in ambient.[1-4] These properties have stimulated intense activity among physicists, chemists and material scientists. Much research has focused on developing routs for the controllable growth of high quality graphene. Historically, graphene can be produced via micromechanical cleavage and SiC decomposition methods.[2, 5, 6] In order to meet even higher requirements, such as good crystallinity, less impurities and large area coverage, the present most successful fabrication technique focuses on chemical vapor deposition (CVD) on copper [5 - 8] Some other graphene growth methods include growth of graphene from solid carbon source, graphene synthesis by ion beam implantation, and graphene formation by decomposition of C_{60}, have also been reported in the recent years.[9, 10] Apart from these, pulsed laser deposition (PLD), which is one of the unique physical vapor deposition (PVD) methods, represents a completely new fabrication way in this field.

For many reasons, PLD is a versatile material fabrication technique. There are many laser parameters regarding sample preparation can be tuned in the PLD system, which have great influence on the sample qualities. First, for laser itself, the laser fluence, wavelength, repetition rate, and pulse duration can be altered. The second aspect includes target to substrate distance, background gas and pressure, and substrate temperature can be varied depending on certain requirements. Since with this technique the energy source, laser, is located outside the vacuum system, it is possible to adopt ultrahigh vacuum (UHV) as well as ambient conditions. It allows growing many kinds of materials; for example, oxides, nitrides, carbides, semiconductors, metals, superconductors, superlattice structures, and even fullerenes and polymers. The pulsed nature of PLD process even allows preparing complex polymer-metal compounds and multylayers. For example, one of the promising realizations of the PLD system is to introduce oxygen to the chamber for oxides based fabrication. It is inevitable for achieving a sufficient amount of oxygen during film growth.

Until now, there have fewer studies on PLD based graphene growth. Indeed, such method used in graphene fabrication originates from the growth of carbon thin film by the PLD system. If a single carbon layer with aromatic ring structure in plane can be obtained, it gives a result of a single layer graphene. For the PLD method, the amorphous C layer can be easily deposited at room temperature. However, in order to fabricate C layer or graphene with a desirable crystalline quality by PLD, some additional conditions are required. For example, the substrate temperatures, vacuum level inside the PLD chamber, appropriate laser operation conditions (i.e. laser fluence and repetition rate) and the choice of catalytic metals. Wang, et. al. have demonstrated that few layer graphene that is bi-layer to multi-layer graphene can be fabricated on catalytic nickel (Ni) thin film by PLD system.[11] The number of graphene layers is found depending on the laser ablation time, and the crystallinity of the graphene depends on the substrate temperature during laser ablation. During the graphene formation, it involves several steps, such as C atoms adsorption, precipitation, segregation and recrystallization. All these processes happen due to the interaction between C atoms and metals. The formation of the graphene on the metal surfaces was first observed during the preparations of platinum (Pt) and ruthenium (Ru) single crystal surfaces. [12, 13]

In fact, the studies of the interaction of the C and the metals has a long history, however, the graphene fabricated by PLD method is realized in the recent years. By comparing with the conventional CVD method which usually involves high processing temperature (>1000 °C) and chemical reactive hydrocarbon gas flow, PLD can reach the same goal at relatively low temperatures and the C target is always in a solid form. Koh, et. al. have demonstrated that few layer graphene can be fabricated at 750 °C on a Ni plate (Figure 1).[14] Apart from the temperature issue, they also showed the cooling rate and laser energy are crucial in fabricating such graphene layers. Afterwards, Wang, et. al. did more systematic studies on the formation of the graphene on the Ni thin film. The significant contribution from their work is the number of graphene layer can be controlled by altering the number of the laser pulses (Figure 2). [11]

Apart from the demonstration of this PLD fabrication technique in this chapter, the characterizations and quantitative studies of graphene mainly rely on fast and non-destructive micro-Raman spectroscopy.[6, 15-17] In the past four decades, it has witnessed that Raman spectroscopy plays an important role in characterizing pyrolytic graphite, glassy carbon, graphitic foams, carbon fibers, nanographite ribbons, carbon nanotubes and fullerences. Owing to the presence of sp^2 bond graphene, Raman spectroscopy gives plenty of inspired information about crystallite size, the introduction of chemical impurities, the magnitude of the mass density, the optical energy gap, the elastic constants, the doping, defects, the crystal disorder, the strain, and the number of the graphene layers. With this respect, the discussion about PLD fabricated graphene here will provide a taste of the power of Raman spectroscopy.

http://dx.doi.org/10.5772/55821

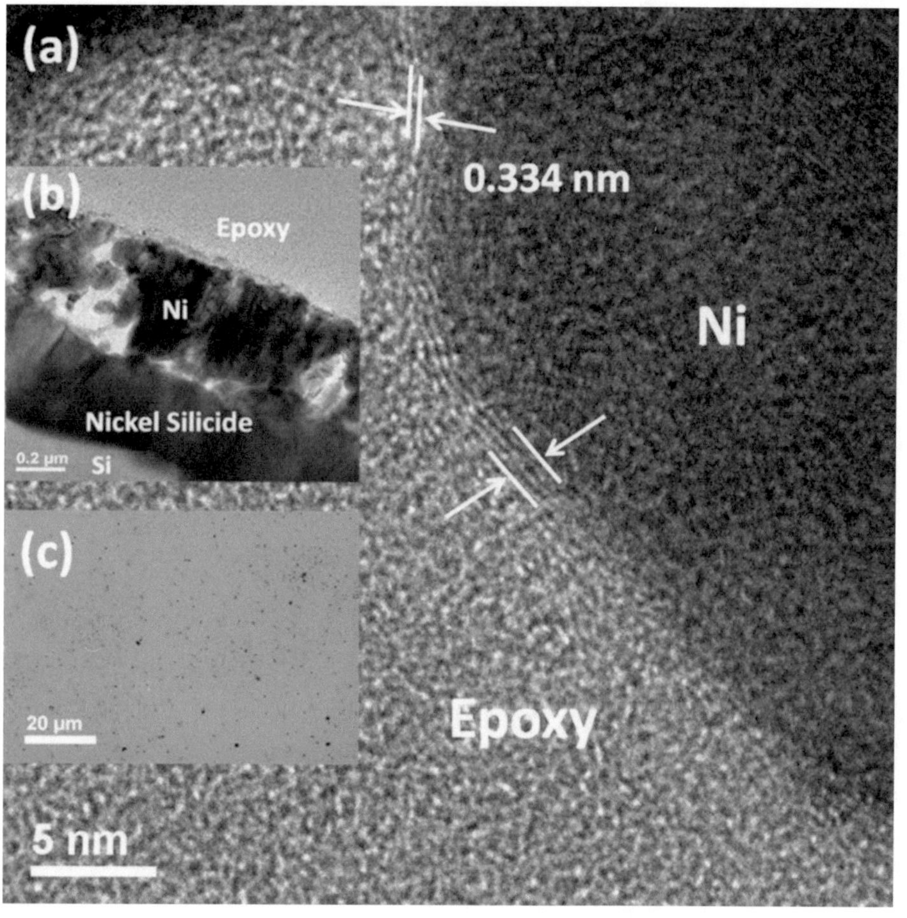

Figure 1. Crossing sectional TEM showing (a) the graphene layers above Ni. (b) Ni/nickel silicide/Si layered structure. Inset c is an overview optical image showing rather uniform coverage [ref. 13. figure 2].

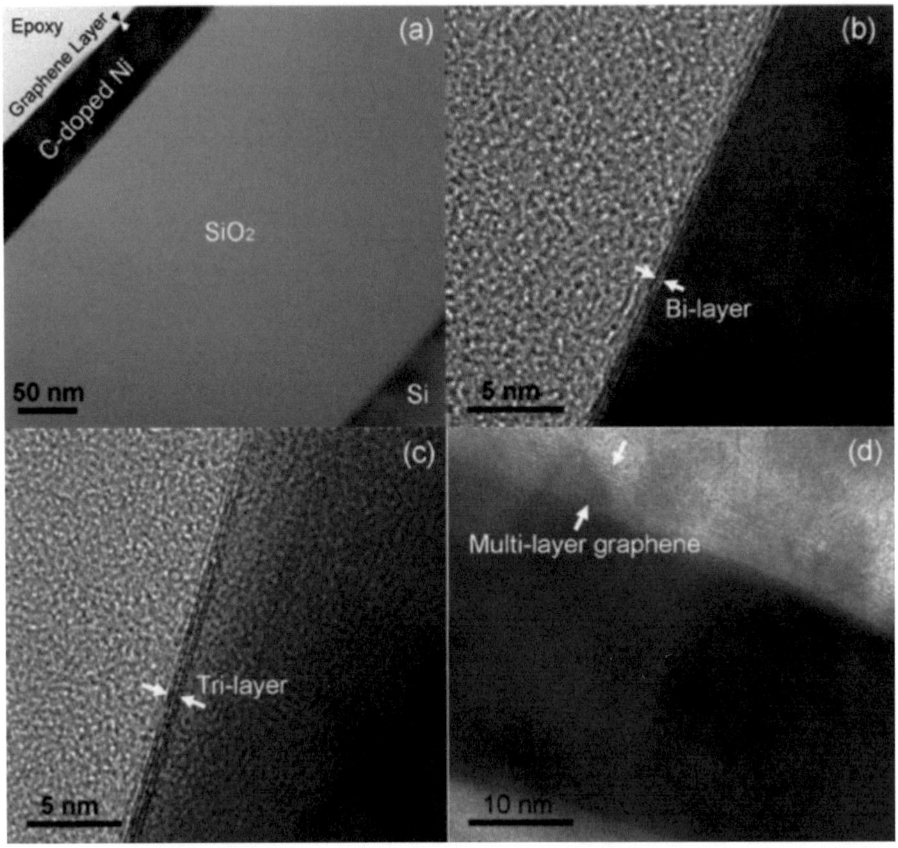

Figure 2. TEM characterization of PLD-grown graphene films. (a) Low magnification TEM image showing PLD-grown graphene on Ni thin film. High resolution TEM images for (b) bi-layer graphene, (c) tri-layer graphene and (d) multi-layer graphene.

2. Metal induced crystallization method

Metal induced crystallization method, also known as metal mediated or metal catalyzed crystallization method, is a crystalline growth technique to fabricate mono- or poly-crystalline materials via the interdiffusion, precipitation, segregation and recrystallization of two materials upon thermal annealing. One of them performs as the catalyst; it is usually a metal, such as aluminum (Al), gold (Au), platinum (Pt), nickel (Ni) and cobalt (Co); [11, 12, 18, 19] while the other is always chosen as a semiconductor material, such as germanium (Ge), and silicon (Si). This technique has been widely explored in the past, in particular for the poly-crystalline Si thin film based solar cell investigation.[20] It has been proposed and experimentally demonstrated in fabricating high crystalline quality and large area polycrystalline Si thin film on various kinds of substrates at low temperature. For both CVD and PVD method based graphene fabrications, metal induced crystallization method acts as a dominating role during the graphitization process at elevated temperature. One way to realize this method in graphene growth is amorphous C layer can be easily changed to crystalline graphene layer based on thermal annealing process. Such process can be understood from figure 3.[21] This process utilizes solid phase sources of C. In this approach, the C is introduced in the amorphous phase with the Ni thin film forming bi-layer stack. Upon high temperature annealing, the C atoms from the a-C layer would dissolve into the Ni layer and be expelled from solution after cooling below the solid solubility limit. By comparing with previous studies about metal-induced Si crystallization, similar mechanism is involved. The driving force for crystallization is thermodynamic stability of the crystalline C and Si phases relative to the amorphous phase.

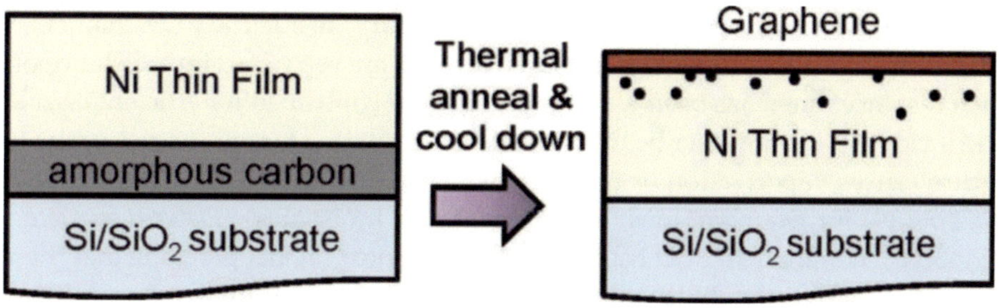

Figure 3. The process schematics for the metal-catalyzed crystallization of amorphous C to graphene by thermal annealing [ref. 19 figure 1].

3. An Introduction to pulsed laser deposition

3.1. Background of PLD

PLD is a thin film fabrication technique using high energy pulsed laser beam to bombard one or more targets at a certain vacuum pressure.[22, 23] The laser shooting areas of the targets experience the transition from the solid to the vapor phase, and subsequently been coated onto a substrate. This growth technique was first used by Smith and Turner in 1965 for the preparation of semiconductors and dielectric thin films.[24] It was further established by Dijkkamp and coworkers on high temperature superconductors in 1987.[25] Afterwards, this technique has been extensively optimized, investigated, and used in oxides, nitrides, carbides, metallic thin films and even organic polymers.[26]

3.2. PLD system setup

PLD is a form of physical vapor deposition (PVD). The system design is somewhat different from other PVD systems such as, thermal evaporation, electron beam evaporation, molecular beam epitaxial growth and magnetron sputter systems, because an external laser source is required. The useful range of laser wavelengths for thin films growth by PLD lies between 200 nm and 400 nm because most materials exhibit strong absorption in this spectral region. Therefore, most photon energies come from laser source can be absorbed by materials. Within this range, there are few commercially available laser sources capable of easily delivering the high energy densities (1 J/cm^2), in relatively large areas (10 mm^2 or larger), which are required for PLD works. A homogenous uniform laser output is also important for high quality thin films fabrications. Most PLD systems work these days use excimer lasers, in which the lasing medium is a mixture of some reactive gases such as krypton (Kr), fluorine (F) and neon (Ne). It is also known that neodymium (Nd): YAG laser can also be used for graphite fabrication on Si substrate, but no one has demonstrated the fabrication of graphene using this laser source so far.

An experimental setup for a typical PLD system is shown in figure 4. It consists of two major components, one is the external laser source and another is the stainless steal vacuum chamber. The vacuum chamber can be placed direct facing the output laser pulse or be set at certain angles. For the latter case, a reflecting mirror is necessary. Inside the PLD chamber, a target and a substrate holder are aligned on the same line but are separated by a distance of 3 cm to 5 cm. Such distance range has been well experimentally confirmed for efficient laser ablation. When the incident focusing laser beam bombards the rotating target, the rise of the localized temperature causes vaporization of the material. It is a feature of plasma plume with high energetic species, for example, ions, electrons, atoms, molecules, clusters, particulates and molten globules. For an ideal PLD based thin films fabrication, people hope that the clusters, particulates and molten globules can be avoided. The film growths depend on several parameters, such as laser fluence, laser repetition rate, substrate temperature and vacuum level. By adjusting the number of laser pulses on the targets, different layers with controllable thicknesses can reach.

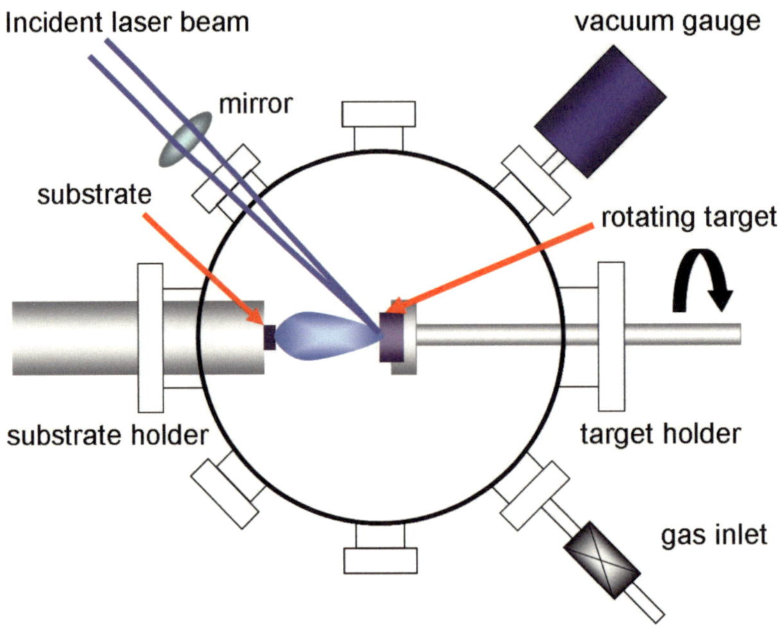

Figure 4. PLD system setup.

3.3. PLD technique in graphene fabrication

In a typical experiment for fabricating graphene layer via PLD, people can select different substrates based on certain requirements. For convenient purpose, we use 300nm SiO_2 coated Si as a typical example to illustrate this fabrication method (figure 5(a)). Furthermore, it is also a good candidate for making bottom gate graphene based field effect transistor (FET) due to the present of sufficient thick SiO_2 layer. Prior to the deposition, the substrate can be cleaned by the conventional chemical means. The fabrications for those metal thin films and C were carried out inside a stainless steal PLD chamber. The SiO_2/Si substrate was firmly attached onto a substrate holder, which is 4 cm in a distance to a PLD target holder. The laser beam can be guided to the target via a focusing lens. More details about the experimental condition can be found from reference 10. A schematic diagram which shows the graphene formation on a thin metal thin film is given in figure 5. The metal thin film is fabricated at room temperature (figure 5(b)). Without breaking the high vacuum, the substrate temperatures were rapidly increased to 650 °C and 600 °C. After the deposition of C (figure 5(c)), the samples were naturally cooled down to room temperature in an ultra-high vacuum. For some PLD systems, the cooling rate can also be controlled, for example by the flow of the liquid nitrogen or pass through the inert gas into the chamber. During this cooling process, the C atom will segregate from the C containing C-metal solid solution and form a continued layer on the topmost surface (figure 5(d)). This is due to the reduction of the solubility of C-metal solution. The metal which locates below graphene layer after fabrication is not a pure thin metal film anymore, because there are still sufficient amount of C atoms which participate the interdiffusion process remain and form C-metal eutectic alloy.

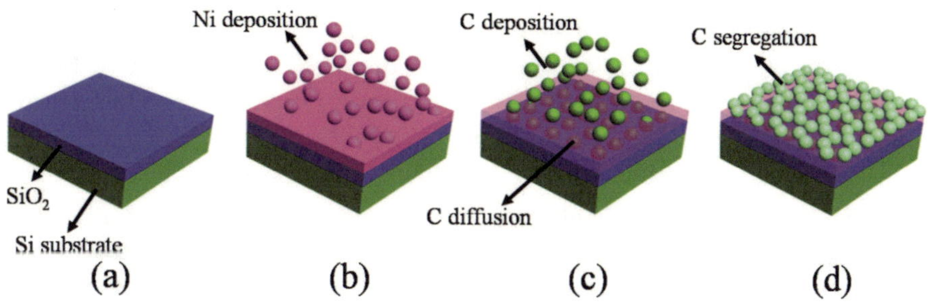

Figure 5. Schematic diagram shows metal-induced graphene formation process by PLD technique. (a) Preparation of SiO_2/Si substrate. (b) Ni thin film deposition at room temperature. (c) C deposition at elevated temperature. (d) Graphene formation at the topmost surface.

4. TEM and Raman spectroscopic studies

Figure 6 shows the Raman spectra for the carbon layers fabricated at 5 difference temperatures, 300 °C, 400 °C, 500 °C, 600 °C and 650 °C respectively on Ni thin film. In figure 6(a), there is no distinct peak can be observed for crystalline C phase. This indicates 300 °C is insufficiently to obtain crystalline carbon layer on Ni thin film. As the deposition temperature for carbon increases, such as shown in figure 6(b) and (c), two remarkable bands tend to emerge at 1350 cm^{-1} and 1550 cm^{-1} respectively. Both of them represent disorder-induced (D) and graphitic (G) bands. For the one locates at 1350 cm^{-1}, its occurrence is due to the breathing mode of sp^2 atoms or A_{1g} symmetry in hexagonal graphitic rings. From both figure 6(b) and (c), a broad bandwidth of such D mode indicates the nonorganized C.[16] Thus, the large full width at half maximum (FWHM) of the D band yields the highly disordered characteristic of the C layer. This mode is forbidden in perfect graphite and become active in the presence of disorder. Furthermore, the D mode is dispersive and varies with photon excitation energy. Therefore, its intensity is strictly related with the presence of sixfold aromatic ring. As the atomic mechanism concerns, the rise of G band is ascribed to the in-plane stretching of the C-C bond in graphitic materials.[17] By comparing with the one of D mode, this one does not require the presence of sixfold ring so that it occurs at all pairs of sp^2 C sites. In the Raman spectra, the G band is due to the doubly degenerate zone center E_{2g} mode. The Raman spectroscopy is very sensitive to this strain effect in sp^2 C based materials. When the interaction between one graphene layer and substrate or another graphene layer is encountered, the induced strain is due to the modification of the C-C bond lengths and angles. As we can see from figure 6(d) and (e), further increase in the deposition temperature of C leads a clear distinguish of D and G bands. The relative intensity ratio of I_D to I_G has been significantly reduced at the deposition temperature of 650 °C for C. It results in the reduction of the crystalline defects. In addition, there is another band gradually become noticeable at 2700 cm^{-1}. It is the second order of zone-boundary phonons, but it has nothing to do with the G band. Because the Raman shift at this point is approximately doubled when comparing with the one of the D band, it is conventionally denoted as 2D or G' band. Owing to the G' band is the second order process associated with a phonon close to the K point in graphene, there has a strong dependence on any perturbation to the electronic and phonon structure of graphene. Therefore, G' band plays an important role for differentiating single, bi- and few- layer graphene upon Raman spectroscopy.[6, 15] The results of the temperature dependence for C deposition suggests that 650 °C is an optimum temperature to fabricate the crystalline C layer on the Ni thin films by the PLD method in this experiment. Apart from Ni, it is also possible to fabricate graphene layer on a Co thin film. Figure 7 shows the temperature dependent graphene fabrication on the Co thin film. By comparing with figure 6, similar results were obtained for these two 3d transition metals.

Figure 8(a) shows the photographic image during graphene layer transferring process. The PLD made graphene transfer method is very similar to the one used in CVD method. A very thin protective poly[methyl methacrylate] (PMMA) layer was initially coated on the top of the graphene/Ni or Co/SiO_2/Si sample by spin coater. The catalytic Ni or Co can be etched away by chemical wet-etching, for instance using an aqueous HCL solution. In figure 8(a), After the Ni thin film was completely dissolved in $FeCl_3$ solution; a 1 × 1 cm^2 few layer graphene coated with the PMMA was detached from the Ni thin film and forms a free-standing layer in the HCl solution. The diluted HCl solution can be further utilized for removing the residual Ni flakes. Afterward, the few-layer graphene can be transferred onto another clean SiO_2/Si substrate and the top as-coated PMMA layer can be dissolve quickly by putting the sample directly into

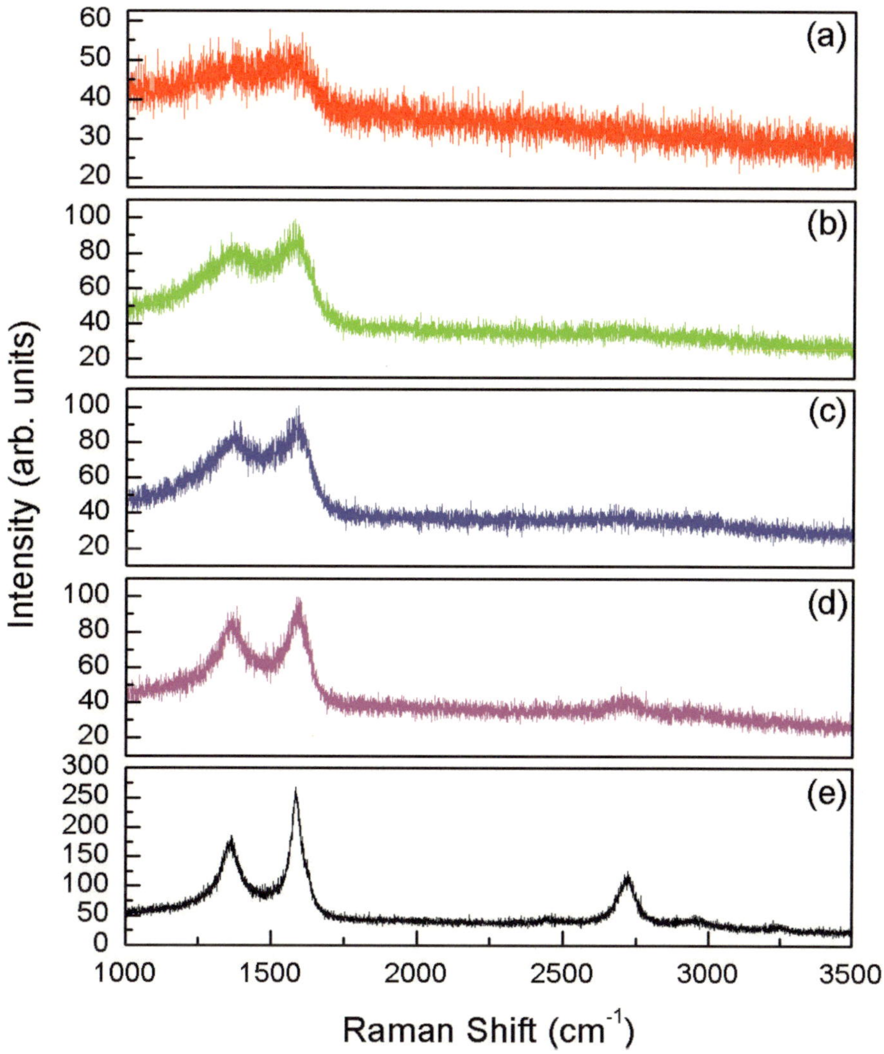

Figure 6. Temperature dependent Ni-induced few layer graphene formation. (a) at 300 °C. (b) at 400 °C. (c) at 500 °C. (d) at 600 °C and (e) at 650 °C.

acetone. The corresponding photographic picture of such graphene layer on the SiO2/Si substrate after successful transfer is shown in figure 8(b). Moreover, the graphene sample preparation method for the transmission electron microscopy (TEM) characterization is slightly different. In this case, after the sample is closely attached to the TEM copper grid, the PMMA can be dissolved by exposing to the acetone vapor for approximately 4 to 5 hours. This transfer process allows maintaining the continuity of the graphene. Figure 8(c) displays the feature of the few-layer graphene and the blue dotted circle denotes the presence of the graphene wrinkles. The formation is primarily due to lattice mismatch between Ni and C and the wrinkles have high possibility to be found at the grain boundaries of as-prepared Ni thin

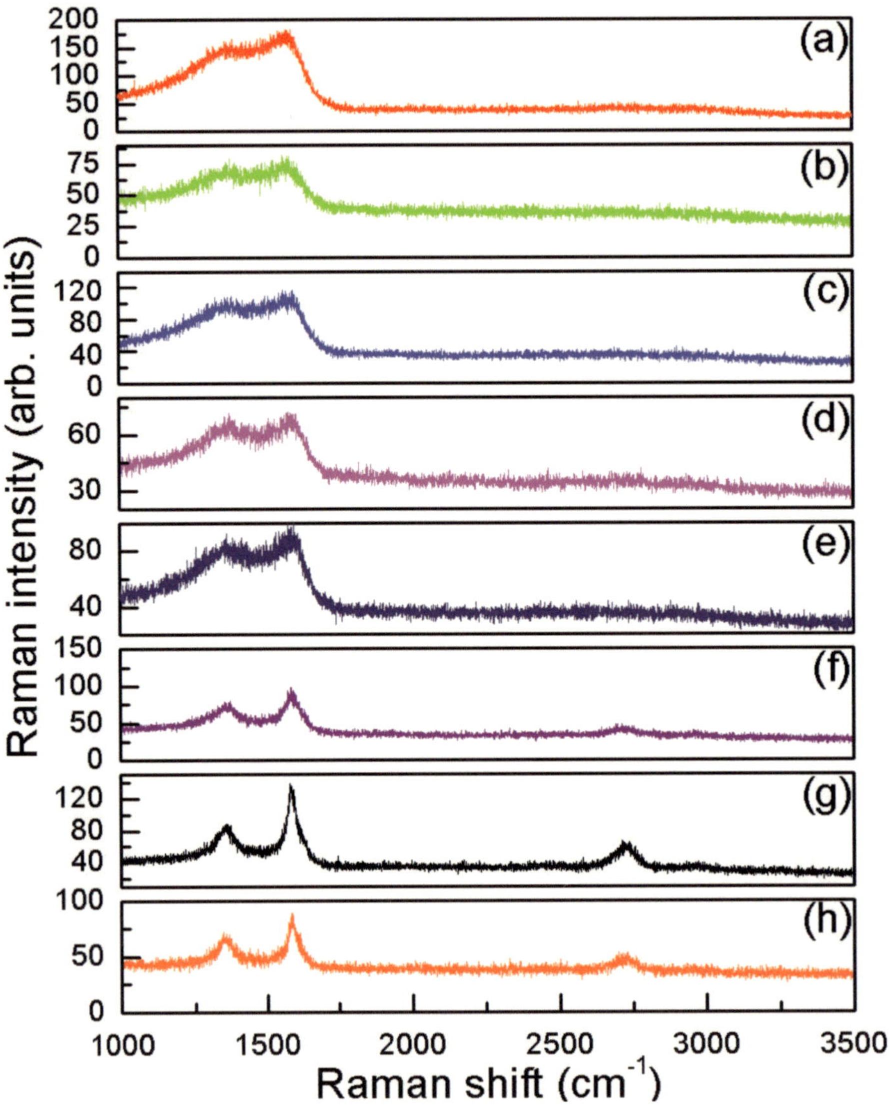

Figure 7. Temperature dependent Co-induced few layer graphene formations. (a) at 300 °C. (b) at 350 °C. (c) at 400 °C. (d) at 450 °C and (e) at 500 °C. (f) at 550 °C. (g) at 600 °C. (h) at 650 °C.

films. Similar with CVD technique, the morphology of fabricated graphene tend to replicate the one of substrate materials. At the edge of the sample which is shown in figure 8(d), the graphene layer by layer feature can be clearly observed. In figure 8(e), the hexagonally symmetric selected area electron diffraction (SADP) for this few-layer graphene was captured indicating the monocrystalline structure of our fabricated few-layer graphene. Two red dotted circles highlight the presence of (0002) lattice plane in the d space. This diffracted spots only appear when few layer or bulk graphite is encountered.

http://dx.doi.org/10.5772/55821

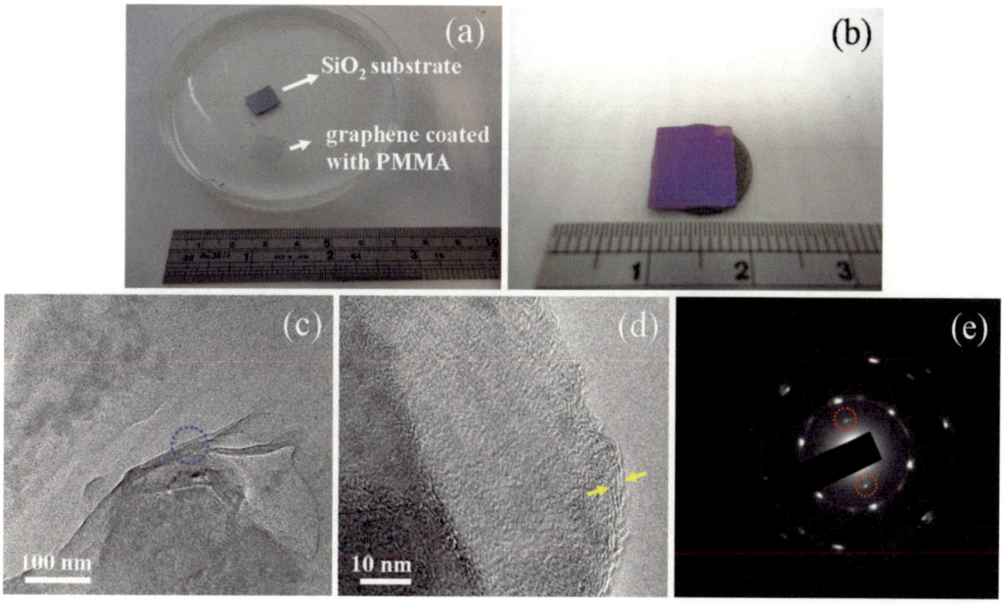

Figure 8. Graphene transfer process after PLD fabrication. (a) Detach of graphene layer from the substrate. (b) Graphene is transferred onto a new piece of SiO_2/Si substrate. (c) Low magnification TEM image shows graphene wrinkles. (d) High resolution TEM image shows few layer graphene edges. (e) The corresponding selected area electron diffraction pattern of few layer graphene.

The high resolution TEM crossing sectional view of bi- and tri- layer graphene are displayed in figure 9(a) and (b) respectively. During the deposition of C at 650 °C, the interdiffusion between C and Ni leads to the formation of the solid solution. Such interdiffusion proceeds until C reaches its saturation status at this temperature. Once the C saturation status reaches, further deposition of C causes a formation of amorphous C layer. Therefore, a well control of the thickness ratio between C and Ni is crucially important in achieving a high quality, bi-, tri- and few layer graphene. When C saturation status reaches at 650 °C, the natural reduction of the substrate temperature to room temperature in high vacuum causes the decrease in the Ni solubility. For this reason, C atoms segregate from the solid solution and form a crystalline and continue graphene layer on the most top of the sample. Figure 9(c) shows the AFM image of the bi-layer graphene. After scratching the sample for depth profile measurement, some graphene parts fold and the relative contrast can be seen from this AFM image. A location across the surface of this piece of graphene was randomly chosen and the corresponding depth profile is shown in figure 9(d). The lateral height which is around 2.5 nm indicates the presence of bi-layer graphene.

Further studies in fabricating graphene by the PLD include the thickness ratio dependence between Ni and C. In this scheme, we found both the number of graphene layers and the D band can be greatly affected. From the Raman spectra in figure 10, C with 3 different thicknesses or laser ablation time were deposited onto 250 nm Ni thin films at 650 °C. The D band is the greatest among those three when the laser ablation time is 120 s (figure 10(a). Moreover,

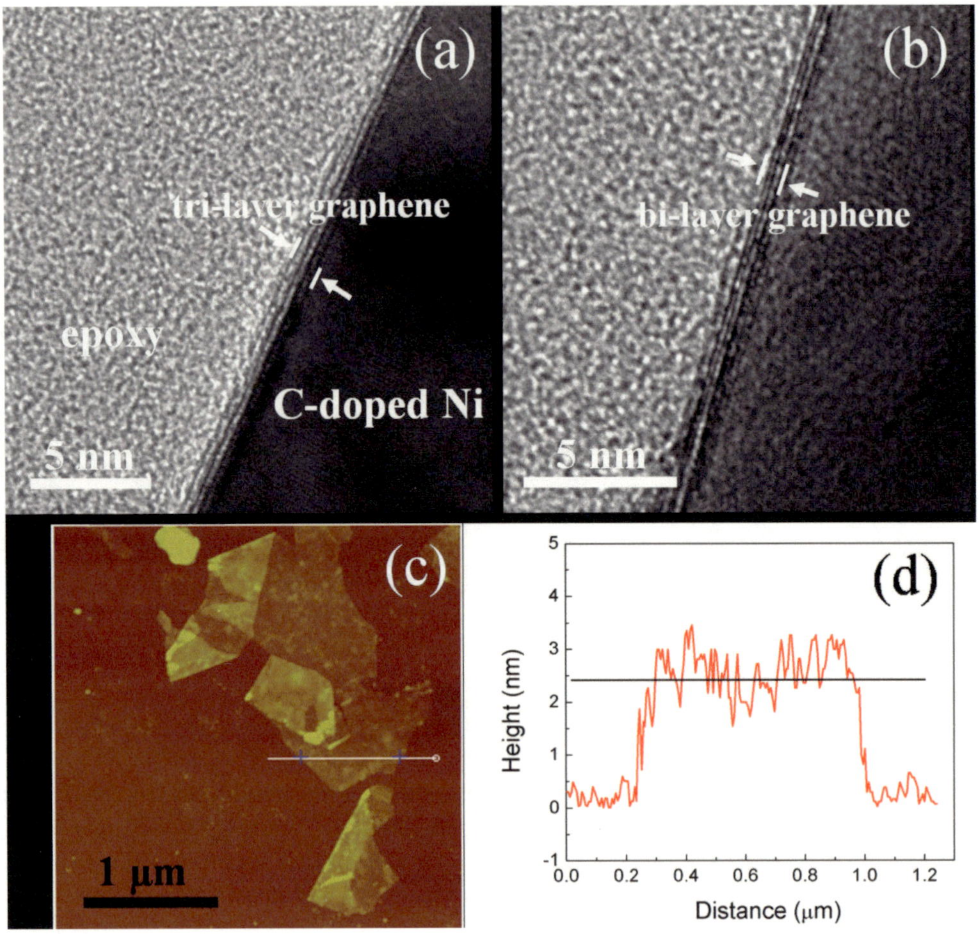

Figure 9. High resolution TEM images show (a) crossing sectional view of tri-layer graphene, (b) crossing sectional view of bi-layer graphene. (c) The AFM image for few layer graphene. (d) The lateral height measurement for few layer graphene.

the intensity ratio of I_{2D}/I_G is 1.24 (Data is summarized in table 1). Thus, both of them imply the poor crystallinity of C layers and the layers. By decreasing the amount of C, such as the resultant Raman spectrum shown in figure 10(b) which corresponds to the laser ablation time of 90 s, the intensity ratio of I_{2D}/I_G is 0.47 which indicates the existence of bi- or tri-layer graphene. Nevertheless, the intensity of the D band is over half of the intensity of the G band. However, when the laser ablation time for C is further decreased to 60 s (figure 10(c)), we observe a significant reduction of the D band Raman intensity and the intensity ratio of I_{2D}/I_G becomes even larger. Similar experiment was done for the graphene fabricated on 250 nm Co thin film. The Raman spectra are shown in figure11 for the different laser ablation times of C, 240 s, 180 s, 120 s, 60 s, and 30 s respectively. The Raman intensities which correspond to each band are summarized in table 2. In contrast to table 1, the graphene which is fabricated on the Co thin film shows similar trend as the laser ablation time for C tends to decrease.

C deposition time (s) / thickness (nm)	D-band position	D-band intensity	G-band position	G-band intensity	2D-band position	2D-band intensity	I_D/I_G	I_{2D}/I_G
120s	1359.30	1011.76	1596.04	814.71	2704.16	130.39	1.24	0.16
90s	1359.30	178.43	1588.48	261.52	2719.27	122.79	0.68	0.47
60s	1346.70	154.17	1580.93	327.43	2688.21	183.99	0.47	0.56

Table 1. Raman intensities for I_D, I_G, I_{2D}, I_D/I_G and I_{2D} / I_G of Figure 10.

C deposition time	D-band position	D-band intensity	G-band position	G-band intensity	2D-band position	2D-band intensity	I_{2D}/I_G
240s	1355.19	358.81	1591.07	403.40	2724.95	97.86	0.24
180s	1355.19	202.91	1581.97	277.00	2724.95	76.17	0.27
120s	1364.30	85.57	1581.97	141.12	2724.95	58.97	0.41
60s	1364.30	65.95	1581.97	113.52	2716.76	55.83	0.49
30s	1364.30	43.17	1581.97	84.08	2707.65	52.29	0.62

Table 2. Raman intensities for I_D, I_G, I_{2D}, I_D/I_G and I_{2D} / I_G of Figure 11.

For the graphene fabricated by the PLD method, the D and 2D bands of bi-layer graphene show some unique characteristics by comparing with few-layer graphene and bulk graphite. As we can see from figure 12(a), the D band of bi-layer graphene possesses a non-symmetric band at 1346.70 cm^{-1} while the few-layer graphene and bulk graphite give symmetric D bands at 1359.30 cm^{-1}. In addition, the Raman spectrum of the asymmetric 2D band of the bi-layer graphene shows red shift relative the ones of few-layer graphene and bulk graphite. Such 2D mode of the bi-layer graphene composes of 4 components, $2D_{1B}$, $2D_{1A}$, $2D_{2A}$, and $2D_{2B}$; in which, $2D_{1A}$ and $2D_{2A}$ have higher intensities than the other two, as shown in figure 12(b). These four components of bi-layer graphene are attributed to the evolution of the electronic band structure. Raman scattering is a fourth order process involving four virtual transitions: [1] a laser induced excitation of an electron-hole pair; [2] electron-phonon scattering with an exchanged momentum; [3] electron phonon scattering with an exchanged momentum; and [4] electron-hole recombination. Based on both TEM and Raman spectroscopic studies, the graphene which is fabricated by the PLD method can be achieved. Nevertheless, there are still some aspects have to be encountered. As we have discussed previously, the number and the quality of graphene layers are decisive by the ratio between C layer and catalytic metals, substrate temperature and laser operation conditions. In order to obtained desirable property of graphene, careful understandings of the graphene and metal interaction, interdiffusion and interface property are crucial.

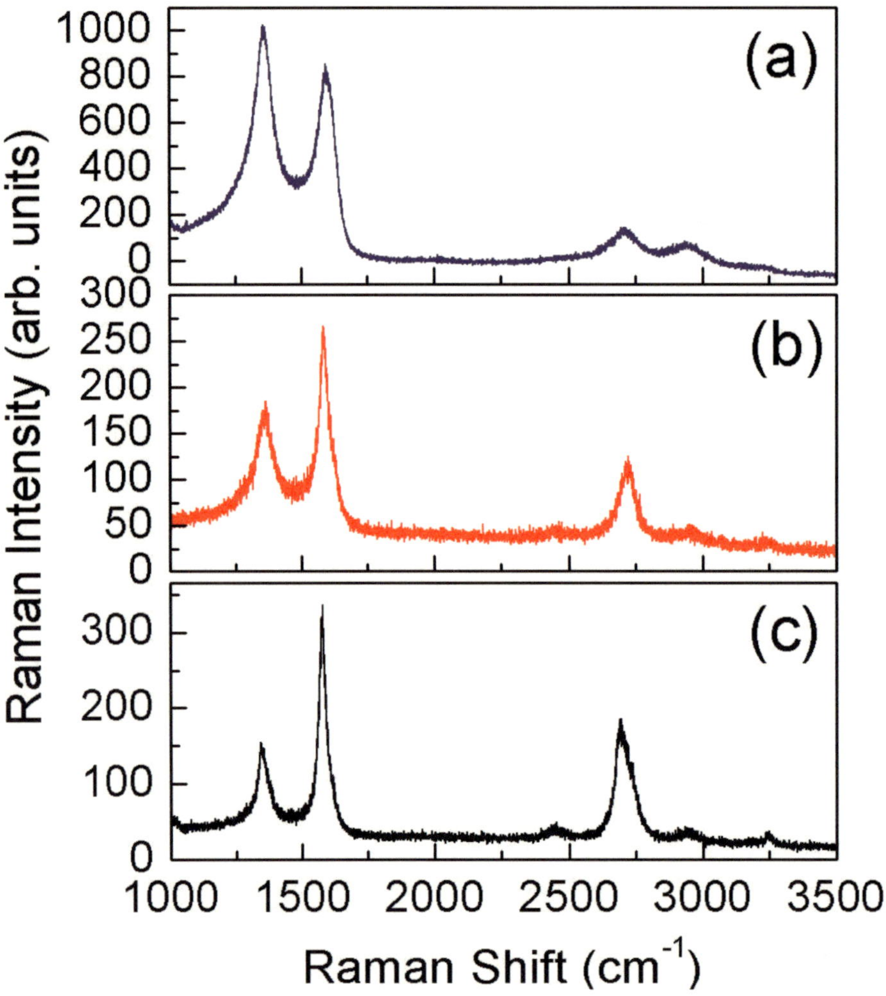

Figure 10. Raman spectroscopic study of different laser ablation time for C deposited onto 250 nm Ni thin film at 650 °C. The corresponding quantities of Raman intensities for each spectrum are summarized in table 1. (a) 120 s laser ablation time. (b) 90 s laser ablation time. (c) 60 s laser ablation time.

http://dx.doi.org/10.5772/55821

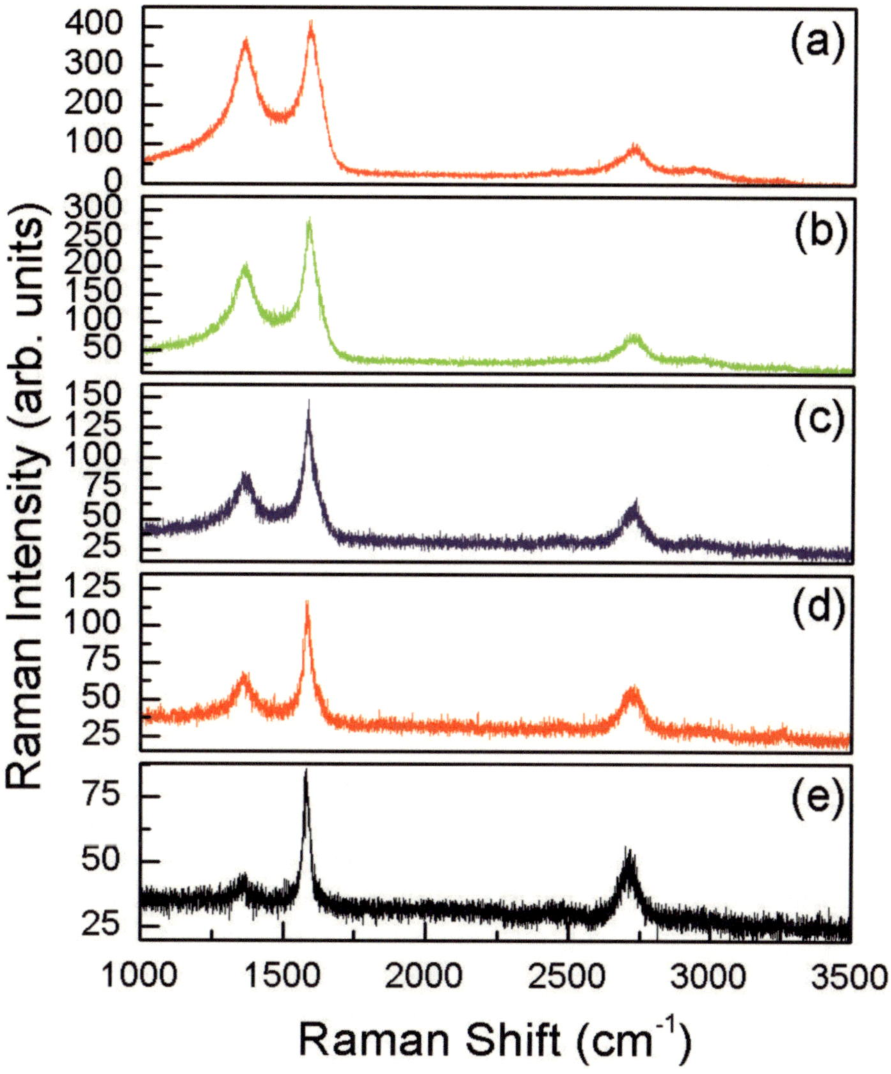

Figure 11. Raman spectroscopic study of different laser ablation time for C deposited onto 250 nm Co thin film at 650 °C. The corresponding quantities of Raman intensities for each spectrum are summarized in table 2. (a) 240 s laser ablation time. (b) 180 s laser ablation time. (c) 120 s laser ablation time. (d) 60 s laser ablation time. (e) 30 s laser ablation time.

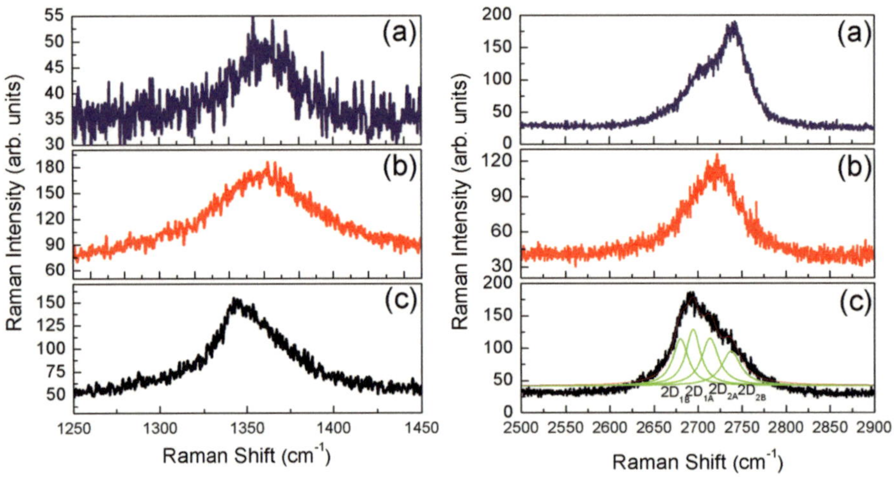

Figure 12. The D and 2D bands for graphene fabricated on Ni thin film by the PLD method.

By the PLD method, other nanostrcutural materials, for instance crystalline Si nanodots, can also be fabricated together with single or multi-layer graphene. Because it is possible to place multi-target within a PLD system, crystalline Si nanodots can then be fabricated via Ni-induced crystallization method as well.[27] Figure 13 (a) and (b) show the SEM images of the crystalline Si nanodots fabricated on top of the graphene layer via the PLD method. Owing to the fact that Si usually required much higher processing temperature (>1000 °C) to be completely crystallized, it has experimentally demonstrated that metal induced crystallization method can reach the same goal with relatively low temperature (< 500 °C). Such kind of Si nanodots structure combined with graphene is very attractive and interesting for studying semiconductor-graphene interface property and eventually reaching the purpose of modern nanoelectronic device design. With regard of such nanostruture, an additional template is necessary in order to define the wide distribution of the Si nanodots. Therefore, a so-called ultra-thin anodic porous alumina (UAPA) template was used in this case. However, the photolithographic and electron beam lithographic techniques can also be applied to graphene. Apparently, it inevitably proves the functionality of PLD based graphene fabrication method in today's graphene research field.

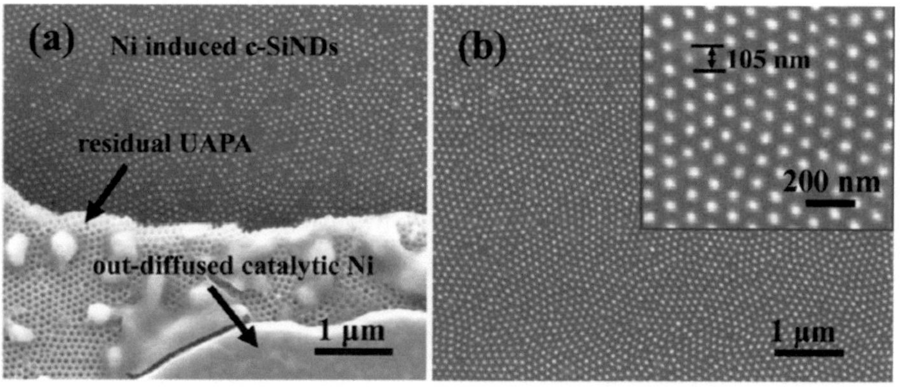

Figure 13. a) Top-view FESEM image of the Ni-induced crystalline silicon nanodots on graphene with remaining UAPA template. (b) The well-ordered nanodots arrays on graphene after lift-off.

5. Conclusion

In this chapter, PLD method has been reviewed for graphene fabrication in this chapter. The foundation of this technique is metal-induced crystallization mechanism. Some catalytic metals need to participate during the crystallization process of C. The advantage of using PLD is to prompt C atoms diffuse into metal thin film at a certain high temperature because PLD can generate high energetic species. After the sample is cooled down, the C atoms segregate from the carbon-metal solid solution and subsequently forming a continuous C thin layer on the top of a carbon containing metal film. The number graphene layer can be controlled by the PLD method, for example by tuning the number of laser pulses, the thickness ratio of C to metal thin film. Besides, we also see the Raman spectroscopy and TEM are both important in characterizing the as-prepared graphene made by the PLD technique. It is expected that PLD based graphene fabrication technique is very promising in current graphene related nano-technology research.

Author details

Kai Wang*

Address all correspondence to: vitalykaiwang@gmail.com

NanoElectronics Group, MESA+ Institute for Nanotechnology, University of Twente, Enschede, The Netherlands

References

[1] I. A. Luk'yanchuk and Y. Kopelevich, Phys. Rev. Lett. 93, 166402 (2004).

[2] K. S. Novoselov, A. K. Geim, S. V. Morozov, D. Jiang, Y. Zhang, S. V. Dubonos, I. V. Grigorieva, and A. A. Firsov, Science 306, 666 (2004).

[3] Y. Zhang, Y.-W. Tan, H. L. Stormer, and P. Kim, Nature 438, 201 (2005).

[4] I. Meric, M. Y. Han, A. F. Young, B. Ozyilmaz, P. Kim, and K. L. Shepard, Nat. Nano. 3, 654 (2008).

[5] X. Li, W. Cai, J. An, S. Kim, J. Nah, D. Yang, R. Piner, A. Velamakanni, I. Jung, E. Tutuc, S. K. Banerjee, L. Colombo, and R. S. Ruoff, Science 324, 1312 (2009).

[6] A. Reina, X. Jia, J. Ho, D. Nezich, H. Son, V. Bulovic, M. S. Dresselhaus, and J. Kong, Nano Lett. 9, 3087 (2009).

[7] R. Hawaldar, P. Merino, M. R. Correia, I. Bdikin, J. Gracio, J. Mendez, J. A. Martin-Gago, and M. K. Singh, Sci. Rep. 2, 1 (2012).

[8] Z. Luo, Y. Lu, D. W. Singer, M. E. Berck, L. A. Somers, B. R. Goldsmith, and A. T. C. Johnson, Chem. Mater. 23, 1441 (2011).

[9] W. K. Kim, Y. M. Jung, J. H. Cho, J. Y. Kang, J. Y. Oh, H. Kang, H. Lee, J. H. Kim, S. Lee, H. J. Shin, J. Y. Lee, Y. C. Kim, I. T. Han, J. M. Kim, J. Yook, S. Baik, and S. C. Jun, App. Phys. Lett. 97, 193103 (2010).

[10] L. s. M. A. Perdigão, S. N. Sabki, J. M. Garfitt, P. Capiod, and P. H. Beton, J. Phys. Chem. C 115, 7472 (2011).

[11] K. Wang, G. Tai, K. H. Wong, S. P. Lau, and W. Guo, AIP Advances 1, 022141 (2011).

[12] J. C. Hamilton and J. M. Blakely, Surf. Sci. 91, 199 (1980).

[13] J. T. Grant and T. W. Haas, Surf. Sci. 21, 76 (1970).

[14] A. T. T. Koh, Y. M. Foong, and D. H. C. Chua, App. Phys. Lett. 97, 114102 (2010).

[15] A. C. Ferrari, J. C. Meyer, V. Scardaci, C. Casiraghi, M. Lazzeri, F. Mauri, S. Piscanec, D. Jiang, K. S. Novoselov, S. Roth, and A. K. Geim, Phys. Rev. Lett. 97, 187401 (2006).

[16] A. C. Ferrari and J. Robertson, Phys. Rev. B 64, 075414 (2001).

[17] F. Tuinstra and J. L. Koenig, J. Chem. Phys. 53, 1126 (1970).

[18] K. Wang and K. H. Wong Semi. Sci. and Tech. 26, 095031 (2011).

[19] J. Wintterlin and M. L. Bocquet, Surf. Sci. 603, 1841 (2009).

[20] K. Wang and K. H. Wong, MRS Online Proceedings Library mrss11-1321-a, 06 (2011).

[21] M. Zheng, K. Takei, B. Hsia, H. Fang, X. Zhang, N. Ferralis, H. Ko, Y. Chueh, Y. Zhang, R. Maboudian, and A. Javey, App. Phys. Lett. 96, 063110 (2010).

[22] M. N. R. Ashfold, F. Claeyssens, G. M. Fuge, and S. J. Henley, Chem. Soc. Rev. 33, 23 (2004).

[23] A. A. Voevodin and M. S. Donley, Sur. Coat. Tech. 82, 199 (1996).

[24] H. M. Smith and A. F. Turner, Appl. Opt. 4, 147 (1965).

[25] D. Dijkkamp, T. Venkatesan, X. D. Wu, S. A. Shaheen, N. Jisrawi, Y. H. Min-Lee, W. L. McLean, and M. Croft, App. Phys. Lett. 51, 619 (1987).

[26] D. P. Norton, in Pulsed Laser Deposition of Thin Films (John Wiley & Sons, Inc., 2006).

[27] G. Tai, K. Wang, Z. Sun, J. Yin, S. M. Ng, J. Zhou, F. Yan, C. W. Leung, K. H. Wong, W. Guo, and S. P. Lau, J. Phys. Chem. C 116, 532 (2011).

Section 3

Properties of Graphene

Chapter 5

Bacterial Interaction with Graphene Particles and Surfaces

Shannon M. Notley, Russell J. Crawford and
Elena P. Ivanova

Additional information is available at the end of the chapter

http://dx.doi.org/10.5772/56172

1. Introduction

Research into the properties and uses of graphene has rapidly expanded over the past decade. Indeed, prior to the seminal paper by Geim and Novoselov in 2004 [1] which eventually led to the 2009 Nobel Prize for physics, the potential of this material was relatively underappreciated. Graphene is a monolayer thick, two dimensional form of carbon atoms linked together in a hexagonal lattice. The sp2 hybridisation of all bonds across the sheet gives rise to its interesting and unique, physical, mechanical, thermal and electrical properties. Thus graphene can be considered to be a 2 dimensional form of its analogue graphite [2]. Importantly, the properties of graphene vary significantly to the bulk material graphite, particularly in terms of electron mobility, and these significant feature differences have driven research in fields as diverse as electronics, materials, energy, defence, security, water and health [3, 4].

Much of the focus of graphene research has been in high value added electronics where miniaturization is a key driver. However to date, the bulk of the material that is produced is geared toward research applications or in composites production. The use of graphene as the load bearing component in composites is highly beneficial [5]. Whilst compatibilization with matrix remains an issue, the high tensile strength of graphene coupled with the ultra-high aspect ratios that are possible using most particulate production methods has led to the rapid uptake of this technology. Furthermore, the potential for using graphene in thin film and coatings applications is increasing, with the goal to produce modified surfaces with improved structural integrity, better heat resistance, conductance, transparence or protection from corrosion [6, 7]. Recent studies have also shown that graphene could also be used in water purification applications due to its high cation exchange capacity and vast available surface area [8]. Other potential applications of graphene include in sensor technology, opto-electronic devices, high electron mobility transistors, supercapacitors, catalysis, photovoltaic (nanocrystal solar cell) and desalination.

Many of these uses for graphene will have significant economic and environmental benefits however it is of great importance that the possible downside effects of incorporating graphene into products which may come into contact with the biosphere are accurately known. A thorough understanding of the interaction with biological material is essential prior to the uptake and utilisation of graphene on a wide scale, particularly if there is significant potential for it to find its way into the environment and human body. Potential biomedical applications for graphene have been suggested. The low surface energy of graphene makes it an attractive substrate for the delivery of hydrophobic drugs. Furthermore, the unique structure could allow its use as a contrast agent in biomedical imaging. The influence of particle size is important in biomedical applications. Single layer graphene may potentially sit in between the leaflets of a lipid bi-layer, it is unlikely though that few layer graphene however could be incorporated. Furthermore, the ability for cells to phagocytise graphene is largely dependent upon particle dimensions. Whilst cytotoxicity is hence of a clear and present concern, graphene as a building block presents a novel opportunity for designing and functionalising systems and products that can potentially transform the manner in which we live. Hence there is a strong push for understanding the interaction of graphene with cells and bacteria. Bacterial interaction with surfaces is ubiquitous in nature. There are also countless examples where biofilm formation leads to significant environmental and health problems. For instance in a marine environment, biofilm formation on ship hulls can lead to corrosion and increased drag as bacteria colonisation can lead to subsequent attachment of other organisms such as barnacles. Whilst this can result in a substantial economic cost, the health problems arising from bacteria attachment to medical implants can be fatal if not properly dealt with in a timely fashion.

This chapter discusses the current knowledge in the area of bacteria – graphene surfaces interactions. A number of graphene production approaches will be also considered along with the brief outline of the graphene surfaces characterisation techniques. A number of recent interesting studies will be reviewed, however at this early stage, no clear conclusions can be drawn on the toxicity of the graphene to bacteria, with some studies suggesting bactericidal activity whilst others suggesting inhibition of attachment and others still no anti-bacterial action. The reasons behind this lack of consistency are therefore of great importance to determine.

2. Production methods and characterisation of graphene

It is useful at this point to define more precisely the term "graphene". The pristine form, as shown in the schematic diagram in Figure 1, contains complete sp2 hybridisation with no defects and only a single layer of atoms. However, in the literature graphene may also refer to few layered material held together through van der Waals forces as well as graphene oxide which is subsequently reduced. It should be noted here, reduced graphene oxide has a significant number of defects which heavily impacts electron mobility and gives rise to a non-

http://dx.doi.org/10.5772/56172

flat (puckered) structure. Few layered graphene also has some of the advantages of graphene such as high aspect ratio yet does not have the same optical and electrical properties of the single layer material.

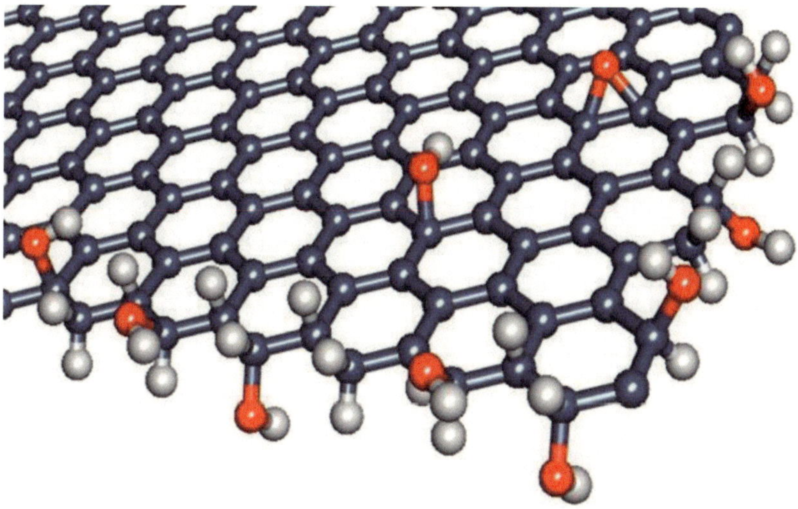

Figure 1. Schematic diagram depicting single layer "pristine" graphene with edge defects (carbon atoms in blue, oxygen in red and hydrogen in white).

It should be apparent that it is not possible for the C atoms at the edges of the particles to be sp2 hybridised. The edges are stabilised through a number of chemical functionalities including epoxy, hydroxyl and carboxylic groups, the latter giving rise to charge in liquids of high dielectric constant such as water. Thus, the smaller the particles, the higher the ratio of edge groups and therefore charges per unit area. Clearly, there are significant differences in the physico-chemical properties of graphene which is influenced ultimately by the production method, these are described below. It is hence not surprising then that the interaction with bacteria would be influenced also by the source of the graphene.

The physical and chemical properties of graphene particles are typically characterised using a variety of techniques including microscopy and spectroscopy. Transmission electron microscopy (TEM) and atomic force microscopy (AFM) imaging are employed to determine the dimensions of the particles. TEM is useful as it can show multiple layers through contrast variation; single layer sheets are relatively transparent to the electron beam however its main advantage is showing the lateral dimensions. AFM is also a technique which has been employed which can give the thickness of the particles. The graphene sheets are typically attached to a substrate prior to imaging, an offset is usually found corresponding to a trapped vapour layer under the particles however for stepped few layer graphene samples, the thickness between layers agrees with literature values. One of the greater advantages of AFM is the ability to combine with Raman spectroscopy to show unequivocally the number of layers of graphene.

Spectroscopic means for characterising graphene are routinely used to give molecular scale chemical properties. Techniques such as Raman and UV-Vis spectroscopy are most common. Raman spectroscopy is considered the gold standard technique for determining the number of graphene layers as well as the presence of any defects [9]. Figure 2 shows a typical Raman spectrum for exfoliated graphene particles as well as graphite, the most prominent peaks are those at 1335 cm^{-1}, 1565 cm^{-1} and 2675cm^{-1} which are termed the D, G and 2D (2D has sometimes been erroneously termed G'). The G peak at 1565cm^{-1} shown in both spectra corresponds to sp2 hybridised carbon and is relatively invariant upon exfoliation. The peak at 1335cm^{1} is sp3 carbon and is introduced through bond cleavage and hence indicates the presence of either edge or basal plane defects. The most important peak though is in the region of 2675cm^{-1} where the position and shape indicate the number of graphene layers. UV-Vis spectrophotometry is also quite useful in determining the presence of defects and has often been used to probe the extent of reduction of graphene oxide to graphene. Strong absorbance across the spectrum indicates a highly conjugated structure and furthermore, the wavelength of intensity maximum provides evidence for the presence of defects.

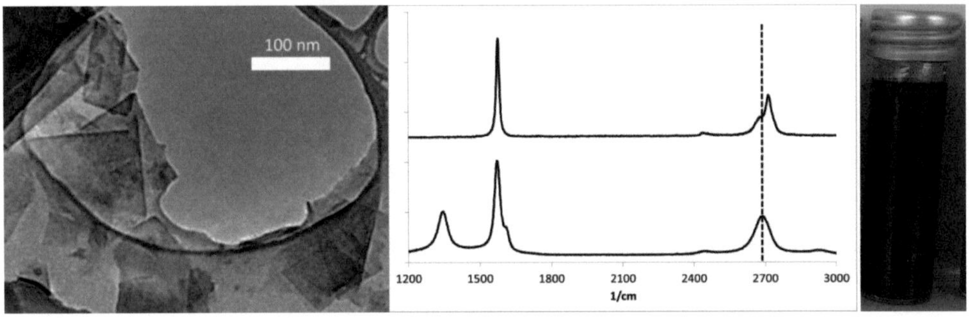

Figure 2. *Left:* TEM of graphene sheets, *middle:* Raman spectra of single layer graphene and bulk graphite, *right:* aqueous suspension of graphene (Reproduced with permission from Reference 30).

The characterisation of graphene is important as it aids in the discussion of differences in the properties depending on preparation method. The most common methods, pertinent to studies involving the interaction of graphene with bacteria are presented below, along with the physical and chemical properties. These include exfoliation of graphite as well the reduction of dispersed graphite oxide and vapour deposition techniques.

2.1. Mechanical exfoliation

This is often referred to as the "Scotch Tape Method" where single and few layer sheets of graphene are removed from a bulk graphite sample through attachment to adhesive tape. This mechanical exfoliation of sheets is considered to be an excellent route to micrometre to sub-millimetre sized graphene particles without the introduction of any defects (apart from those at the edges). Thus, the sheets produced using this method are considered "pristine" and hence have excellent electrical conductivity. Graphene prepared through mechanical exfoliation is used mainly in research applications where low quantity, high purity samples are required.

Most often, the samples are fixed to a supporting substrate exposing the conjugated basal planes. Thus, the surface for bacterial study is "atomically" smooth with lateral dimensions in excess of the typical size of a bacteria cell.

2.2. Epitaxial growth and Chemical Vapour Deposition (CVD)

Both techniques of epitaxial growth and chemical vapour deposition rely on a suitable substrate for supporting the deposited graphene [10]. Typically these substrates are metal or metal-like (eg silicon) as process parameters usually involve high temperatures precluding deposition onto polymer surfaces. Epitaxial growth of graphene was first demonstrated on a silicon carbide substrate [11-13]. Temperatures in excess of 1300 °C and ultra-high vacuum are normally required to produce a graphene multilayered materials through sublimation of silicon, usually with thicknesses in the range of 5-100 sheets but over large (wafer scale) areas. However some studies have had success with moderate vacuum conditions in controlled environments. Epitaxial growth of graphene shows great promise as lithographic techniques may be used in conjunction to produce patterned surfaces. It should be noted though, that epitaxially grown graphene has significantly different properties to the exfoliated material which is largely due to interfacial effects of the underlying SiC substrate. For example, the quantum Hall effect is absent with epitaxial graphene and furthermore, slight differences in periodicity and electron diffraction patterns are observed. Thus electron mobility is not of the same order as exfoliated graphene.

Aside from SiC substrates, other materials such as transition metals have been employed to produce graphene layers. CVD has been employed with iridium [14], ruthenium [15], platinum and nickel [16]. Nucleation and growth occurs after exposure of the substrate to a hydrocarbon gas such as methane or ethylene at low pressures (often ultra high vacuum conditions) and high temperatures. The condensed gas is converted to a carbon layer upon cooling of the substrate. This typically produces a multilayered graphene surface, the properties of which are similar to those detailed for epitaxially grown graphene.

Both techniques of epitaxially growth and CVD are capable of producing large area graphene with high transparency and conductance which are suitable for studies of bacteria. They also have the added advantage of ease of patterning which can be beneficial for potential investigations of directed growth of cells [16]. There is control though over the thickness (ie number of layers) which is also advantageous and the roughness of such graphene surfaces is somewhat conformal to the underlying substrate [17].

2.3. Reduced Graphene Oxide (RGO)

One of the most common processes for producing industrial scale quantities of graphene is based upon a method first developed by Hummers for oxidising graphite [18]. Briefly, the modified Hummers method involves the oxidation of graphite powder with for example, sulphuric acid and potassium permanganate to produce graphite oxide. This product is then dispersed in water and subjected to sonication to split apart the bulk material into single and few layer graphene oxide [5, 19, 20]. The suspension is relatively stable due to the high charge (zeta potential greater than – 60 mV) on the exfoliated graphene oxide sheets. Subsequently, the graphene oxide is reduced, often with hydrazine to re-form the C-C bonds with most, but importantly not all of the conjugation maintained. Whilst possible to produce on an industrial scale, the Hummers method does have some significant drawbacks including the highly toxic nature of some reactants as well as high cost of oxidisers. Furthermore there is a somewhat complicated post-production purification treatment. However one of the greatest disadvantages is the introduction of a significant level of defects which reduces the conductivity of the particles substantially as well as creating a degree of roughness [10]. Hence RGO is most often used in composites applications where conductivity is not as important and roughness increases the surface area. In terms of potential interaction with bacteria, the reduced conductivity and increased roughness are not necessarily detrimental and studies involving graphene oxide and RGO have been reported.

A similar procedure using carbon nanotubes as the starting material has also been reported which can result in the production of graphene. Often termed "unzipping", carbon nanotubes have been treated with strong oxidizers to cleave bonds allowing the tube to unroll. The product is subsequently reduced to repair defects and regenerate the sheet conjugation. The graphene particles produced from unzipping of nanotubes are hence of limited dimensions which also means they have an inherently larger edge defect ratio and charge per unit area than other preparation techniques.

2.4. Liquid phase exfoliation

One promising technique for the production of relatively chemically unmodified graphene in large quantities is liquid phase exfoliation, sometimes in the presence of surfactants and polymers [21-32]. This method for the preparation of graphene has been performed in water, organic solvents as well as ionic liquids [33, 34] however most often the concentration produced is low (< 1mg/mL). Typically, the interfacial energy of the liquid phase is matched to the surface energy of graphite promoting exfoliation under ultrasonication. In other words, the cohesive energy of the liquid phase and solid phase are comparable. This is achieved with appropriate choice of organic solvent, for example, N-methylpyrrolidone (NMP), N,N-Dimethylacetamide (DMA), γ-butyrolactone (GBL) and 1,3-dimethyl-2-imidazolidinone (DMEU) or ionic liquid with a surface tension of the order of about 41 mJ/m^2. In aqueous suspension, surface tension is reduced to the optimum energy range through the addition of a surfactant in order to separate the graphene sheets beyond the range of the strong attractive van der Waals forces. The surfactant also has a secondary role in preventing re-aggregation through adsorption to the graphene surface resulting in a repulsive energy between sheets.

A variety of surfactants have been employed in aqueous solution processing through sonication exfoliation. Monomeric surfactants, typically with cationic charge have been extensively used. Through adsorption, the graphene produced in this manner has a significant positive charge preventing re-aggregation. The surfactant is not adsorbed irreversibly so changes in solution conditions (eg through dialysis or dilution) can result in de-stabilisation of the suspensions. Polymeric type surfactants such as Pluronics have also been used which tend to improve stability due to the strongly bound nature of the hydrophobic components [30].

http://dx.doi.org/10.5772/56172

Clearly, the presence of any bound surfactant to the surface of graphene produced in this manner will have a significant influence on the interaction with bacteria. Furthermore, depending on processing parameters such as sonication power, sonication time, presence of salt and surfactants and centrifugation, the particles produced can be chemically varied. High sonication power leads to cavitation which can cleave bonds and oxidise the sheets as well as introduce defects. However, gentle sonication gives relatively pristine sheets suitable for both composites and electronics applications. Centrifugation is also important to separate single and few layer graphene from the unexfoliated bulk material. Importantly, liquid phase exfoliation always produces a distribution of graphene thickness and hence careful control of sedimentation is required to maximise the proportion of single layer graphene. Sample thickness may also influence the interaction with bacteria as will be discussed in later sections.

2.5. Surfaces prepared from graphene particles

Graphene surfaces have been prepared in a number of ways using particles as a starting material. These include "paper" like surfaces, supported and free standing, thin films deposited from solution onto a substrate as well as the development of thicker interfaces using the layer-by-layer technique. These methods are described in greater detail below.

One convenient method for producing macroscopically large areas of graphene oxide and graphene is to form a "paper" [20, 35]. Typically, a suspension of the particles is vacuum or pressured filtered, the directional flow gives rise to the alignment of particles to form a lamella structure as demonstrated by x-ray diffraction. Upon drying of this graphene particulate filter cake, the attractive forces between the sheets leads to a very strong material which can be peeled away from the underlying membrane substrate. The resultant material is often referred to as paper as the process is very similar to that involving wood fibres. In contrast though to conventional paper surfaces, graphene paper has a higher tensile strength and stiffness. The alignment of the particles means that the surface topography is flat and furthermore, the thickness of the carbon film can be tuned simply by increasing the concentration of graphene in the suspension prior to filtering. Many studies have presented cross-sectional images of the paper using SEM where the lamella structure is confirmed. Depending on thickness of the graphene paper, such cross-sections could also be used for bacterial studies. It is also possible to create hybrid paper structures through reactions within water layers in the graphene film or through simple addition of other materials [36, 37].

The direct deposition of graphene particles onto a planar surface has also been extensively used to create areas of graphene suitable for further studies. Attempts have been made to "stitch" together such particles however many of the important properties such as high conductivity were lost. Thus, in most cases, a mosaic like structure of single and few layer graphene particles is present. Imaging of these surfaces using AFM profiling, combined with Raman spectroscopy is arguably the best way of characterising the graphene particles but it should also be apparent that these type of surfaces present a convenient way of producing graphene modified surfaces. Depending on the source of such particles, the areas covered by graphene is variable but is usually of the order a square micron.

In order to achieve a complete coverage, the layer-by-layer technique [38, 39] has been used. This involves the alternate deposition of positively and negatively charged components, using polyelectrolytes, however with a native negative charge, graphene oxide and graphene with edge defects have been introduced into such assemblies [20, 29, 40]. The thickness of the film containing graphene can be easily selected by determining the desired number of bilayers. Figure 3 shows the structure of the multilayer assembly incorporating cationically stabilised graphene and poly(acrylic acid) as a function of the number of layers [29].

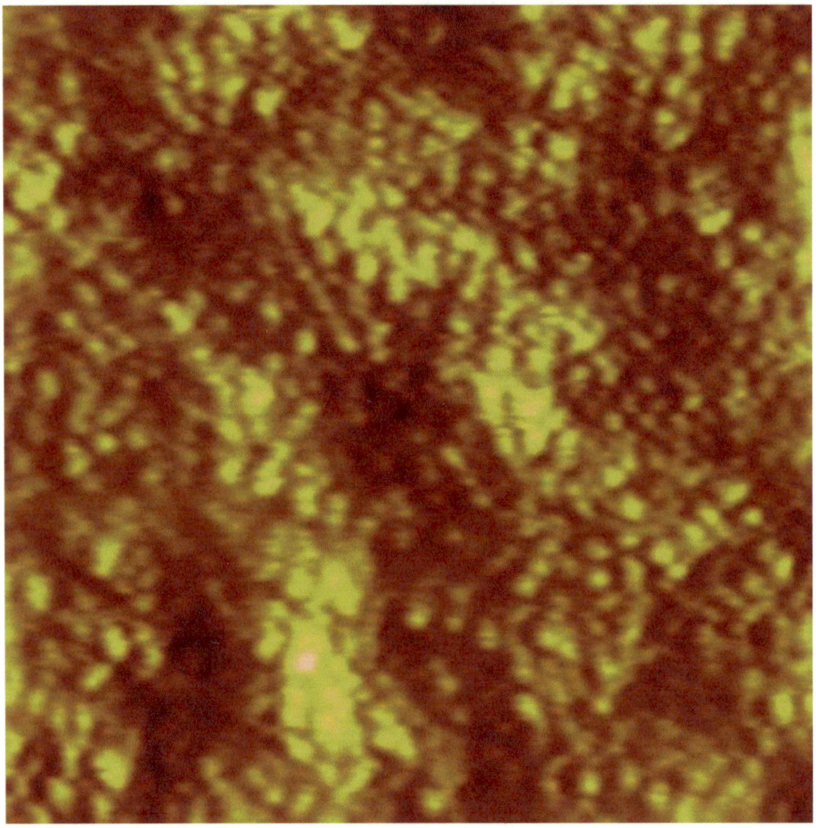

Figure 3. Surface structure of a graphene thin film prepared through the alternate deposition of cationically modified graphene and polyacrylic acid probed using atomic force microscopy imaging, the scale is 500 nm × 500 nm. (Reproduced with permission from Reference 29).

3. Some insights on the nature of bacterial attachment to surfaces

In a thermodynamic sense, adhesion can be defined as the energy required to create new surface area upon separation in a defined medium (such as air or water). As this suggests, the effective surface energies of the two interfaces are hence of critical importance. Thus the surface chemistry of the bacteria as well as the substrata to which the bacteria may attach must be well understood and considered as one of the possible parameters that may either promote or inhibit bacterial attachment and subsequent biofilm formation. The physical surface structure is also a critical factor influencing the interaction with bacteria. These details are discussed further below.

The chemistry of the substrate has been shown to influence the adhesion of bacteria and the build-up of the biofilm. Typically, bacteria have shown greater affinity for hydrophobic surfaces [41]. This has been hypothesised to be due to hydrophobic interactions although this is certainly dependent on the type of bacteria studied. Furthermore, an effective strategy for the creation of anti-bacterial surfaces has been to modify the surface to give a hydrophilic chemistry with polymeric materials such as polyethylene glycol (PEG) which are highly hydrated. In order for attachment to occur, the water molecules must be displaced, which is a thermodynamically unfavourable result. At high ionic strengths, the PEG chains can dehydrate and the efficacy for preventing bacterial adhesion of such surfaces is weaker [42]. It is commonly believed that adsorption on hydrophobic surfaces is rapid with strong binding forces, while adhesion to hydrophilic surfaces followed the model of reversible and irreversible adhesion proposed by Marshall [43] and can be described by DLVO theory (Derjaguin, Landau, Verwey, and Overbeek) [44, 45]. Initially, a weak and reversible stage of the adhesion was observed at separation distances of several nanometers, at which point the bacterium can be removed by shear forces or desorb spontaneously. At a later stage, this attachment can be converted into irreversible adhesion by synthesis of extracellular biopolymers or by stabilization of conformational changes in existing polymers. These polymers bridge separation distances of less than 1 nm, displacing the adsorbed water and /or neutralising the electrostatic repulsion. Another factor that may influence the adhesion of bacteria is the presence of surface charge. Many surfaces carry a significant net charge due to ion dissolution, ion adsorption or surface ionisation, the latter effect gives rise to the potential for either net positive or net negative charge depending on pH. This tends to be less important for polymeric surfaces which are inherently hydrophobic. It is interesting to note that graphene, in a similar sense to graphite has an intermediate wettability and depending on the method of production, may have a varying level of charge.

At this point, it is useful to introduce some definitions to make a distinction between surfaces which inhibit attachment and surfaces which effectively destroy bacteria on contact [46]. Both types of surfaces are often classed as "anti-bacterial" however the latter should more correctly be classed as bactericidal. Some chemistries such as those involving quaternary ammonium groups have shown strong bactericidal action through inducing permeability of the membrane. It has been demonstrated that surface morphology can also be used to kill bacteria. Typically, such structured surfaces have significant roughness or peaks with high aspect ratios which can cause the bacterial cell wall to rupture under its own weight upon contact [46]. Clearly in the case of graphene, the edges also fulfil this latter criterion however the basal planes can be considered atomically smooth. Therefore there is an expectation then that the interaction of bacteria with graphene will be highly dependent on the orientation of the surface relative to the bacteria.

3.1. Surface chemistry of bacterial cells

The bacterial cell outer layer is composed of complex biological molecules; therefore at the time of the cell-surface interaction the cell surface chemistry is heterogeneous and depends on various physicochemical and biological parameters. For example it was shown that on the surface of the cyanobacterium *Calothrix* sp. KC97 that the three main functional groups (carboxyl, phosphoryl, and amine) present are present in several discrete binding sites [47]. Carboxyl groups are found in abundance in cell wall constituents such as diaminopimelic and glutamic acids in the peptide cross-linkages of peptidoglycan molecules; or carboxylated teichoic acids, covalently bound to peptidoglycan, or other phosphoryl-containing polymers in the cell envelope which include LPS. Lipids present in the cell wall may also exhibit phosphoryl groups of the phosphatidylglycerol and phosphatidylcholine. It was also reported that the negatively charged carboxyl and positively charged amine groups, present at natural pH, usually occur in approximately equal proportions. The slight increase in the negative charge with increasing pH was suggested to be predominantly a result of deprotonation of carboxylic and phosphoric groups to form negatively charged species but may also be contributed to by deprotonation of amine groups to form electroneutral species.

In addition to the inherent heterogeneity of cellular outer layers, bacteria secrete a variety of extracellular polymeric substances (EPS), including polysaccharides, proteins, and nucleic acids, that vary in molecular mass and structural properties [48]. Extracellular polysaccharides (EPSs) are present in many forms, including capsular polysaccharides, which are also referred to as 'cell-bound extracellular polymeric substances' which remain connected to cell surfaces by means of a covalent attachment to phospholipid or lipid A molecules present at the bacterial surface [49]. While some capsular polysaccharides may be released into the growth medium (*i.e.* become 'free') as a consequence of the low stability of the phosphodiester linkage between the polysaccharide and the phospholipid membrane anchor, other tightly attached capsular polysaccharides form a distinct structural layer (the capsule) which encloses the cell and serves as a protective layer. This layer acts as a shield on the cell surface, affording the cell protection from major bacterial pathogens [50]. Most significantly, EPS play a major role in mediating the bacterial colonization of surfaces, biotic and abiotic, by enabling cell adhesion and co-aggregation via dipole interactions, covalent or ionic bonding, steric interactions, and hydrophobic association, making the target surface more attractive for bacterial attachment. The temperature, solution pH, electrolyte and macromolecule concentration, and adsorbent surface chemistry will directly influence the chemical composition and structure of the EPS substances that are responsible for the surface conditioning.

3.2. Interaction of bacteria with "graphene" modified surfaces

Only a relative few studies have investigated the interaction of bacteria with graphene or graphene-like surfaces in comparison to other materials such as minerals, metals and plastics. The unique structure of graphene, in particular the physical dimensions which are characterised by the high aspect ratio, provide opportunities for creating surfaces with tailored properties which may inhibit or promote bacterial attachment as well as potentially inducing cell death. The varied preparation procedures for producing single and few layer graphene

http://dx.doi.org/10.5772/56172

materials also provide the ability to tune the interaction with bacteria. These are perhaps the reasons why there is no consistent picture regarding the interaction of bacteria with such surfaces modified in this way [51]. However, there are clear circumstances where the build-up of the biofilm is inhibited. The available studies are summarised below.

There is great interest in producing graphene modified surfaces for anti-bacterial applications. Studies have suggested that such surfaces are indeed anti-bacterial yet others show the promotion of cell attachment. The first study undertaken for the investigation of bacterial interaction with a graphene-like surface was by Akhavan and Ghaderi in 2010 that used Gram-negative (*E. coli*) and Gram-positive (*S. aureus*) bacteria [52]. In this study, single and few layer graphene oxide and reduced graphene oxide deposited onto a stainless steel substrate was used. The orientation of particles was such that a significant number of edges were exposed. A substantial loss in cell viability for both *E. coli* and *S. aureus* was observed. However within this investigation, some trends were reported. For example, the cell viability was less for the Gram positive bacteria *S. aureus* and furthermore, the reduced graphene oxide surface had a greater ability to inhibit attachment as well as kill the bacteria. This study showed that membrane damage induced through contact with the graphene oxide and reduced graphene oxide particles occurred through measurement of the efflux of the cytoplasmic materials which supports the observation of a greater toxicity toward Gram-positive bacteria. The authors suggested in addition that the reduced graphene oxide particles have sharper edges than the oxidated counterparts leading to increased potency however this theory is not well supported by the direct imaging of particles, nor the extensive literature in this area [1, 2].

Hu *et al* also investigated the interaction of *E. coli* with graphene oxide and reduced graphene oxide surfaces prepared through vacuum filtration into a paper like surface [53]. In this instance, the particles lay relatively flat on the surface with only a few edges oriented away from the interface. An airborne bacteria test was used in this study with the viability of cells subsequently determined. Similar to the work of Akhavan and Ghaderi, a significant loss of viability was observed for both graphene oxide and reduced graphene oxide; however in contrast they determined the greatest loss for the graphene oxide. Again, the toxicity mechanism was suggested to be due to membrane damage induced through contact with the particulate edges which was confirmed through scanning electron microscopy imaging.

In contrast to the above mentioned studies where graphene oxide has shown anti-bacterial activity, the study of Ruiz *et al* suggested that there is no detrimental effect to bacteria and indeed that bacterial growth rates were higher [54]. Graphene oxide paper surfaces were used in this study, the material prepared using a modified Hummers method together with *E. coli* as the model bacteria. Growth rates of bacteria were up to 3 times as fast on membranes modified with a high surface coverage of graphene oxide with a preferential attachment to areas with high particle density. No high resolution images of the paper surface were presented however it is likely that using this vacuum filtration preparation procedure that the graphene oxide particles are lying flat on the membrane substrate with few exposed edges. Previously it has been suggested that the cytotoxicity is due to membrane damage caused by rupture due to particle edges [51]. The oxidised basal planes which are however exposed in this case may confer an increased wettability on the particles which in turn allows a stronger interaction with the bacterial LPS through hydrogen bonding.

The interaction of bacteria with composite surfaces of graphene materials and other polymers has also been investigated. Many exfoliation techniques including the Hummers method and liquid phase exfoliation lend themselves to easy modification of the surface properties of the graphene particles through binding with polymers in either a physisorbed or chemisorbed sense [28]. This is due to the introduction of reactive sites or more simply charged groups which may lead to electrostatic adsorption of oppositely charged (cationic) polyelectrolytes. The combination with inherently anti-bacterial polymers such as those containing quaternary ammonium groups has proven highly useful. Sreeprasad *et al* produced composite films of graphene oxide (and reduced graphene oxide) and lactoferrin or chitosan [55]. *E. coli* was used as the bacteria in this study. Without surface modification through protein or polymer adsorption, the anti-bacterial activity of graphene oxide and reduced graphene oxide was weak. A significant increase in cytotoxicity was observed through incorporation of lactoferrin however the greatest loss in cell viability was due to the combination of lactoferrin and chitosan. There was no statistical difference in the graphene material observed.

Another study involving graphene oxide composite films with included polyvinyl-N-carbazole, a polymer with inherent antibacterial characteristics [56, 57]. The efficacy of these composite surfaces against two Gram-negative bacteria (*E. coli* and *C. metallidurans*) and two Gram-positive bacteria (*B. subtilus* and *R. opacus*) was investigated. The anti-bacterial activity of the composite material was higher than the unmodified graphene oxide particles and surfaces. Not only did the surfaces inhibit build-up of the biofilm but it also inactivated a significant proportion of bacteria. Again, there was a greater loss in viability of Gram-positive bacteria exposed to the composite film. This study also showed that whilst anti-bacterial activity was observed, no significant cytotoxicity toward eukaryotic cells (human fibroblast) was observed suggesting that graphene oxide-polyvinyl-N-carbazole composites could potentially be used in clinical applications.

More recently another graphene oxide-polymer composite material also showed anti-bacterial activity as well as the promotion of the growth of human cells (adipose-derived stem cells) [58]. Poly-l-lysine (PLL) is a cationic polyelectrolyte which can adsorb irreversibly to graphene oxide through an electrostatic interaction. PLL has been shown to have anti-microbial activity but has also been used for the immobilization of cells to negatively charged surfaces such as silica hence Some *et al* used functionalised graphene oxide with PLL to investigate cytotoxicity of *E. coli*. In addition to the graphene oxide composite, the particles were subsequently reduced and the anti-bacterial activity studied. This study showed that there was no anti-bacterial activity of unmodified graphene oxide, as with previous studies, the surfaces were essentially arranged in a flat conformation and hence edges were not exposed. However, the graphene oxide-PLL composite showed potent anti-bacterial activity toward the Gram-negative bacteria *E. coli*. The relatively highly cationically charged surface was speculated to be the mechanism inhibiting attachment. Some *et al* also modified the composite to boost the amount of cationic charges through covalent attachment of a diazonium salt which resulted in a decrease in cell adherence. Staining experiments demonstrated that cell death was also induced.

http://dx.doi.org/10.5772/56172

In the current literature, all studies involving bacteria have been conducted with either graphene oxide or reduced graphene oxide particles which have subsequently been formed into a surface for attachment and viability studies. There is still no consensus view in regard to the inherent anti-bacterial nature of such surfaces. The information available suggests there is some difference toward cytotoxicity depending upon the orientation of the particles attached to the surface. Graphene oxide and reduced graphene oxide sheets with the sharp edges exposed to contact with the bacterial membrane seem to induce rupture. In contrast, the studies which have shown no anti-bacterial activity tend to involve surfaces with the basal planes at the interface. As outlined in the previous sections, the materials used to date in studying the interaction with bacteria are quite distinct from what could be called "pristine" graphene or graphene produced through epitaxial growth. That is, no studies involving the highly conjugated, defect free and single layer of sp2 hybridised carbon have been reported.

4. Interaction of graphene particles with bacteria in suspension

Many of the studies outlined above also investigated the interaction of bacteria with graphene oxide and reduced graphene oxide particles in suspension with comparable results to those involving substrates produced from these particles [54, 55, 58, 59]. These studies usually investigated cell viability as a function of time and in general terms, greater loss in viability was observed for Gram-positive bacteria. In all of these studies, membrane rupture was proposed to be the toxicity mechanism. In suspension, the orientation of the graphene oxide or reduced graphene oxide particles may interact with bacteria in a manner of edgewise contact. Undoubtedly the evidence supports the suggestion of membrane rupture inducing cytotoxicity; however no molecular scale mechanism has yet been proposed.

In addition to the studies reported in the previous section, there have been others which have investigated the cytotoxicity of graphene and similar materials toward a range of bacteria in suspension. Such studies are highly important as there are many potential uses where graphene and graphene oxide have been proposed to be employed from clinical applications to water purification and coatings. In these uses it could reasonably be expected that the particles may interact with a range of cellular materials including bacteria. A recent study chronicled the antibacterial activity of a family of related materials dispersions including graphite, graphite oxide, graphene oxide and reduced graphene oxide toward the model Gram-negative bacteria *E. coli* [59]. All of these materials showed some anti-bacterial activity using a colony counting method. The materials with smallest thickness, that is graphene oxide and reduced graphene oxide showed a greater cytotoxicity with a preference for the reduced form. These authors noted that membrane rupture occurred however also demonstrated that oxidative stress was also a potential mechanism for the observed cell death.

The antibacterial efficiency of reduced graphene oxide nanosheets was also studied by Krishnamoorthy *et al* [60]. The authors studied the minimum inhibitory concentration (MIC) of the particles against both Gram-positive (*E. faecalis* and *B. subtilis*) and Gram-negative (*E. coli* and *S. typhimurium)* bacteria. The MIC of reduced graphene oxide particles was observed to be significantly lower for the Gram-negative bacteria. This was hypothesised to be due to the much thinner peptidoglycan layer of these types of bacteria. This is in direct contrast to other studies which suggest that the presence of the secondary cell membrane of Gram-negative bacteria provides a better resistance to membrane induced damage in the presence of these particles. The authors also provided evidence that enhanced lipid peroxidation occurred in suspensions containing reduced graphene oxide.

Many studies have investigated the difference in activity between graphene oxide and reduced graphene oxide yet to date only one has looked into the particle size dependence [61]. Liu *et al* demonstrated that larger sheets had a stronger antibacterial action than smaller graphene oxide particles. Different sized sheets were prepared through varying the sonication time in the modified Hummers method used to prepare the graphene oxide suspension. The lateral size difference was up to 2 orders of magnitude. The loss of viability of *E. coli* cells increased with increasing size and was due to the ability of larger sheets to effectively isolate and wrap bacterium cells preventing proliferation.

Akhavan *et al* have investigated the potential of "wrapping" of bacteria with graphene particles in order to reduce bioactivity [62]. The model Gram-negative bacterium *E. coli* was used in this study. The authors observed no significant inactivation of the bacteria in the presence of graphene oxide suspension or reduced form of graphene. However, the study was also conducted in the presence of melatonin, a reductant, which resulted in the aggregation of bacteria and functionalised graphene particles. AFM showed that the bacteria were ensconced by particles which resulted in a decrease in active cells.

Additional work by Akhavan and Ghaderi has demonstrated that *E. coli* was able to reduce graphene oxide under acidic conditions [63]. The oxygen containing functional groups were reduced by as much as 60% over a period of two days as the graphene oxide particles were supposed to be biocompatible. Interestingly, upon conversion to reduced graphene oxide, antibacterial action was shown, due in large part to the existing attached bacteria to the graphene surfaces as well as growth inhibition of these particles as previously discussed.

Another successful strategy that has been employed for producing increased anti-bacterial action of graphene oxide has been to functionalise the surfaces with nanometric silver particles. Zhang *et al* used gelatin as a "green" reducing agent for Ag^{2+} ions to form silver nanoparticles of size about 100 nm [64]. These were also stabilized with gelatin and subsequently mixed with a graphene oxide suspension. The positively charged amino groups of the gelatin stabilised silver particles were then electrostatically attracted to the negatively charge groups on the graphene oxide surface. These composite particles showed concentration dependence for the growth inhibition of *E. coli* with as little as 10 ppm required to reduce growth by 99.9% (the corresponding non-functionalised graphene oxide particles reduced growth by 38%). Combining graphene oxide with silver nanoparticles was also undertaken by Das *et al* to investigate the interaction with the Gram-negative bacteria *E. coli and P. aeruginosa* [65]. Sodium borohydride as a reducing agent was added to a suspension of graphene oxide particles in the presence of $AgNO_3$ under vigorous stirring to produce decorated sheets. The size of the Ag nanoparticles was of the order of 5 - 25 nm. The authors showed an increasing growth delay of both bacterial strains with increasing nanoparticle concentration.

A variation on the surface modification of graphene oxide sheets with nanoparticles is to create a hybrid material which is photocatalytic. This approach was taken by Liu *et al* where the surface was functionalised with Ag_3PO_4 nanoparticles to create a material with strong absorbance in the visible region [66]. These particles show enhanced photocatalytic activity as demonstrated by the degradation of a model dye AO7 through the production of reactive oxygen species. The authors also observed strong antibacterial action against *E. coli* of the composite graphene oxide-Ag_3PO_4 to a similar extent of synthesised Ag_3PO_4 nanoparticles. The mechanism of action was suggested to be due to oxidative stress induced disinfection upon exposure to visible light.

5. Outlook

Whilst much work has been undertaken toward understanding the interaction of bacteria and graphene (and related materials) few clear unambiguous trends can currently be drawn on bacteria properties or physico-chemical properties of the particles. The overwhelming majority of studies have been conducted with graphene oxide prepared using the Hummers method or with the subsequently reduced graphene oxide particles. As discussed, many of the unique properties of graphene such as high electron mobility are significantly diminished upon oxidation and reduction. Hence in order to adequately state the bacterial interaction with graphene and related materials, further studies using differently sourced materials must be undertaken. Such studies should involve exfoliated graphene which retains the extended conjugation or alternatively, epitaxially grown graphene materials.

The orientation of the graphene sheets immobilised at the interface has a clear influence on the observed bactericidal action. Single layer graphene and graphene oxide have a thickness of about 0.3 nm, or atomic thickness, and hence have been speculated to be able to rupture the cell membrane. This was shown with both Gram-positive and Gram-negative bacteria which suggest the ultrastructure of prokaryotic cell walls is not a factor in the observed bactericidal activity. Furthermore, the studies which had an orientation of particles parallel to the interface may have inhibited bacterial attachment in some cases however cell death was not evident. In addition, the lateral size of particles had an influence on cell viability which supports the idea that the edge density is important. Surfaces patterned to a greater extent, particular on the length scale of the size of the bacteria will improve understanding of the role of particle orientation. Many techniques are described in the literature which may allow a greater degree of control over the orientation of particle surfaces through lithography. Epitaxially grown graphene surfaces combined with photolithography may prove useful to achieve this control.

The bacteria species studied to date are predominantly common types such as *E. coli* or *P. aeruginosa*. The variety of bacteria available is vast with diverse properties. It would be of great interest to expand the knowledge of the interaction with graphene and graphene oxide materials beyond these systems to others of high importance in clinical applications such as *S. aureus* or those with substantially different surface chemistries including highly hydrophobic bacteria. The studies detailed above have also used bacteria with simple geometries such as spheres and rods but other more complex geometries are common.

The majority of studies have focussed on anti-bacterial applications of graphene oxide and reduced graphene oxide derivatives. Whilst this is undoubtedly an important avenue of research, there are many applications where promoting bacterial growth on surfaces is advantageous. Controlling particle orientation as well combining with biocompatible polymer materials may lead to the potential for utilising graphene particles in biomedical applications.

The increased potential use of graphene, graphene oxide and reduced graphene oxide surfaces and particles in technology demands a thorough understanding of cytotoxicity. Nanoparticles are becoming more prevalent in biomedical applications. Examples include the use of particles in medical imaging. Another prominent example is the use of gold nanorods which may induce local heating once incorporated into tissue for novel anti-tumour applications. Porous nanoparticles have also been suggested for drug delviery. Bacteria form an important component of the biomass and important function in the environment. Hence it is timely and essential that all potential benefits and disadvantages of the use of this new class of materials are appreciated prior to the widespread uptake and use of graphene in novel applications.

Acknowledgements

SN acknowledges financial support from the Australian Research Council under the Future Fellowships scheme.

Author details

Shannon M. Notley*, Russell J. Crawford and Elena P. Ivanova

*Address all correspondence to: snotley@swin.edu.au

Faculty of Life and Social Sciences, Swinburne University of Technology, Hawthorn VIC, Australia

References

[1] Novoselov, K. S, et al. Electric field effect in atomically thin carbon films. Science, (2004). , 666.

[2] Novoselov, K. S, et al. Two-dimensional atomic crystals. Proc. Natl. Acad. Sci. USA, (2005). , 10451.

[3] Geim, A. K. Graphene: Status and prospects. Science, (2009). , 1530-1534.

[4] Geim, A. K, & Novoselov, K. S. The rise of graphene. Nat. Mater., (2007). , 183.

[5] Stankovich, S, et al. Graphene-based composite materials. Nature, (2006). , 282-286.

[6] Singh RamanR.K., et al., Protecting copper form electrochemical degradation by graphene coating. Carbon, (2012). , 4040-4045.

[7] Prasai, D, et al. Graphene: corrosion-inhibiting coating. ACS Nano, (2012). , 1102-1108.

[8] Li, D, & Kaner, R. B. Graphene-based materials. Science, (2008).

[9] Ferrari, A. C, et al. Raman Spectrum of Graphene and Graphene Layers. Phys. Rev. Lett., (2006). , 187401.

[10] Allen, M. J, Tung, V. C, & Kaner, R. B. Honeycomb carbon: A review of graphene. Chem. Rev., (2010). , 132-145.

[11] De Heer, W. A, et al. Epitaxial graphene. Solid State Communications, (2007). , 92-100.

[12] Berger, C, et al. Electronic confinement and coherence in patterned epitaxial graphene. Science, (2006). , 1191-1196.

[13] Berger, C, et al. Ultrathin epitaxial graphite: 2D electron gas properties and a route twoard graphene-based nanoelectronics. J. Phys. Chem. B, (2004). , 19912-19916.

[14] Diaye, N, et al. Structure of epitaxial graphene on Ir(111). New Journal of Physics, (2008). , 043033.

[15] De Parga, A. L. V, et al. Periodically rippled graphene: Growth and spatially resolved electronic structure. Phys. Rev. Lett., (2008). , 056807-056804.

[16] Reina, A, et al. Large are, few layer graphene films on arbitrary substrates by chemical vapour deposition. Nano Lett., (2009). , 30-35.

[17] Rasool, H. I, et al. Continuity of graphene on polycrystalline copper. Nano Lett., (2011). , 251-256.

[18] Hummers, W. S, & Offeman, R. E. Preparation of graphitic oxide. J. Am. Chem. Soc,, (1958). , 1339-1339.

[19] Stankovich, S, et al. STable aqueous dispersions of graphitic nanoplatelets via the reduction of exfoliated graphite oxide in the presence of poly(sodium-4-styrenesulfonate). J. Mater. Chem., (2006). , 155-158.

[20] Li, D, et al. Processable aqueous dispersions of graphene nanosheets. Nature Nanotechnology, (2008). , 101-105.

[21] Coleman, J. N, et al. Two-dimensional nanosheets produced by liquid exfoliation of layered materials. Science, (2011). , 568-571.

[22] Hernandez, Y, et al. Measurement of multicomponent solubility parameters for graphene facilitates solvent discovery. Langmuir, (2010). , 3208-3213.

[23] Hernandez, Y, et al. High-yield production of graphene by liquid-phase exfoliation of graphite. Nat. Nanotechnol., (2008). , 563.

[24] Lotya, M, et al. Liquid phase production of graphene by exfoliation of graphite in surfactant/water solutions. J. Am. Chem. Soc,, (2009). , 3611-3620.

[25] Lotya, M, et al. High-concentration, surfactant-stabilized graphene dispersions. ACS Nano, (2010). , 3155-3162.

[26] Smith, R. J, et al. Large-scale exfoliation of inorganic layered compounds in aqueous surfactant solutions. Adv. Mater., (2011). , 3944-3948.

[27] Smith, R. J, Lotya, M, & Coleman, J. N. The importance of repulsive potential barriers for the dispersion of graphene using surfactants New Journal of Physics, (2010). p. art (125008)

[28] Griffith, A, & Notley, S. M. pH dependent stability of graphene suspensions with adsorbed weakly ionisable cationic polyelectrolyte. J. Colloid Interface Sci., (2012). , 210-215.

[29] Notley, S. M. Adsorption of polyelectrolyte modified graphene to silica surfaces: monolayers and multilayers. J. Colloid Interface Sci., (2012). , 35-40.

[30] Notley, S. M. Highly concentrated aqueous suspensions of graphene through ultrasonic exfoliation with continuous surfactant addition. Langmuir, (2012). , 14110-14113.

[31] Seo, J, et al. High-concentration aqueous dispersions of graphene using nonionic, biocompatible block copolymers. J. Phys. Chem. Lett., (2011). , 1004-1008.

[32] Liang, Y. T, & Hersam, M. C. Highly concentrated graphene solutions via polymer enhanced solvent exfoliation and iterative solvent exchange. J. Am. Chem. Soc,, (2010). , 17661-17663.

[33] Nuvoli, D, et al. High concentration few-layer graphene sheets obtained by liquid phase exfoliation of graphite in ionic liquid. J. Mater. Chem., (2011). , 3428-3431.

[34] Wang, X, et al. Direct exfoliation of natural graphite into micrometre size few layers graphene sheets using ionic liquids. Chem. Comm., (2010). , 4487-4489.

[35] Dikin, D. A, et al. Preparation and characterization of graphene oxide paper. Nature, (2007). , 457-460.

[36] Hu, T, et al. Flexible free-standing graphene-TiO2 hybrid paper for use as lithium ion battery anode materials. Carbon, (2013). , 322-326.

[37] Liu, F, et al. Folded structure graphene paper for high performance electrode material. Adv. Mater., (2012). , 1089-1094.

[38] Decher, G. Fuzzy Nano-assemblies:Toward layered polymeric multicomposites. Science, (1997). , 1232-1237.

[39] Notley, S. M, Eriksson, M, & Wågberg, L. Visco-elastic and adhesive properties of adsorbed polyelectrolyte multilayers determined in situ with QCM-D and AFM measurements. J. Colloid Interface Sci., (2005). , 29-37.

[40] Szabo, T, Szeri, A, & Dekany, I. Composite graphitic nanolayers prepared by self-assembly between finely dispersed graphite oxide and a cationic polymer. Carbon, (2005). , 87-94.

[41] Doyle, R. J, & Rosenberg, M. Microbial cell surface hydrophobicity, in American Society for MIcrobiology(1990). Washington.

[42] Liu, G, et al. Protein resistance of (ethylene oxide)n monolayers at the air-water interface: effects of packing density and chain length. Phys. Chem. Chem. Phys., (2007). , 6073-6082.

[43] Marshall, K. C. Biofilms: an overview of bacterial adhesion, activitiy and control at surfaces. ASM News, (1992). , 202-207.

[44] Bruscher, H. J, & Van Der Mei, H. C. Physico-chemical interactions in initial microbial adhesion and relevance for biofiom formation. J. Dent. Res., (1997). , 24-32.

[45] Weerkamp, A. H, Uyen, H. M, & Busscher, H. J. Effect of zeta potential and surface energy on bacterial adhesion to uncoated and saliva coated human enamel and dentin. J. Dent. Res., (1998). , 1483-1487.

[46] Ivanova, E. P, et al. Natural bactericidal surfaces: mechanical rupture of Pseudomonas aeruginosa cells by cicada wings. Small, (2012). , 2489-2494.

[47] Phoenix, V. R, et al. Characterization and Implications of the Cell Surface Reactivity of Calothrix sp. Strain KC97. Appl. Environ. Microbiol., (2002). , 4827-4834.

[48] Nazarenko, E.L., R.J. Crawford, and E.P. Ivanova, The structural diversity of carbohydrate antigens of selected gram-negative marine bacteria. Marine Drugs, 2011. 9: p. 1914-1954

[49] Deng, L, et al. Characterization of the Linkage between the Type III Capsular Polysaccharide and the Bacterial Cell Wall of Group B Streptococcus. J. Biol. Chem., (2000). , 7497-7504.

[50] Dong, C, et al. Wza the translocon for E. coli capsular polysaccharides defines a new class of membrane protein. Nature, (2006). , 226-229.

[51] Jastrzebska, A. M, Kurtycz, P, & Olszyna, A. R. Recent advances in graphene family materials toxicity investigations. J. Nanopart. Res., (2012). , 1320-1328.

[52] Akhavan, O, & Ghaderi, E. Toxicity of graphene and graphene oxide nanowalls against bacteria. ACS Nano, (2010). , 5731-5736.

[53] Hu, W, et al. Graphene-based antibacterial paper. ACS Nano, (2010). , 4317-4323.

[54] Ruiz, O. N, et al. Graphene oxide: A nonspecific enhancer of cellular growth. ACS Nano, (2011). , 8100-8107.

[55] Sreeprasad, T. S, et al. Transparent, luminescent, antibacterial and patternable film forming composites of graphene oxide/reduced graphene oxide. ACS Applied Materials and Interfaces, (2011). , 2643-2654.

[56] Mejias CarpioI.E., et al., Toxicity of a polymer-graphene oxide composite against bacterial planktonic cells, biofilms, and mammalian cells. Nanoscale, (2012). , 4746-4756.

[57] Santos, C. M, et al. Chem. Commun., (2011). , 8892.

[58] Some, S, et al. Dual functions of highly potent graphene derivatives-poly-l-lysine composites to inhibit bacteria and support human cells. ACS Nano, (2012). , 7151-7161.

[59] Liu, S, et al. Antibacterial activity of graphite, graphite oxide, graphene oxide and reduced graphene oxide: Membrane and oxidative stress. ACS Nano, (2011). , 6971-6980.

[60] Krishnamoorthy, K, et al. Antibacterial efficiency of graphene nanosheets against pathogenic bacteria via lipid peroxidation. J. Phys. Chem. C., (2012). , 17820-17827.

[61] Liu, S, et al. Lateral dimension-dependent antibacterial activity of graphene oxide sheets. Langmuir, (2012). , 12364-12372.

[62] Akhavan, O, Ghaderi, E, & Esfandiar, A. Wrapping bacteria by graphene nanosheets for isolation from environment, reactivation by sonication and inactivation by near-infrared irradiation. J. Phys. Chem. B, (2011). , 6279-6288.

[63] Akhavan, O, & Ghaderi, E. Escherichia coli bacteria reduce graphene oxide to bactericidal graphene in a self-limiting manner. Carbon, (2012). , 1853-1860.

[64] Zhang, D, Liu, X, & Wang, X. Green synthesis of graphene oxide sheets decorated by silver nanoprisms and their anti-bacterial properties. J. Inorg. Biochem., (2011). , 1181-1186.

[65] Das, M. R, et al. Synthesis of silver nanoparticles in an aqueous suspension of graphene oxide sheets and its antimicrobial activity. Colloids Surf. B, (2011). , 16-22.

[66] Liu, L, Liu, J, & Sun, D. D. Graphene oxide enwrapped Ag3PO4 composite: towards a highly efficient and stable visible-light induced photocatalyst for water purification. Catal. Sci. Technol., (2012). , 2525-2532.

Chapter 6

Porous Nanocarbons: Molecular Filtration and Electronics

Boyang Wang, Artem Baskin and Petr Král

Additional information is available at the end of the chapter

http://dx.doi.org/10.5772/56247

1. Introduction

Graphene is an atomically thick monolayer formed by hexagonally arranged carbon atoms with sp^2 planar covalent bonds [1]. It has unique mechanical [2], electronic [3], optical [4], and transport properties [5], leading to many potential applications [6]. Its low-energy electron excitations are 2D massless chiral Dirac fermions [3], moving 300 times slower than light [7]. These fermions have nearly ballistic transport and high mobility at room temperature [3, 5, 7–9]. In magnetic fields, their motion is associated with several quantum Hall effects, observable at room temperature [10, 11]. Many methods have been proposed to fabricate large-area semiconducting graphene materials, such as chemical [12–14], sonochemical [15], and lithographic [16, 17] methods. Graphene is a great candidate for electronic applications [6], especially upon further functionalization [18–20]. Graphene was modified by wrapping [21–24], oxidation [25–27], etching [12, 28, 29], covalent [20, 30–34], noncovalent [20, 35], and biological [36] functionalization.

Despite the fact that graphene has excellent transport characteristics, its applicability in semiconductor industry is limited, since it lacks a significant band gap $\gg kT$ [3]. Therefore, graphene conductance can not be easily modified by doping and electric fields, as in silicon [37]. However, a band gap can be opened in geometrically restricted graphene [5, 38–42]. The simplest restrictions originate in edges [43], giving graphene unique structural [44–46], electronic [39, 41] and chemical properties [47]. Graphene edges have been modified [48, 49] by selective chemical bonding [50] and etching [29, 51]. They have been controlled by electrochemical [52] and optical [53] means, and characterized spectroscopically [54–56]. When graphene is restricted by two parallel edges, a graphene nanoribbon (GNR) is formed with armchair (AGNR) or zig-zag edges (ZGNR). Metallic or semiconducting GNRs can be obtained by cutting graphene [57], carbon nanotubes or by purely organic means [58]. While tight-binding calculations showed that AGNRs appear to be metallic at certain widths [59],

DFT calculations predicted that all AGNRs are semiconducting, with a band gap roughly inversely proportional to the GNR width [59–62], but all ZGNRs are semimetallic.

Alternatively, graphene can be restricted by perforation with nanopores, in analogy to a two-dimensional Swiss cheese [63, 64]. Graphene nanopores can be functionalized to create materials with rich applications in physics [65], chemistry [66, 67], biology [68] and various technologies [69, 70], such as molecular separation [63, 71–75] and electronics [6, 62, 76]. Chemical functionalization of graphene nanopores can improve their separation ability [63, 77–83], and detection [74, 84]. For example, oxidation of packed multilayered graphene sheets could allow selective ultrafast water flows through these layers, and prevent the passage of helium, nitrogen, argon, and hydrogen [80]. Nitrogen substitutions of carbon atoms in the pore rims are important for a quick separation of ^{3}He and ^{4}He nuclei [77]. Diamond-like carbon nanosheets with nanopores can selectively filtrate organic solvents [81], such as ethanol and azobenzene.

2. Preparation of porous nanocarbons

The porous graphene can be prepared by top down and bottom up approaches. In the top down approach, nanopores can be drilled in graphene by electron beams [73, 85–87] or by oxidative [64, 75, 88] and steam [89] etching. In Fig. 1 (left), an electron-beam drilled nanopore in graphene is shown [73]. This pore has been used to detect the passage of ionic solutions and DNA molecules [73, 86, 87]. Graphene nanopores prepared by oxygen plasma etching [64] have homogeneous distribution of sizes, which leads to opening of an effective energy gap of 100 meV in the graphene material. Steam etching of graphene by water vapor in a seal vessel at 200 K temperature is a rather slow process [89], but with a better control of the porosity by the etching time.

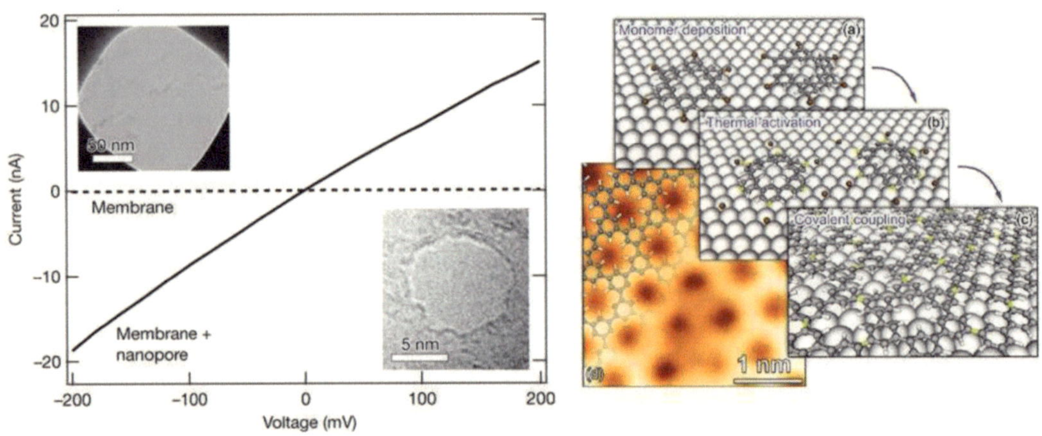

Figure 1. (left) Experimentally prepared graphene nanopore by electron beam drilling. From Ref. [73]. (right) Synthetic growth of porous graphene from organic molecules covalently bound on the surfaces of metals. From Ref. [90].

Precise graphene nanopores could also be prepared by bottom up approaches, where porous graphene can be grown from conjugated aromatic molecules covalently interconnected on the surfaces of other materials (metals). In Fig. 1 (right), we show graphene nanopores with precisely functionalized rims experimentally prepared in this manner [90]. This procedure

http://dx.doi.org/10.5772/56247

uses surface-assisted coupling of molecular precursors into linear polyphenylenes and their subsequent cyclodehydrogenation. The porous graphene was synthesized by depositing hexaiodo-substituted macrocycle cyclohexa-m-phenylene (CHP) on Ag (111) surface, and annealing to temperatures above 570 K initiates the polymerization reaction. Six-fold aryl-aryl coupling of the CHP monomers resulted in a 2D polyphenylene network. This contained periodic nanopores in a hexagonal pattern in the polymerized graphene layer, also with atomic precision. Another bottom up approach used the self-assembly of close-packed polystyrene nanospheres to synthesize semiconducting porous graphene networks [62]. Highly ordered networks of carbon atoms based on non-covalent interactions, such as hydrogen bonding, metal coordination, or aromatic interactions have also been grown on various single crystalline substrates [58, 90, 91].

Although both the bottom up and top down approaches create porous graphene with a band gap, the amount of imperfections is significantly smaller in the first approach. This is of importance for electronic and optical applications [6]. Field effect transistors made of porous graphene fabricated in a top down way demonstrate a high hole mobility at room temperature [64]. In a functionalized and modified graphene, spectroscopic methods are used to characterize the electronic states [51, 92] and chemical structures before [92–98] and after [99–101] modifications. These methods include Raman [99, 102–105], resonance-Raman [51], surface-enhanced Raman [106], infrared [95, 107–109], plasmon [110], nuclear magneto-resonance [111], scanning tunneling [95, 112], photoemission [113] and Auger electron spectroscopies [114].

3. Ionic and molecular separation by graphene nanopores

Biological ionic channels play a key role in many cellular transport phenomena [115]. The large selectivity of these proteins is realized by precisely arranged arrays of charged amino acids that can efficiently recognize and guide the fast passing ions [116]. Much simpler non-biological ionic and molecular channels are made from zeolites [117], carbon [118], silica [119] and other materials [120, 121]. For many industrial applications, these channels are often not selective and transparent enough. Therefore, ion and molecular channels with novel structures and recognition principles need to be designed.

Molecular dynamics studies have predicted [122] and experiments have demonstrated fast transport of gases [123, 124] and liquids [125–128] through carbon nanotube (CNT) membranes. Hydrated ions could also pass through CNTs of large enough diameters [129]. Recently, graphene [1] and boron-nitride [130] monolayers have been prepared and intensively studied [131–133] for their many potential applications. The first is a semimetal that can screen free charges (screening length of 4 Å [134]), while the second is a semiconductor with a large band gap that screens charges very little.

3.1. Ionic and atomic separation by graphene nanopores

Note that graphene layer is impermeable to even small atoms, such as helium [135]. However, recently, functionalized nanopores in graphene monolayers have been theoretically designed and have shown that they could serve as ionic sieves of high selectivity and transparency [63], in analogy to natural functional proteins [136–138].

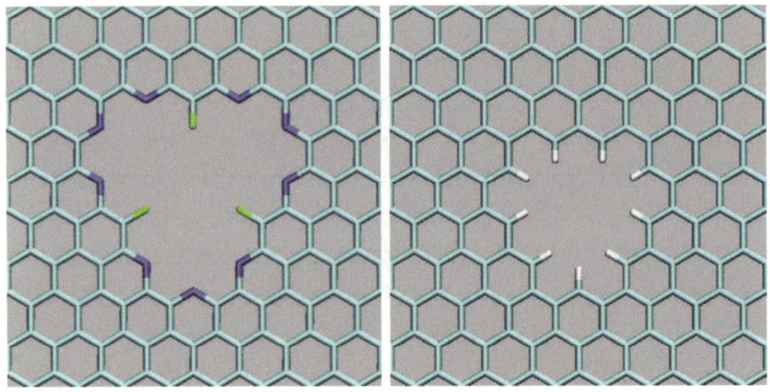

Figure 2. Functionalized graphene nanopores. (Left) The F-N-terminated nanopore. (Right) The H-terminated nanopore. From Ref. [63].

In Fig. 2, we display the studied chemically modified graphene nanopores. The F-N-pore (left) is terminated by negatively charged nitrogens and fluorines, favoring the passage of cations. The H-pore (right) is terminated by positively charged hydrogens, favoring the passage of anions. These nanopores could be formed in the graphene monolayers by ion etching [139], and their chemical modifications could be realized by local oxidation, similarly as on graphene edges [140, 141].

We modeled the passage of hydrated ions from the first and seventh periods through these tiny nanopores. Our molecular dynamics (MD) simulations [142–144] were performed with the NAMD package [145], based on the CHARMM27 force field [146]. The model graphene sheets (sizes of 23x23 Å) have chemically modified nanopores in their center. The atomic charges of these sheets are obtained ab initio with B3LYP density functional [147], using the Gaussian03 package [148]. These calculations show that when the ion passes through the graphene nanopore, the semimetallic sheet becomes highly polarized.

Therefore, a screening charge of the same size but opposite sign to that of the ion goes to the pore rim and an opposite charge remains spread on the rest of the sheet. In the presence of water the screening charge should be more delocalized at the pore entrance, due to water polarization. In the MD simulations, we model these effects by homogeneously spreading the screening charge for the studied ion on the first atomic layer of the pore and the atomic ligands attached to it. The screening charge is kept there during the whole simulations, since its presence is little affecting the dynamics when the ion is far away from the pore. The opposite polarization charge on the sheet is neglected, for simplicity, so the combined system of the ion and the sheet is kept neutral.

In the simulations, the sheet is placed in a periodic unit cell with 10 Å of water (320 waters in total) on each side. One ion is placed in the cell and driven by an electric field E applied in the direction perpendicular to the graphene sheet. Each simulation run typically lasts 100-500 ns. It turns out that the F-N-pore is only passed by the Li^+, Na^+ and K^+ ions, while the H-pore is only passed by the Cl^- and Br^- ions. This selectivity of the nanopores, even in the presence of large driving field of E = 0.1 V/nm, is caused by the Coulombic coupling between the ion and the functionalized nanopore rim. At this field, the passage rates of Li^+, Na^+ and K^+ have the ratio of 9:14:33, while those of F^-, Cl^- and Br^- have the ratio of 0:17:33. Therefore, the nanopores are also highly selective to the sizes of the ions. The Li^+ and F^- ions have the lowest passage rates, because of their small radii and large coupling to

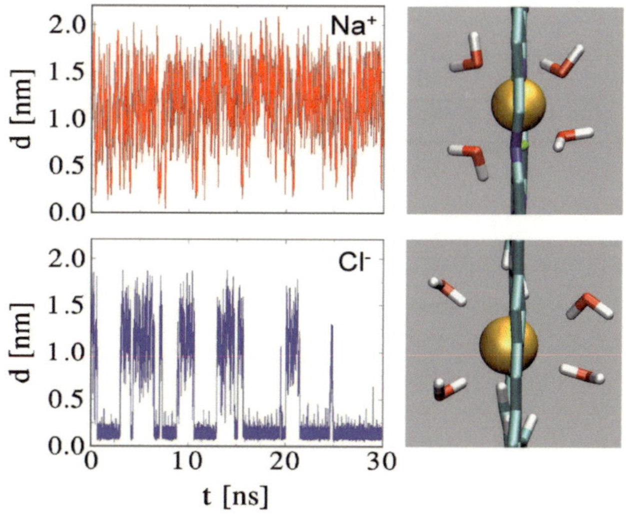

Figure 3. (left) The time-dependent distance d between the Na^+ and Cl^- ions and the centers of the F-N-pore and H-pore, respectively, at the field of E = 6.25 mV/nm. The ion passages through the two nanopores have very different dynamics. (right) The passing ions are surrounded by two water half-shells, but only Cl^- has relatively stable binding to the pore. From Ref. [63].

the water molecules in their hydration shells. On the contrary, K^+ and Br^- have large radii, so it is relatively easy to break their hydration shells in the pores.

In Fig. 3 (right insets), we also show the configurations of the selected Na^+ and Cl^- ions passing through the two nanopores. The polar and charged rims replace some of the water molecules from the first hydration shell of the ions passing through the nanopores. Due to the atomic thickness of the graphene layer, the passing ion is surrounded by two separated first hydration half-shells at its both sides. The close contact between the pore rim and the ion leads to the high selectivity of the ion passage.

Fig. 3 (left insets) displays the time-dependent distance d between the ions and their pore centers, obtained at the low field of $E = 6.25$ mV/nm. The character of the obtained fluctuations reveals that the ion passages through the two nanopores have very different dynamics [149]. The Na^+ ion passes the F-N-pore rather fast, without binding with it. It rarely gets closer than 5 Å to the pore center and stays most of the time in the water region ($d > 10$ Å). The ion does not pass through the pore center, but rather through one of the three smaller holes in the nanopore with the C_3 symmetry. This asymmetric hole can not easily break the hydration shell of the Na^+ ion. The K^+ ion binds weaker to its hydration shell and it starts to bind to the pore.

The Cl^- ion has more stable binding to the symmetric H-pore, where it already stays for about 70% of the time. The electric field decreases the ion-pore binding, which is reflected in shorter time periods spent by the ion in the pore. At the field of $E = 0.1$ V/nm, the Cl^- ion already stays in total only 30% of the time inside the pore. The unidirectional ion flow in the presence of electric field is reflected in the slight left-right asymmetry in the number of entrances of the ion in the nanopore and exits out of it.

Figure 4 shows the results of the ion flow through the system at different fields E. We display the field dependence of the average number of Na^+ and Cl^- ion passages, r_{Na} and

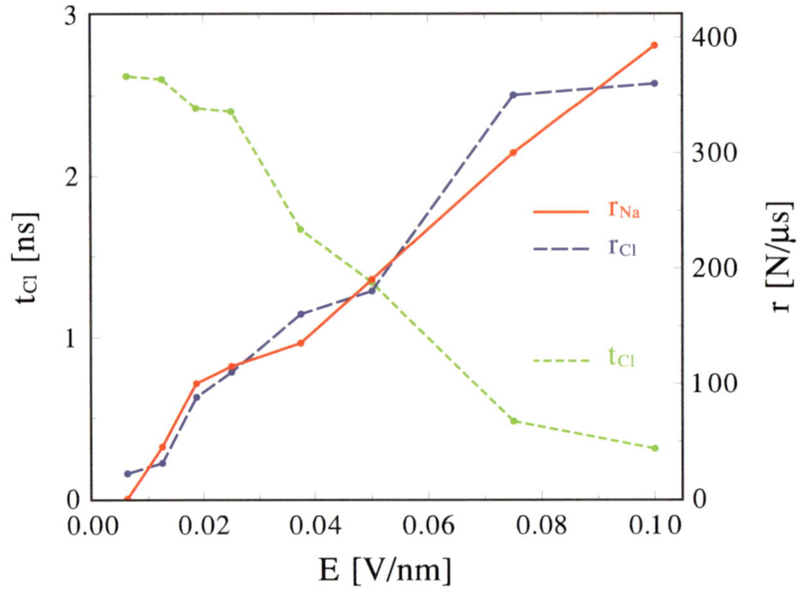

Figure 4. (right axis) Dependence of the r_{Na} and r_{Cl} flow rates of the hydrated Na^+ and Cl^- ions through the F-N-pore and H-pore, respectively, on the applied electric field E. (left axis) The average time t_{Cl} spent by the Cl^- ion inside the H-pore. From Ref. [63].

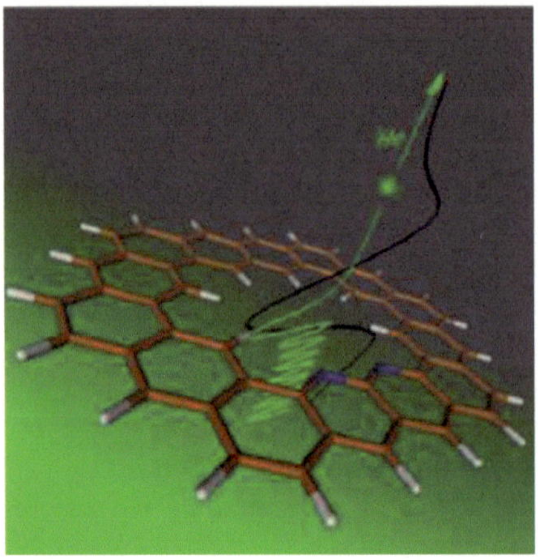

Figure 5. A graphene nanopore functionalized by nitrogen atoms, used in the separation of ^3He and ^4He isotopes. From Ref. [77].

r_{Cl}, through the F-N-pore and the H-pore, respectively. Both rates monotonously grow with the field, showing the accelerated passage of the ions through the pore and the surrounding water layers. We also show the field-dependence of the average residence time t_{Cl} of the Cl^- ion in the H-pore, defined as the time during which the ion stays within the 5 Å distance from the pore center. We can see that t_{Cl} decreases monotonously with the field that helps the ion to overcome the Coulombic binding to the nanopore.

http://dx.doi.org/10.5772/56247

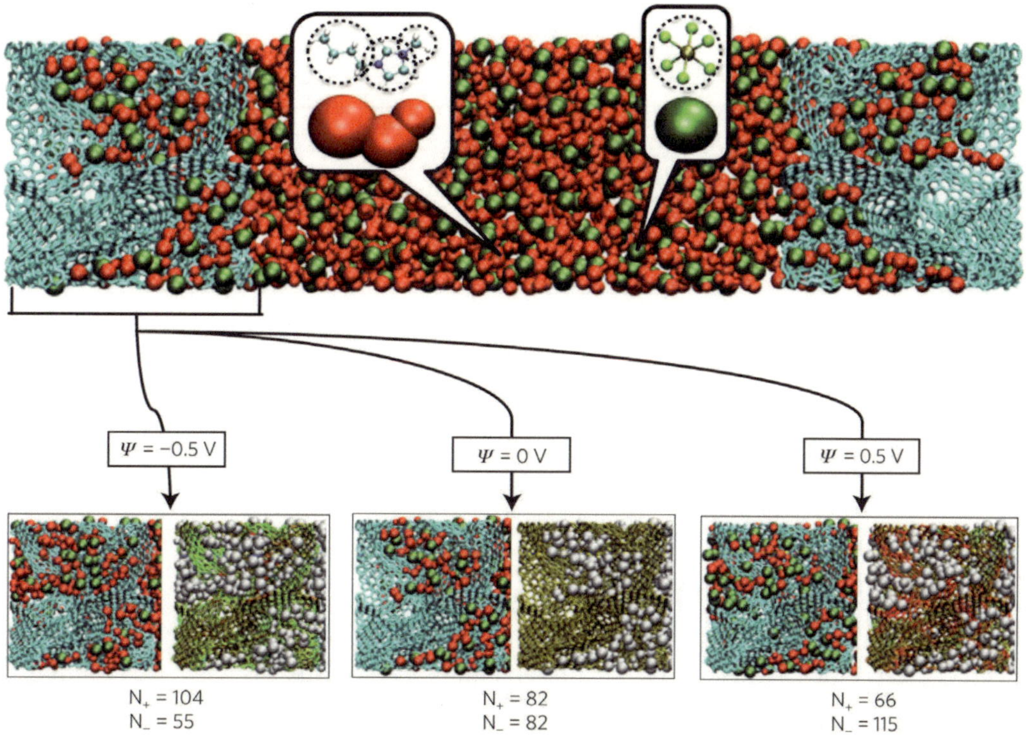

Figure 6. The EDLC simulation cell. Upper panel: the simulation cell consists of a BMI-PF_6 ionic liquid electrolyte surrounded by two porous electrodes (CDC-1200) held at constant electrical potentials (blue: C atoms, red: the three sites of BMIC and green: PF_6 ions; a coarse-grained model is used to describe these ions). Lower panel: structure of the electrode for various voltages. For each value, the same snapshot is shown twice: The ionic distribution is shown on the left. The degree of charging of the electrode atoms is shown on the right, where the carbon atoms are coloured according to the charge q they carry (green: $q < 0$, red: $q > 0$ and yellow: $q = 0$). From Ref. [151].

The above results have shown that chemically functionalized nanopores in graphene monolayers can have a high selectivity for the passage of hydrated ions. These ionic and molecular nanopores could be optimally designed [150] by choosing the type of monolayer material, the pore sizes and the structure/number of ligands attached to their rims. These ultrasmall nanopores could be applied in molecular separation, desalination and battery technologies.

Selective atom passage through graphene nanopores was also studied by first principle techniques [77]. In Fig. 5, we show the configuration of one of the studied systems with a chemical structure possibly matching the separated ^{3}He and ^{4}He isotopes. The tunneling rates of the isotopes through the different nanopores were calculated. The barriers for ^{3}He to pass through the three nanopores were found to be very different one from another. The hydrogen, half-nitrogen (one side), and full-nitogen (both sides) terminated nanopores had the barriers, $\approx$ 0.078 eV, $\approx$ 0.025 eV, and $\approx$ 0.0076 eV, respectivelly. To achieve a high flux through the nanopore, while keeping a large difference in the transimission probabilities of ^{3}He and ^{4}He isotopes through the nanopore, the one-sided nitrogen doped pore (the second pore) was chosen, as shown in Fig. 5. The transmission ratio between ^{3}He and ^{4}He was 2 at 20 K, and raised with decreasing temperature to 19 at 10 K.

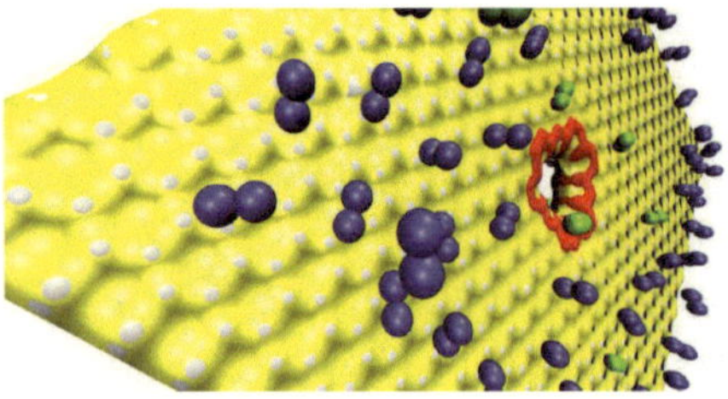

Figure 7. The graphene nanopore used in separations of hydrogen and nitrogen gases. The graphene sheet is represented by a yellow plane with white points referring to carbon atoms, and the pore shape is colored by red; hydrogen and nitrogen molecules are colored by green and blue, respectively. From Ref. [152].

Porous graphene could also be used as a unique material for electrodes in supercapacitors and batteries, as shown in recent simulations with polarizable force fields [151]. In Fig. 6 the studied system is shown. Upon charging the porous nanocarbon electrodes, ionic liquid electrolytes could enter into them through the nanopores. The anode retains more anions than cations, and the cathode retains more cations than anions. The supercapacitor formed could store charges with a very high efficiency.

3.2. Molecular separation by graphene nanopores

In the separation of neutral molecules by graphene nanopores, van der Waals interactions are crucial, since size exclusion is the main separation mechanism. Recently, for example, the separation of nitrogen and hydrogen gases in graphene nanopores have been examined by MD simulations [152], as shown in Fig. 7. In the minimal graphene nanopore size studied here, hydrogen molecules could pass through the pore and nitrogen molecules were totally excluded. When the graphene pore size became larger, N_2 passed through the nanopore much faster than H_2 due to a different mechanism. It was found that N_2 was largely adsorbed on the graphene surface, due to strong van der Waals attraction, and it diffused on it, before reaching the pore. On the other hand, H_2 was much less adsorbed on the graphene, and it had to diffuse in the 3D space before reaching the nanopore.

It was also shown by MD simulations that graphene nanopore could enhance the separation of carbon dioxide from nitrogen molecules [153]. Separation of hydrogen molecules from methane molecules based on the size exclusion mechanism was also investigated by ab initio methods [71]. The potential energy surfaces of a hydrogen and methane molecules passing through a chemically modified graphene nanopore were obtained by density functional methods. A dramatic difference between these two potential energy surfaces was found, which caused a highly efficient separation of hydrogen molecules from methane molecules. It was also shown that methane could be separated from air by graphene nanopores [154], which could serve as atmospheric nanofilters [155].

The flow of water through graphene nanopores was also simulated [72] in a configuration shown in Fig. 8. The passage of a single water molecule through such a small graphene nanopore necessitates it to break most of its hydrogen bonds, which significantly slows down the water flux through the nanopore (two water molecules that are separated by the graphene nanopore have one hydrogen bond in common). As the pore size increase so does the flux. Graphene nanopore of a diameter $d > 0.8$ nm were proven to have higher water flux than a carbon nanotube of the same diameter. Therefore, porous graphene could have significant applications in desalination [156].

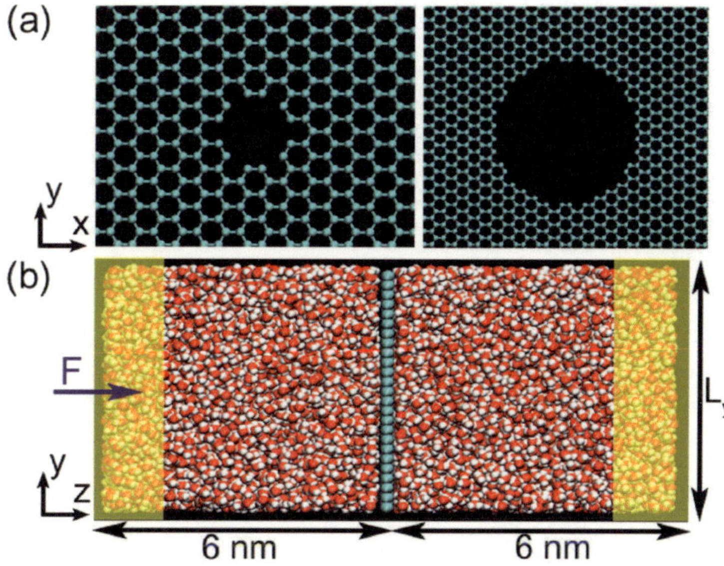

Figure 8. (a) Graphene membrane with a nanopore of diameter $d_{avg} = 0.75$ nm (left) and $d_{avg} = 2.75$ nm (right). (b) Simulation setup. Cyan color represents carbon atom, red color and white color represent the oxygen and hydrogen atoms of a water molecule, respectively. Two water reservoirs are attached to each side of the porous graphene membrane. $L_y = L_x = 4$ nm when the pore diameter is 0.75 nm, and $L_y = L_x = 6$ nm when the pore diameter is 2.75 nm. In the shaded region ($z = 1$ nm), external forces are applied on water molecules to create a pressure drop across the membrane. From Ref. [72].

It was shown that graphene with nanopores could be a good candidate for water desalination [156, 157]. Traditional water desalination industry applied polymeric reverse osmosis membranes, which had high salt rejection, but with a very low water flux rate. Graphene nanopores could selectively reject salt [157], and with a water flux several orders of magnitude higher than polymeric reverse osmosis membranes. The salt rejection mechanism is based on size exclusion [157], since salt ions have hydration shells that are much bigger than single water molecules. The maximal diameter of the hydrogen-terminated nanopore that could reject salt is 5.5 Å. The salt rejection rate decreases at higher pressures, since the ion hydration shells break. Since, large areas of graphene can be grown on copper foils [158], it is likely that graphene membrane can have industrial applications for water desalination in the future.

Recently, other graphene-based filtration membranes were studied as well. For example, small pieces of graphene oxide (GO) were packed to form multilayered membranes [80], shown in Fig. 9 (a-b). In each layer of the membrane, the GO pieces were disconnected, with empty space between the pieces, as shown in Fig. 9 (c). When water was adsorbed on the membrane, it found empty spaces in each GO layer and protruded toward the next layer. In this way, water was able to penetrate through the whole membrane. However, such water permeable GO membranes prevents the leakage of He gas, as shown in Fig. 9 (d).

Graphene nanopores could also be used for DNA sequencing [73, 86, 87, 159]. Previous sequencing nanopores were relatively thick and retained multiple DNA base pairs inside the nanopore during the sequencing process, so the detection of these bases were not efficient. On the contrary, graphene nanopores can provide detection within an atomically thick

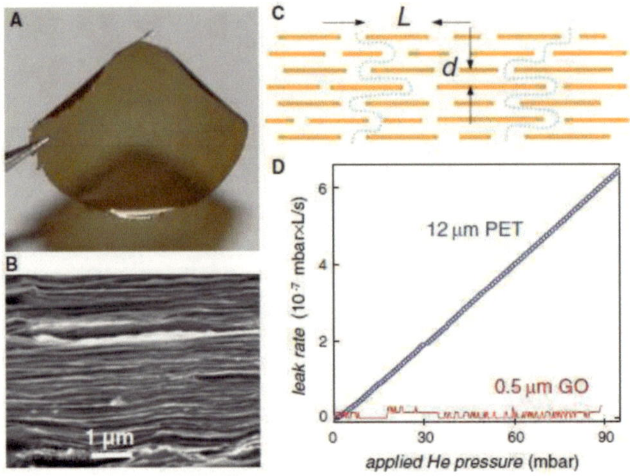

Figure 9. He-leak through GO membranes. (A) Photo of a 1-mm-thick GO film peeled off of a Cu foil. (B) Electron micrograph of the films cross section. (C) Schematic view for possible permeation through the laminates. Typical L/d is 1000. (D) Examples of He-leak measurements for a freestanding submicrometer-thick GO membrane and a reference PET film (normalized per square centimeter). From Ref. [80].

monolayer. The detection of DNA basis might be possible to deduce from the fluctuations of ionic currents passing simultaneously with the DNA through the graphene pores.

In the experiments [73], double stranded DNA molecules and solvated ions were electrophoretically passed through a 5 nm graphene nanopore, shown in Fig. 1 (left). Ions could pass smoothly through the nanopore, and their electric current is proportional to the applied voltage. However, other molecules present in the nanopore reduce the ion current. When a DNA molecule passes through the nanopore, it causes electric current blocade, and reduces the current for a certain time, which can reflected the size and conformation of the passing DNA. If the DNA molecule is folded, then the blocade duration is short (e.g. on the order of 10 microseconds). When the DNA is unfolded, it passes the nanopore as a linear chain, and the current blocade lasts for a longer time (e.g. 200 microseconds). In this way, details about the DNA molecule and its passage through the graphene nanopore could be detected.

The passage of DNA through a graphene nanopore was also investigated by MD simulations in the presence of electric fields [74, 160]. To make sure that DNA translocation through the nanopore completes within a time scale that is accessible for molecular dynamics, a bias voltage of $\phi = 2.1$ V was applied to the system. It turned out that additional charges at the nanopore also influenced the field-driven DNA passage through the nanopore. A positively charged nanopore allowed faster transport of DNA through the nanopore than negatively charged ones, since DNA alone is negatively charged. Figure 10 shows a trajectory of DNA conformations during the translocation through the nanopore within 15 ns in the electric field. The partially folded DNA had various conformations while passing through pore, and the electric current fluctuated with these conformations. When part of the DNA was inside the nanopore, the current had different peaks signifying the translocation events. It was shown that poly(A-T)20 and poly(G-C)20 duplexes passing through the nanopore produce different ionic current profiles. This means that A-T and G-C base pairs could be distinguished. Molecular and atomic species passing through graphene nanopores

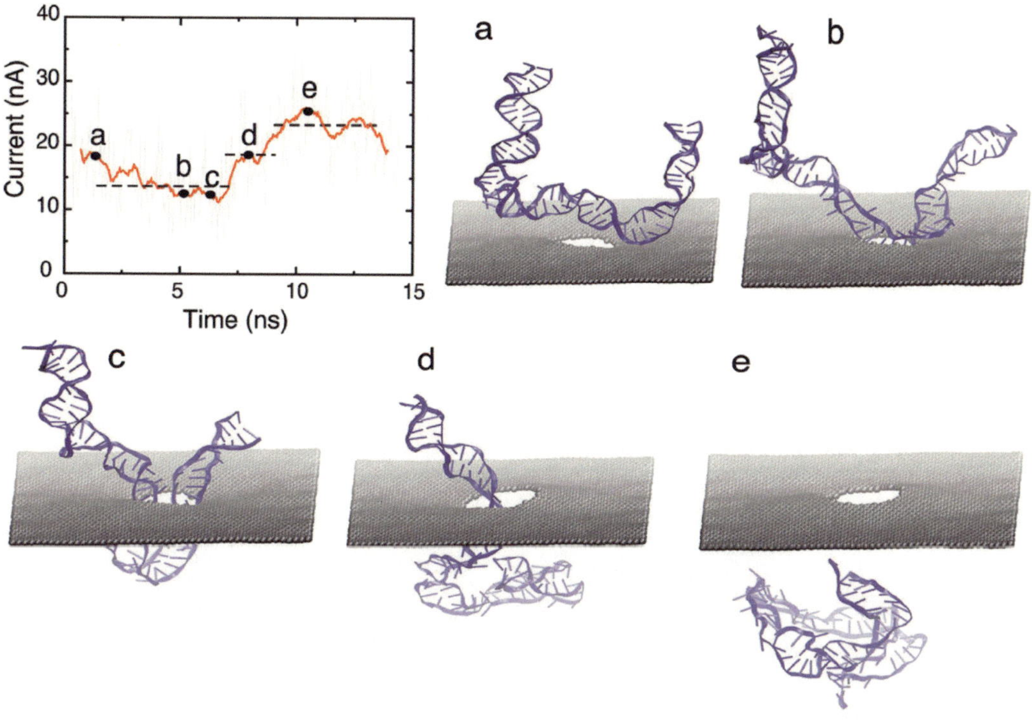

Figure 10. Translocation of partially folded DNA. The three dotted lines correspond to plateaus in ionic current signature. Snapshots of DNA conformation during translocation is shown in (a-e). (a) Initial conformation of DNA; (b) DNA captured by pore mouth; (c) both chains of folded DNA in the pore; (d) one chain leaves pore; (e) DNA exits the pore completely. The diameter of the pore is 3 nm and the bias voltage was 2.1 V. From Ref. [74].

could in principle be detected electronically from the change of the electronic structure of porous nanocarbons in the presence of passing molecules [71]. For example, the electron transmissivity of porous nanocarbons may significantly vary when the particular molecule passes through the pore [161–163].

4. Electronic Structure/Transport in Porous Nanocarbons

Given the discussed possibilities of molecular separation and detection in graphene nanopores, it is important to understand the electronic structure of porous graphene in vacuum and in the presence of molecular fields. Therefore, we review the electronic properties of *porous nanocarbons* (PNCs), as shown in Fig 11. In general, PNCs are derived from graphene, carbon nanotubes (CNT), and nanoribbons (GNR) by introduction of functionalized pores. These nanopores allow tuning of their electronic structures, which can be useful in a wide range of electronics applications [164–169].

4.1. Electronic structures of porous nanocarbons

The main effects that the pores cause in pristine nanocarbons are: 1) the energy band gap opening; 2) the emergence of midgap energy states; 3) flat and quasi-flat bands at zero and low non-zero energies. Although, partial results for the electronic structures of porous

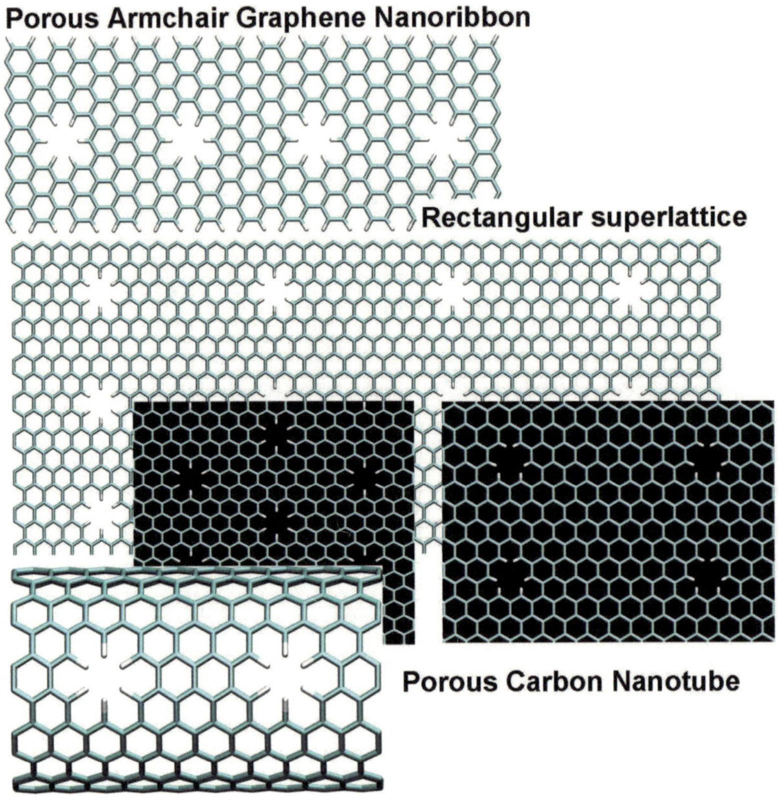

Figure 11. Examples of PNCs.

graphene [167–169] and other PNCs have been obtained [5, 59, 170–173], general principles that would unify their electronic structures are missing.

Graphene and its derivatives were studied theoretically by a variety of methods, such various DFT and tight-binding (TB) approaches, using semiempirical parametrization of electron-electron interactions in model Hamiltonians [168, 169]. The pseudo-relativistic behavior of electrons close to the Fermi level in graphene can be modeled by a Dirac equation (DE) [167]. Semiempirical approaches were also used, such as the Hückel molecular orbital method [170, 173], crystal orbital methods [174], or mean-field resonating-valence-bond theories designed to treat unpaired π-electrons in benzenoid carbon species [175]. The principal problem for any theoretical approach dealing with PNCs is to determine the correlation between geometrical structure of PNCs and their electronic structures. Therefore, we will focus on this correlation and describe it separately for different PNCs systems.

Porous graphene nanoribbons (PGNRs)

Graphene and its nanoribbons can be directly grown on material surfaces. Therefore, graphene-based materials could have direct applications in nanoelectronics. In contrast to pristine 2D graphene sheets and CNTs, GNRs have (armchair, zigzag, or mixed) edges.

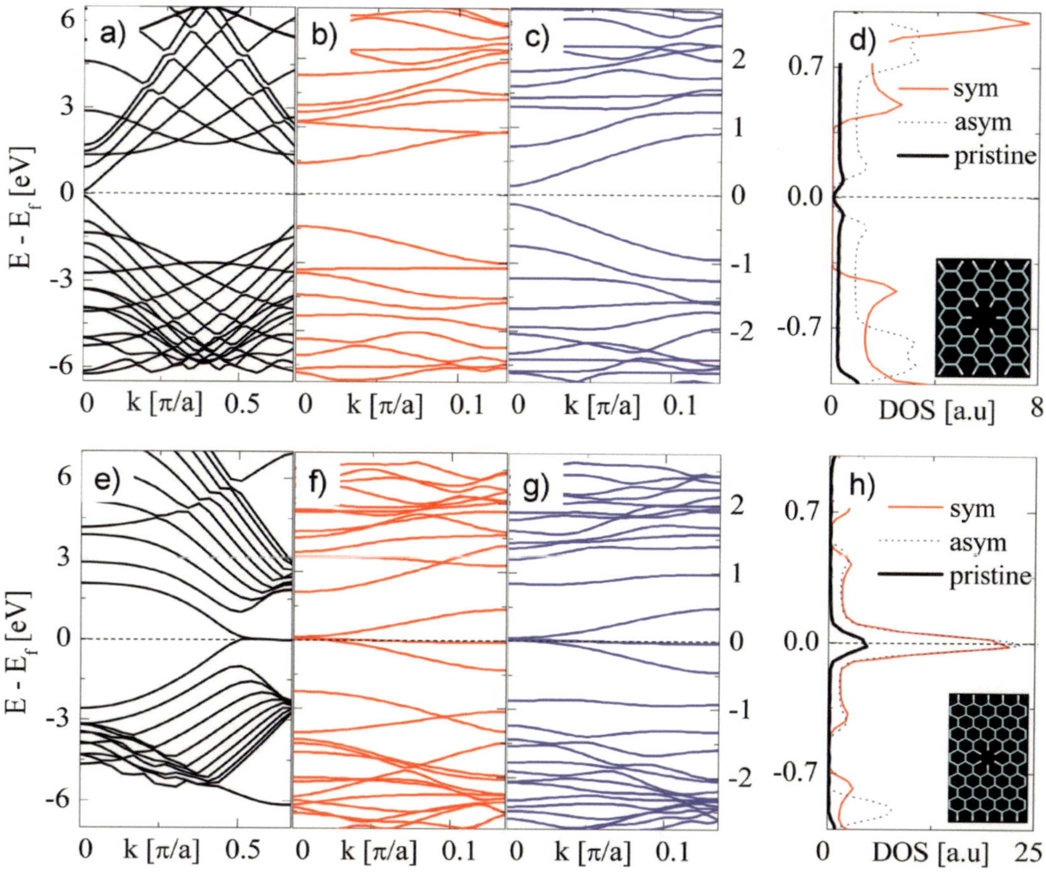

Figure 12. Band structure of: a) pristine 11-AGNR, b) 11-AGNR with centered SP, c) 11-AGNR with shifted SP, e) pristine 10-ZGNR, f) 10-ZGNR with centered SP, g) 10-ZGNR with shifted SP. The energy scales for (b, c, f, g) cases are the same. Density of states: d) centered and shifted SP in 11-AGNR, h) centered and shifted SP in 10-ZGNR. Insets: Unit cells for 11-AGNR and 10-ZGNR with the standard pore (*a* is the lattice constant of the supercell). From Ref. [76].

Although pristine GNRs can be metalic or semiconducting, their metallicity can be tunned by local changes of their morphology and edge chemistry [14, 38, 59–61, 164, 171, 176–181].

We begin our review by examining the band structures of porous armchair (AGNR) and zigzag (ZGNR) graphene nanoribbons, which are studied by large scale *ab-initio* DFT calculations [76]. We have used SIESTA 3.0 [182] in supercells of (> 40 atoms), and neglected (in most cases) spin degrees of freedom. The calculations were done with the Perdew-Zunger LDA functional [183] and pseudopotentials with the cutoff energy of 400 Ry (double-zeta basis and polarization orbitals, 13 and 5 orbitals for C-atom and H-atom, respectively). See other details in Ref. [76].

All the GNRs are classified by the number of carbon dimers, N, that form the ribbon (N-GNR) [5]. As an example, we have studied the porous 11-AGNR and 10-ZGNR (both metallic when pristine), with the distances between nanopores of 12.78 Å and 12.3 Å, respectively, and elementary cells are shown in Figs. 12 (d, h). We have described PNCs perforated with pores of different shapes, sizes, and locations. Most of the results are obtained for arrays of honeycomb-shaped pores, called the "standard pore" (SP) [90, 155], where in each pore six C atoms (one benzene ring) are excluded from the studied nanocarbon and the dangling bonds are terminated by hydrogens.

In Figs. 12 (a, b), one can see that the introduction of a periodic array of SPs in the center of 11-AGNR causes a significant band-gap opening ($0.15 - 0.92$ eV). On the other hand, both pristine and porous (same pores) 10-ZGNR have no energy band gap, as seen in Figs. 12 (e, f). When the SPs are displaced by one honeycomb cell towards the GNR-edge, the band gap in 11-AGNR shrinks ≈ 3.75 times, while 10-ZGNR remains metallic, as shown in Figs. 12 (c, g). We also checked that the PGNR band structures monotonously approach their pristine form as the separation between adjacent pores is increased. The densities of states (DOS) for the band structures presented are displayed in Figs. 12 (d, h).

The energy band gaps in ZGNRs and AGNRs are known to arise from a staggered sublattice potential and a quantum confinement [38, 60], respectively, and they depend on the ribbon width and its functionalization [166, 184, 185]. The wave functions of the HOMO and LUMO bands in AGNRs, which contribute directly to the area near E_f, are localized at the center of the ribbons, keeping their edges chemically stable [166]. In ZGNRs, these wavefunctions are localized at the ribbon edges. Consequently, when the SPs are positioned at the center of 11-AGNR a band gap opens in its band structure, while 10-ZGNR remains metallic. The metallicity of ZGNRs is caused by flat bands present at E_f, originating from highly localized states formed at the zigzag edges [166], as discussed below. When the pore is closer to the edge of 11-AGNR, its band structure is influenced less, while no significant change is observed in 10-ZGNR. These results can be generalized for other PGNRs provided that the pore is honeycomb like.

Porous graphene (PG)

Porous graphene (graphene antidot lattice) consists of a periodic arrangement of holes in a graphene sheet [167]. PG has been studied in a number of works [186–188], but in most studies hexagonal lattices of near-circular holes are considered [167–169]. Here, we use the $\{L, R\}$ PG superlattice classifaction [167] where L and R are measured in units of the graphene lattice constant $a = 2.46$ Å (see Fig. 13).

Various approaches have been used to describe the band structure of PGs. In TB approaches, a single-electron Schrödinger equation for an effective potential is solved. In the simplest version of TB, the Bloch wavefunctions are expanded over only localized atomic π orbitals of the carbon lattice atoms calculated with LCAO method and only nearest-neighbor overlap matrix elements are retained. These overlap matrix elements are regarded as emperical parameters fitted to the experimental data (e.g. the Fermi velocity). The DE can improve the bands close to their linear crossing points. In contrast to TB calculations, DFT approaches do not imply particle-hole symmetry around the Fermi level, as seen in Fig. 14.

The energy band gaps calculated in PG lattices by DE, TB and DFT methods are summarized in Table 15. The study of many different graphene antidot lattices disclosed a simple scaling law between their pore sizes and band gaps, $E_g = K\sqrt{N_{hole}}/N_{cell}$, where N_{hole} is the number of carbon atoms that have been removed from the pristine unit cell containing N_{cell} carbon atoms and $K \approx 25$ eV [167, 168]. Therefore, the band gaps increase with the increasing hole size and decreasing lattice constant.

http://dx.doi.org/10.5772/56247

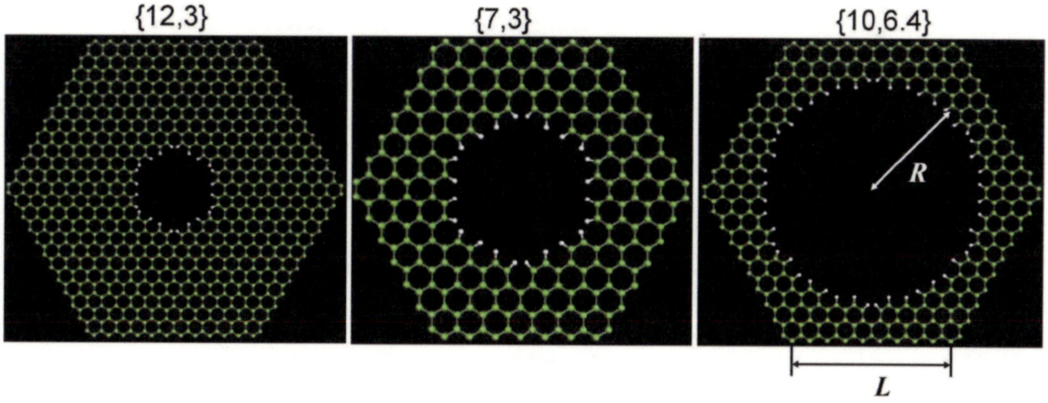

Figure 13. Unit cells of three hexagonal graphene antidot lattices with different side lengths L and hole radii R. The structures are denoted as L, R with both lengths measured in units of the graphene lattice constant $a \simeq 2.46$ Å. The edges of the holes have been hydrogen-passivated (hydrogen shown as white atoms). From Ref. [167].

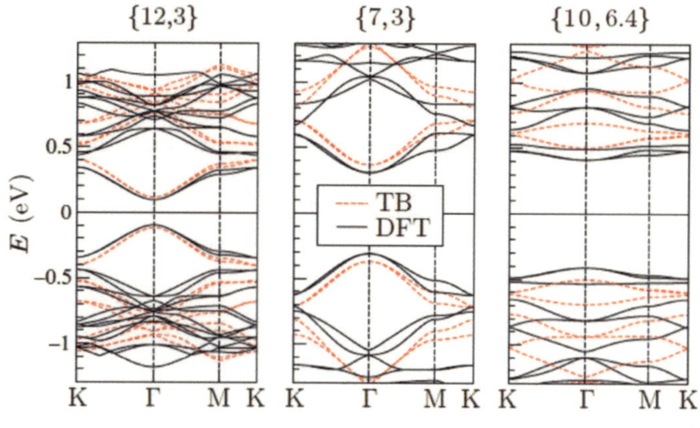

Figure 14. Band structures of three representative graphene antidot lattices. Full lines indicate results obtained using DFT, while TB results are shown with red dashed lines. From Ref. [167].

	{12, 3}		{7, 3}		{10, 6.4}	
	eV	$\Delta_{\{12,3\}}$	eV	$\Delta_{\{12,3\}}$	eV	$\Delta_{\{12,3\}}$
DE	0.54 (0.29)	1	1.27 (0.82)	2.35 (2.83)	1.53 (1.22)	2.83 (4.21)
TB	0.23	1	0.74	3.22	1.01	4.39
DFT	0.19	1	0.61	3.21	0.82	4.32

Figure 15. Band gaps of three representative graphene antidot lattices obtained by solving the DE, via TB, and using DFT. Values in parentheses are obtained using the DE. The band gaps are given in eV as well as in dimensionless values relative to the size of the band gap for the $\{L, R\} = \{12, 3\}$ structure [167].

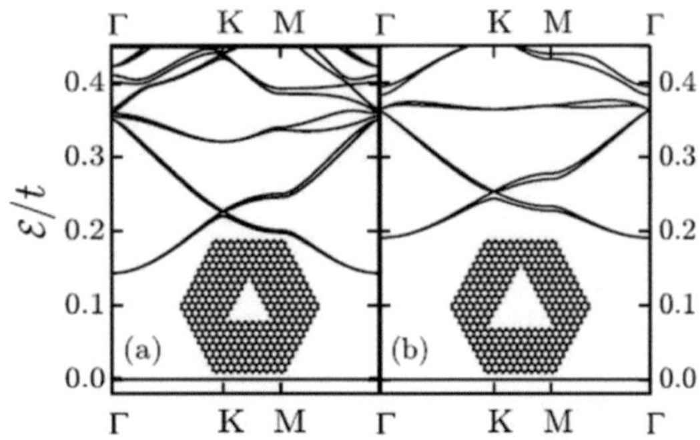

Figure 16. Band structure of an antidot lattice with triangular antidots calculated with TB method with $t \approx 2.8$ eV. Only bands above the Fermi level $\mathcal{E} = 0$ are shown. Γ, K, M stand for the high-symmetry points in the Brillouin zone. The flat band at the Fermi level is a) sixfold degenerate and b) ninefold degenerate. Adapted from Ref. [169].

The emergence of midgap energy states and corresponding quasi-flat bands around the Fermi level is another well pronounced effect related to the modification of the intact graphene lattice with nanopores. These states are relatively isolated, they have a minimal overlap of their wavefunctions, and thus they form flat bands. Graphene is known to be a bipartite lattice that can be divided into two sublattices A and B. According to the Lieb's theorem [189, 190], in a system where N_A and N_B are the total number of sites on the respective sublattices there should be $N_A - N_B$ zero energy states (so called, a global sublattice imbalance). The corresponding total magnetic moment per unit cell is just a half of this difference [167, 169].

In antidot graphene lattices, flat bands form at E_f due to the sublattice imbalance along the edges (zigzag GNRs) and pore rims. The electronic states corresponding to the zero energy flat bands (see Fig. 16) are predominantly localized in the vicinity of the pore rim (e.g. triangular pore). These localized states at zigzag rims were also studied in different types of pores, called "anti-molecules", where a set of simple rules was shown to link the net number of unpaired electrons with the degeneracy of flat bands [164, 167, 169, 174, 191]. It was shown that even a small number of zigzag-type of sites (local sublattice imbalance) per unit cell is enough to generate flat bands and related magnetism [176].

In previous studies, it was assumed that the perforation of graphene by nanopores opens up the energy band gap. The only exclusion from this trend was related to triangular pore shapes, due to their global sublattice imbalance. We have performed extensive *ab initio* calculations of electronic structures of PG-superlattices with different arrangements of the SPs [76]. SPs, largely used in our study, keep the global sublattice balance of the unit cell, but their short zigzag-like rims break the local balance and may still give rise to flat bands.

In Figs. 17 (a-d), we show the band structures of (edgeless) PGs with one honeycomb and three different rectangular SP-superlattices defined by N_A, N_Z, and displayed in Figs. 17 (e, f). Fig. 17 (g) also shows the studied rectangular superlattice with triangular-shape pores. The honeycomb SP-superlattice has flat bands around $E \approx -1.5$ eV (Fig. 17 a), which are absent in the rectangular SP-superlattices (Figs. 17 (b-d)). According to [169], such quasi-flat bands at non-zero energy might be ascribed to the local sublattice imbalance (globally $n_A = n_B$). However, the local sublattice imbalance due to SPs can not be the origin of flat bands in the band structure of the honeycomb SP-superlattice, since it has the same pores as the other

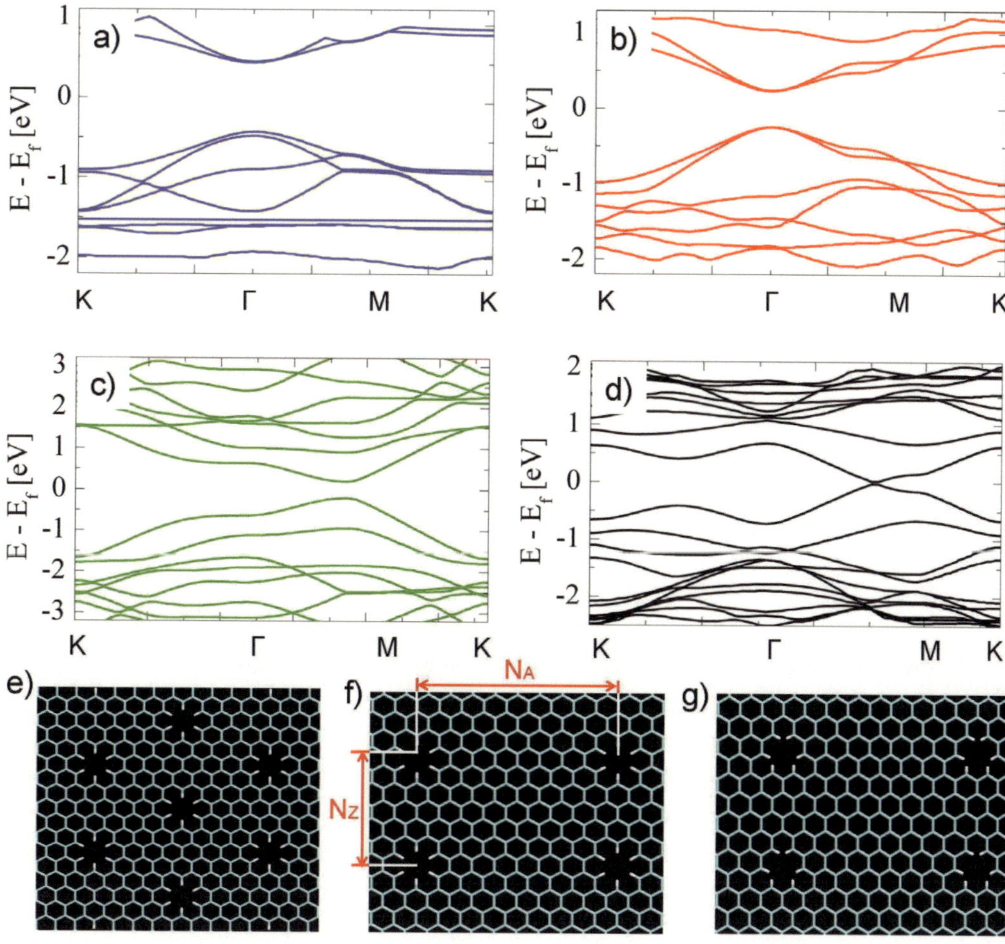

Figure 17. (top) Band structure of a) a honeycomb SP-superlattice in e) and three rectangular SP-superlattices with b) $N_A = 15$, $N_Z = 4$, c) $N_A = 7$, $N_Z = 2$, and d) $N_A = 7$, $N_Z = 8$. (bottom) e) honeycomb SP-superlattice, f) rectangular SP-superlattice characterized by N_A, N_Z, g) rectangular superlattice with triangular-shape pores. From Ref. [76] (see Fig. 16)

structures with no flat bands. Therefore, the flat bands are most likely related with the honeycomb SP-arrangement, which might produce differently localized states. These results show that the small number of zigzag sites at the very short SP-rim can not generate flat bands at the Fermi level or at its vicinity. In contrast, even the smallest triangular-shape pores break the global sublattice balance and generate an unbalanced π-electron density [175], associated with the appearance of rim-localized states and flat bands at E_f (compare with Fig. 16).

With these observations, we can now relate the band structures of PGNRs and PGs. Since SPs do not break the global sublattice balance in the PGNRs, their presence does not generate any new flat bands at E_f (see Fig. 12); on the contrary, we found that when AGNRs are perforated with triangular-shape pores, their band structure always contains flat bands at E_f. Even though the SPs do not create flat bands at E_f in GNRs, they still influence their band structure. While the band structures of AGNRs (Figs. 12 (b, c)) is influenced a lot, the

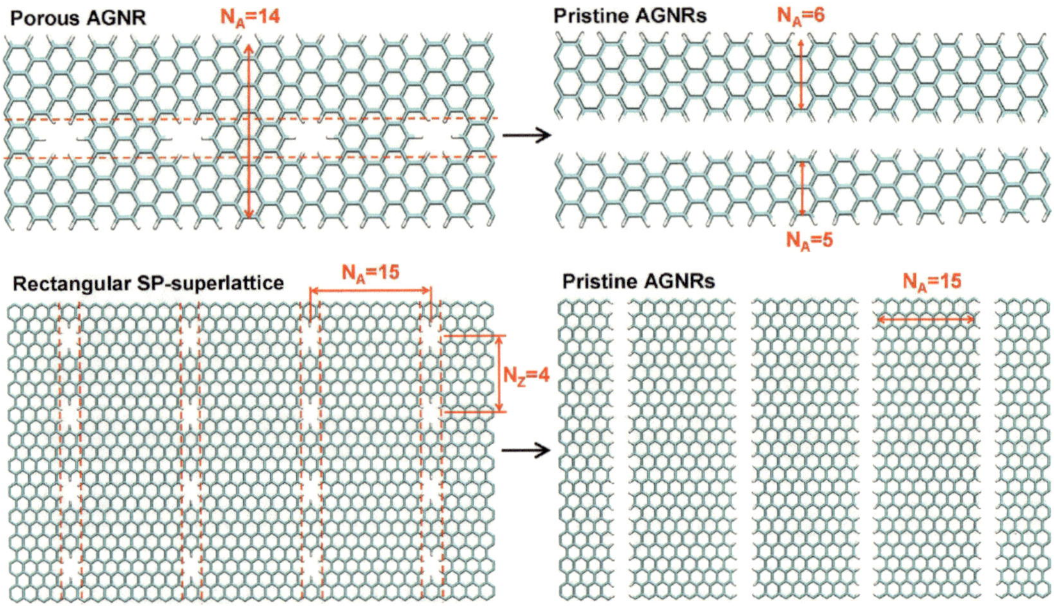

Figure 18. (top) Effective replacing of porous N-AGNR by "daughter" pristine N_1- and N_2-AGNR ($N = 14$, $N_1 = 5$, $N_2 = 6$). (bottom) Effective replacing of rectangular SP-superlattice with $N_A > N_Z$ by set of pristine AGNRs ($N_A = 15$, $N_Z = 4$). From Ref. [76].

band structures of ZGNRs with flat bands at E_f (local sublattice imbalance caused by zigzag edges) can not be significantly modified by SPs. Therefore, we conclude that all porous ZGNRs are metallic (checked by calculations).

Interestingly, we found that rectangular PG-superlattices perforated with SPs and larger pores of honeycomb symmetry can be both metallic and semiconducting [76]. It turns out that we can generalize these observations into a hypothesis that, in the first approximation, the electronic structure of these superlattices has the same type of conductivity as many parallel AGNRs or ZGNRs (of effective widths N_A or N_Z), depending on the ratio $\rho = N_A/N_Z$ (see Fig. 18 bottom). If $\rho \gg 1$, one can see the superlattice as being "cut" into separated N_A-AGNRs, while for $\rho \ll 1$ the same is true for separated N_Z-ZGNRs.

The above hypothesis was largely confirmed by our follow up calculations. For example, the conductivity in the PGs with $N_Z = 2$ is the same as in the corresponding AGNRs: metallic for $N_A = 5, 11, 17...$ and semiconducting for $N_A = 7, 9, 13, 15...$ (see also Table. 1). If we continue with the $N_A = 9, 15, 21, ...$ semiconducting AGNRs and increase the initially small $N_Z = 2$, we find that the PGs remain semiconducting for (roughly) $N_Z < N_A$, with the band gap shrinking with increasing N_Z, signaling the transition to the ZGNR-dominated metallic conductivity. If we continue with the the $N_A = 11$ metallic AGNRs, the PGs become metallic for all the N_Z, since the ZGNRs that take over at $N_Z > N_A$ are all metallic. Finally, when we continue from $N_A = 7, 13...$, the metallicity appears abruptly at $N_Z \geq 4$. In other cases, we expect that the transition between the AGNR and ZGNR-type of behavior occurs somewhere around the "diagonal", $\rho = 1$. Our calculations also show that the metal-semiconductor transitions predominantly occur in two regions of the Brillouin zones, as seen in Figs. 17 (b-d), and the bands can be partially flat in the k_y direction ($N_A = 7, 13, ..a.$).

Porous nanotubes

Unlike porous graphene and porous nanoribbons, the electronic structure of porous carbon nanotubes (PCNTs) was studied much less [173, 192]. However, there is a tight connection between CNTs and GNRs and clear correspondence between the electronic structures of pristine CNTs and structurally analogous GNRs. Moreover, the electronic structures of PNCs are similar to those of pristine GNRs [76].

We follow the Ezawa's notations [178] in description of GNRs and CNTs. While ZGNRs indexed by $\langle p,0\rangle$ are metallic for all p, only one third of AGNRs indexed by $\langle p,1\rangle$ are metallic. Since, only one half of AGNRs can be rolled into ZCNTs, only one third ZCNTs are metallic [178]. We found that this mapping can be generalized to the point that every metallic/semiconducting CNT corresponds to a metallic/semiconducting GNR. The unambiguous relationship can be expressed in terms of the chirality index, p, and the number of GNRs carbon dimers, N, as follows

$$\begin{aligned} &\text{ZCNT}\langle p,0\rangle \Rightarrow N\text{-AGNR}, \quad \text{where } N = 2p+2, \\ &\text{ACNT}\langle p,p\rangle \Rightarrow N\text{-ZGNR}, \quad \text{where } N = 2p+1. \end{aligned}$$

Intuitively, we can look at this CNT-GNR correspondence as a consequence of CNT "cutting" (Figs. 19 (top, middle)), which preserves the type of conductivity. The AGNRs and ZGNRs that do not match any CNTs are all semiconducting and metallic, respectively, as summarized in Table. 1.

AGNR <p,1>	dimers, N	7	8	9	10	11	12	13	14	ZGNR <p,0>	5	6	7	8	9	10	11	12
	p-index	4	5	6	7	8	9	10	11		4	5	6	7	8	9	10	11
	cond.	s	m	s	s	m	s	s	m		m	m	m	m	m	m	m	m
	geometry	s	a	s	a	s	a	s	a		a	s	a	s	a	s	a	s
ZCNT<(p+1)/2,0>	cond.		m		s		s		m	ACNT <p/2, p/2>	m		m		m		m	

Table 1. Correspondence rule for GNRs and CNTs. N - number of dimers forming GNR, p - index for the chirality vector in GNR and CNT; cond. - conductivity property (semiconducting (s) or metallic (m)); geometry defines the symmetry with respect to the mirror plane perpendicular to the ribbon and containing its axis: symmetric (s), asymmetric (a). Examples of AGNRs for $N = 7, \ldots 14$ and ZGNRs for $N = 5, \ldots 12$ are illustrated. From Ref. [76]

Analogously, porous CNTs might have band structures similar to porous GNRs. Tight-binding calculations predicted [173] that a line of SPs (separated by $\sim$ 12.8 Å) should cause band gap opening in ACNTs, whereas porous ZCNTs should be semiconducting regardless of the pore shape. These results are in contradiction with our *ab-initio* calculations, which show that the metallicity of pristine ACNTs (Fig. 20 a) is preserved in the porous ACNTs (Fig. 20 e), even for triangular-shape pores with clear zigzag-like rims. In metallic ZCNTs, the SP-perforation causes band gap opening, as shown in Figs. 20 (b, f), while in semiconducting ZCNTs, it causes band gap shrinking, as seen in Figs. 20 (c, g).

The above observations allow us to build a *unified model* that can predict the type of conductivity in porous nanocarbons perforated with SPs and other pores that do not break the global sublattice balance. The model is based on the assumption that when nanocarbons (NCs) are perforated by a line of relatively close SPs, the type of conductivity in these PNCs is the same as in the (daughter) systems obtained from these nanocarbons by removing all C atoms within a stripe going in the direction of the pores and having the same width as the pores (all dangling C bonds are H-terminated). We call these modified NCs the daughter

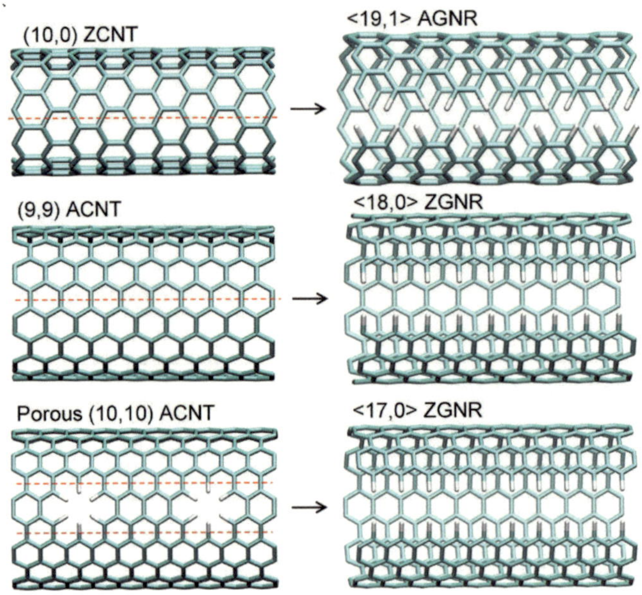

Figure 19. Cutting of: (top) (10,0)ZCNT into ⟨19,1⟩AGNR and (middle) (9,9)ACNT into ⟨18,0⟩ZGNR (opening of the GNRs is schematically shown). (bottom) Removal of atoms from porous (10,10)ACNT leading to ⟨17,0⟩ZGNR. From Ref. [76].

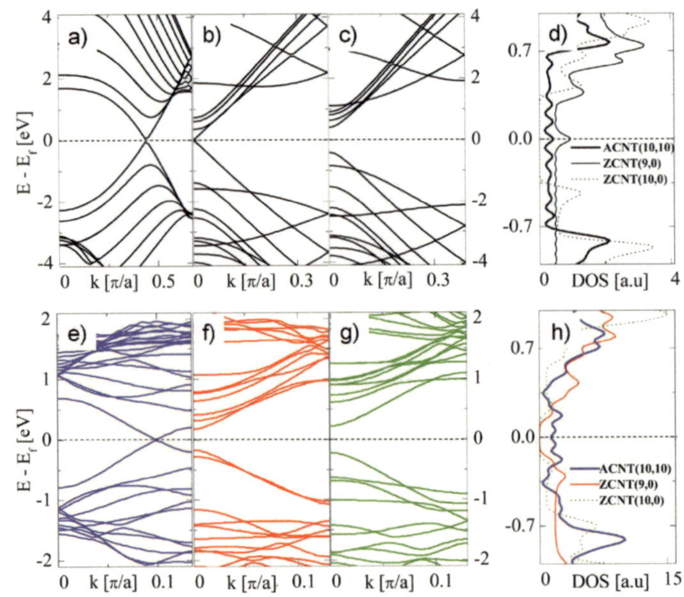

Figure 20. Band structures in pristine CNTs: a) ACNT (10,10), b) ZCNT (9,0), c) ZCNT (10,10); d) DOS. Porous CNTs: e) ACNT (10,10), f) ZCNT (9,0), g) ZCNT (10,10); h) DOS. The energy scales for (b,c) and (f,g) cases are the same. From Ref. [76].

http://dx.doi.org/10.5772/56247

systems of the original NCs (two GNRs for PGNRs, one GNR for PCNTs, and a GNR-lattice for PG). This rule predicts that: (1) Porous ACNTs are metallic as the (daughter) ZGNRs; Fig. 19 (bottom) shows effective replacing of porous ACNT by pristine ZGNR. Perforating the resulting ZGNRs (and the other half of ZGNRs that can not be rolled up into CNTs) gives two metallic ZGNRs, preserving the ZGNR-metallicity. (2) Porous ZCNTs may give semiconducting or metallic AGNRs. Cutting all the AGNRs may give pairs of AGNRs with any conductivity. These results were confirmed by *ab initio* calculations.

We now use these rules to predict metallicity in porous AGNRs with SPs. We assume that their band gaps are $E_{BG} \simeq \min(E_{1BG}; E_{2BG})$, where E_{1BG}, E_{2BG} are band gaps of their two daughter AGNRs (see Fig. 18 (top)). With this inference, we can derive an analytical expression describing the dependence of the band gap on the width of the porous AGNRs. For simplicity, we consider SPs positioned in the middle of AGNRs of the width of $W = a\sqrt{3}(N-1)/2$, where a is the C-C distance and N is the number of dimers. By evaluating the widths of the pristine daughter AGNRs, we find that porous N-AGNRs are potentially metallic if the number of C-C dimers is given by at least one of these equations

$$N = \left(6k + 11 + (-1)^k\right)/2, \quad N = 6k + 3 + 2(-1)^k,$$
$$N = 12k + 9, \quad (k = 0, 1, 2...), \tag{1}$$

i.e., if $N = 5, 6, 7, 8, 9, 12, 14, 17, 18, 19, 20, ...$ In Fig. 21 (top), we compare the *ab-initio* energy band gaps in pristine [59, 180] and porous AGNRs to validate the above model. In contrast to the pristine ribbons, where the metallic points emerge with the period of 3, ($N_{met} = 3m + 2$), the band gaps of porous AGNRs have a more complex dependence. Nevertheless, the positions of the band gap minima agree with Eqn. 1.

We can extend the assumptions used in Eqn. 1 to PNCs perforated with larger and shifted honeycomb-like pores. Their presence may still be reduced to removing from the AGNRs a layer of atoms of the width given by the pore size, where the minimum band gap of the two resulting AGNRs can determine the band gap of the porous AGNR. For example, when the SP is shifted in the 11-AGNR by one honeycomb from the ribbon center, the two daughter pristine 4-AGNRs are replaced by 2-AGNR and 6-AGNR (all semiconducting). This should lead to a band gap shrinkage, in agreement with our *ab-initio* calculations, presented in Figs. 12 (b, c). Alternatively, we can replace the SP by a double-size hexagonal pore with 24 C-atom excluded. If the 11-AGNR and 12-AGNR are perforated by such pores, they become semiconducting, since their cutting leads to semiconducting 2-AGNRs and 2- and 3-AGNR, respectively. These results are in agreement with *ab-initio* calculations, giving in 11-AGNR and 12-AGNR the band gaps of 1.1 eV and 1.18 eV, respectively. We have also tested the triple-size hexagonal pore (54 C-atoms excluded) in order to check how its long rims affect band structure of GNRs. Our calculations show that no additional features (e.g. flat bands at the Fermi level) appear when the GNR are perforated by these pores. In contrast, when these AGNRs are perforated with SPs in the ribbon center, only the 11-AGNR is semiconducting.

Finally, we discuss porous ZCNTs that can have any conductivity. In Fig. 21 (bottom), we present the energy band gap of porous ZCNTs in dependence on the chirality index, p. It exhibits similar periodicity as in the pristine ZCNTs. However, the model is not reliable in porous ZCNTs. For example, the porous ZCNT(7,0) and ZCNT(8,0) have band gaps similar

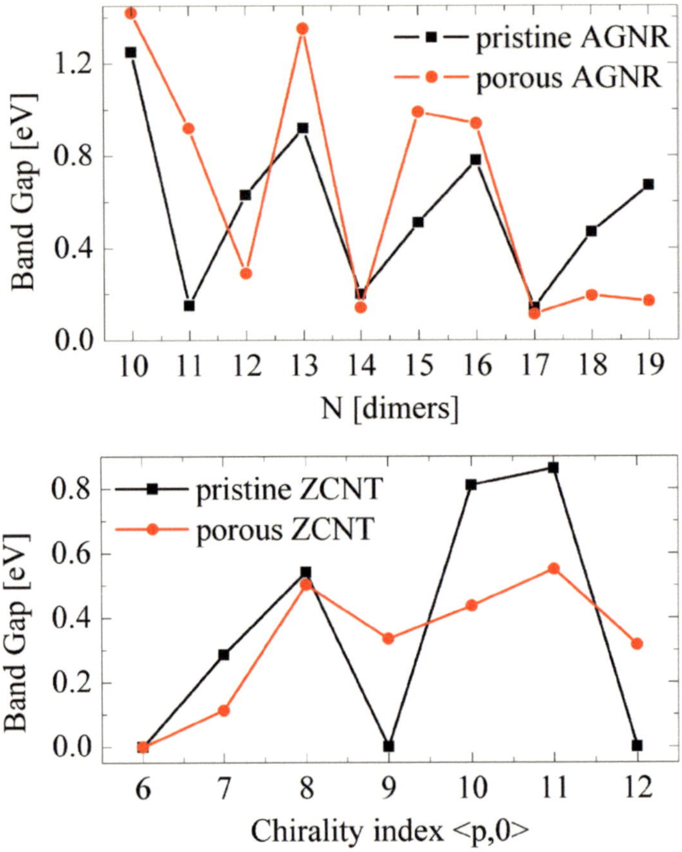

Figure 21. (top) Dependence of the band gap in the pristine and porous AGNR on the number of dimers (the central position of the SP). (bottom) The same dependence in ZCNT on the chirality index. From Ref. [76].

to the daughter 11-AGNR and 13-AGNR, respectively. But the same is not true for the porous ZCNT(9,0) and ZCNT(10,0) paired with the daughter 15-AGNR and 17-AGNR, respectively. In principle, this failure might be caused by the fact that the AGNRs are not calculated deformed as the corresponding daughter ZCNTs [193, 194]. However, our calculations show that the bent 17-AGNR has almost the same band gap as the pristine 17-AGNR. Therefore, a more quantitative approach needs to be used here.

It is of interest to see if other types of periodic modifications can also be used to tune the band structures of nanocarbons. To briefly examine this idea, we have replaced SPs by Stone-Wales 55-77 defects [195]. In Fig. 22, we show the band structures of 11-AGNR, 10-ZGNR, and graphene superlattices modified in this way. The periodic array of SW 55-77 defect in 11-AGNR leads to small bang gap opening, as shown Fig. 22 (a), in analogy to 11-AGNR with SPs (Fig.12 b). The band structure of 10-ZGNR, shown in Fig. 22 (b), is not sensitive to this perturbation, as in the SP-perturbations (Fig.12 f). On the other hand, when we replace in graphene superlattices SPs with the SW 55-77 defects, we can obtain qualitatively different band structures. In particular, the band structure of graphene modified by SW 55-77 defects in the array with $N_A = 7$ and $N_Z = 4$ (Fig. 22 c) is similar to that of the SP-superlattice with the same N_A and N_Z, but here we also observe opening of a small band gap. In Fig. 22 (d),

http://dx.doi.org/10.5772/56247

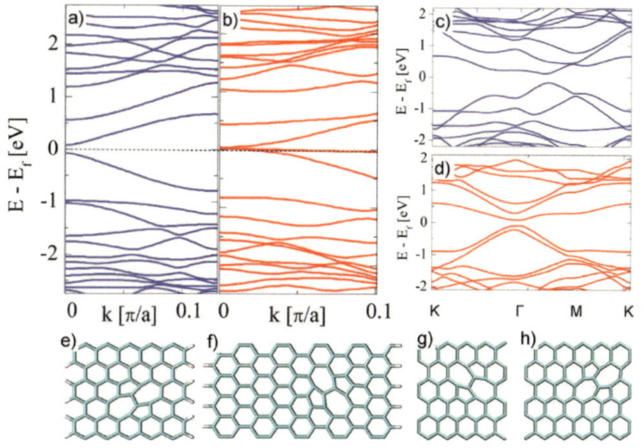

Figure 22. Band structures of nanocarbons with periodic 55-77 Stone-Wales defects: a) 11-AGNR, b) 10-ZCNT, c) graphene superlattice with $N_A = 7$ and $N_Z = 4$, d) graphene superlattice with $N_A = 9$ and $N_Z = 4$. Figures (e-h) show unit cells for the respective cases. From Ref. [76].

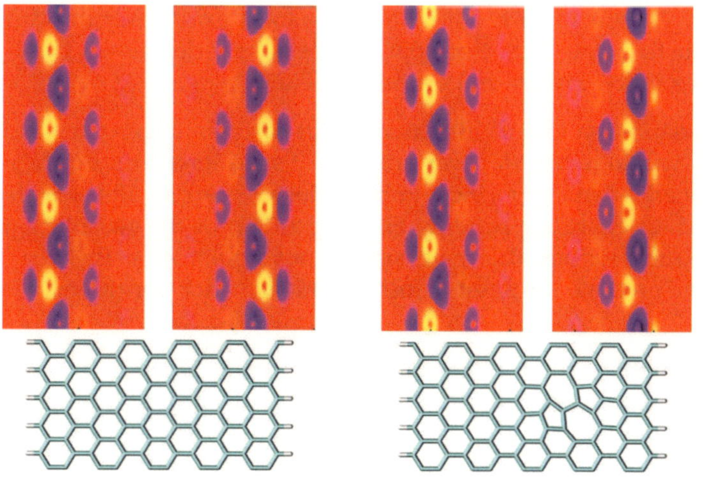

Figure 23. 10-ZGNR edge magnetism. The alternation of spin imbalance as result of introduction of 55-77 SW defects. From Ref. [76].

we show the band structure of the SW-graphene superlattice ($N_A = 9$ and $N_Z = 4$), which is semiconducting as the PG-superlattice with the same N_A and N_Z. These observations show that periodic defects could also be used to tune band gaps in nanocarbons, but the rules might be slightly different.

For completeness, we have recalculated some of the above structures including spin polarization (we used a set of LDA and GGA functionals). It turns out, the band structures of nanocarbons can be modified by the spin polarization (zigzag edges) [164], but the presence of SPs does not introduce additional magnetic features beyond the changes described already in the non-magnetic calculations. Interestingly, the presence of arrays of SW (55-77) defects in ZGNRs can alternate mutual orientation of the magnetic moments localized at the opposite edges, due to topological changes in the sublattices, as shown in Fig. 23.

We have shown that the formation of nanopores in PNCs can lead to the appearance of band gaps, midgap states and quasi-flat bands around the Fermi level. We have discussed a unified picture of electronic structures in PNCs, where the formation of periodic arrays of nanopores can be regarded as an effective cutting. We have also seen that nanopores can induce a local spin imbalance related to a non-zero magnetic moment. All these effects show that the electronic parameters of PNCs are highly tunable, which can be used in numerous potential applications.

Additional control of the electronic structure in PNCs

The electronic structure of PNCs can be further influenced and controlled by the chemistry of the edges and pore rims, various defects, mechanical deformations, etc. For example, it was shown that hydrogen adsorbate structures on graphene are thermodynamically stable well above the room temperature, and the resulting band gap opening is stable against a weak disorder [165, 185]. Extensive studies of GNRs and PGNRs with chemically functionalized edges were carried on [166, 184]. It was shown that edge or bulk functionalization or atomic substitution in PNCs (n/p doping) is an efficient mean to change electronic characteristics of nanocarbons. In particular, chemical functionalization of zigzag-type of edges breaks the spin degeneracy, as shown in Fig. 24 [166]. It promotes a metal-semiconductor transition or a half-semiconducting state with two spin channels (spin-dependent band gap or opposite spin polarizations in the valence and conduction bands).

An edge functionalization of AGNRs does not lift the spin degeneracy and it can generate electronic states within a few eV away from the Fermi level in the valence band with no effect on their energy band gap [166]. Substitutions of carbon atoms with N and B atoms can produce various effects, depending on the position of the substitutional site [166]. In particular, edge substitutions at low density do not significantly alter the band gap, while bulk substitutions promote the onset of metal-semiconductor transitions, as shown in Fig. 25 [166].

Mechanical deformations, such as application of uniaxial or shear strains and out-of-plane deformation or bending, can also modify the energy band gap. Compared to a graphene sheet where the band gap remains close to zero even if a large strain is applied, the band gap of GNRs and PGNRs depends strongly on their edge shape [194]. For AGNRs, a weak uniaxial strain changes the band gap in a linear fashion, whereas a large strain results in periodic oscillation of the band gap. In contrast, the shear strain always reduces the band gap. In ZGNRs, a strain changes the spin polarization at the edges, and thereby modulate the band gap. Out-of-the-plane radial deformations can cause a semiconductor-metal transition in GNRs, due to the hybridization of σ and π orbitals and modifications of orbital interactions [193]. Mechanical deformations may also be related with other interesting phenomena, such as a piezoelectricity in a boron-nitride analogue of graphene [196].

4.2. Electron transport in porous nanocarbons

Electronic structures in graphene-based systems determine their electronic transport properties. Specifically, electronic transmission through such systems under bias (current-voltage characteristics) strongly depends on their energy band gap, densities of states, bands lineup, spin states, and the reciprocal positions of an Fermi energy level of their electrodes with respect to their HOMO and LUMO bands [197–199].

http://dx.doi.org/10.5772/56247

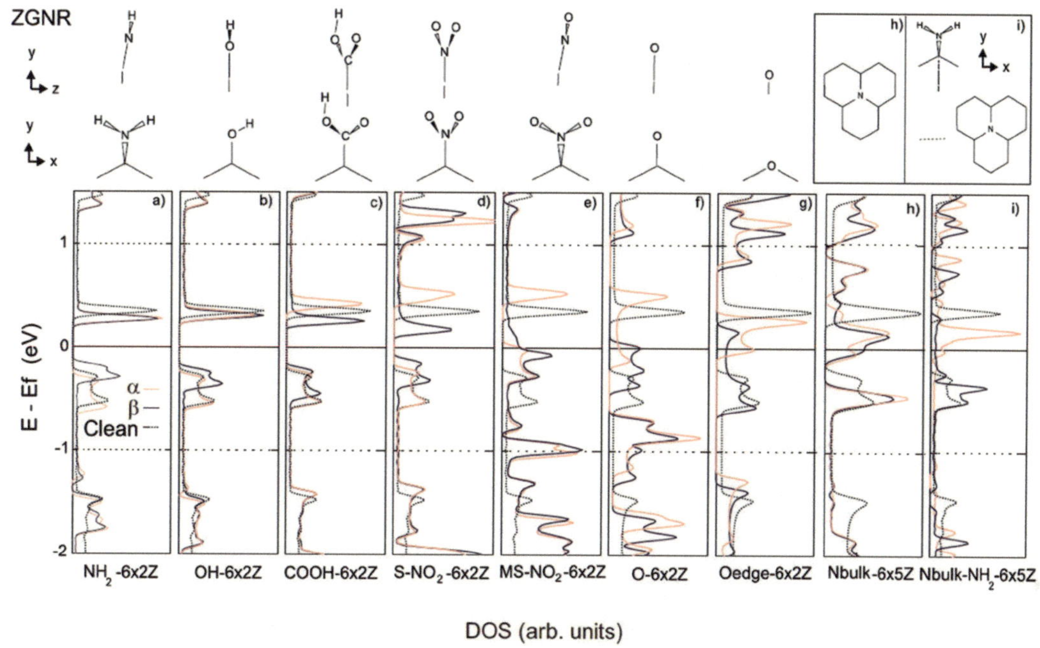

Figure 24. Spin density of α (red solid line) and β (black solid line) for functionalized ZGNRs in comparison with a clean ribbon (black dotted line). 6-ZGNR functionalized with a) NH_2, b) OH, c) COOH, d) stable-NO_2, e) metastable-NO_2 and f) O radicals in one of the ribbon edges, resulting in (a)-(f) lifting of the spin degeneracy, (a)-(d) spin-selective band gap, or (e) and (f) semiconductor-metal transition. g) Oxygen edge substitutional atom favoring a semiconductor-metal transition. (h) 6-ZGNR with a nitrogen bulk substitution in the center of the ribbon. i) Nitrogen bulk substituted ribbon shown in h) with an extra single NH_2 at the edge of the opposite carbon sublattice, with the maximum distance between N and NH_2 in the periodic direction, resulting in a semiconductor-metal transition. From Ref. [166].

The quantum transport in biased graphene nanoribbons (GNRs) was extensively studied [48, 200–202]. For example, it was shown that spin-valve devices based on GNRs can exhibit magnetoresistance values that are thousands of times higher than previously reported experimental values [203]. GNR doping was shown to significantly affect their transport properties [48, 199, 204]. Modeling of electron transport in GNRs was done using the Landauer-Buttiker formalism [205], non equilibrium Green's function (NEGF) techniques [206], and other methods.

In Fig. 26, we show the calculated quantum conductances of two pristine armchair GNRs of different widths and a pristine zigzag GNR in anti-ferromagnetic and ferromagnetic states [200]. One can see that the energy band gap of armchair GNRs depends on the ribbon widths. The step-like behavior of the transmission spectrum is related with the available conductance channels (bands) [201]. In zigzag GNRs, the electron transport is dominated by edge states which are spin-polarized. Depending on the edge spin configuration (↑↓ or ↑↑) the zigzag GNRs are predicted to be semiconducting or semi-metallic, which is reflected in the presence or absence of the energy interval of zero transmission, respectively.

We have calculated by NEGF techniques the electron transmission through porous armchair GNRs. The system setup and the electron transmission spectra are shown in Fig. 27. As one can see, the perforation of 11-AGNR with a pore can significantly affect its conductance.

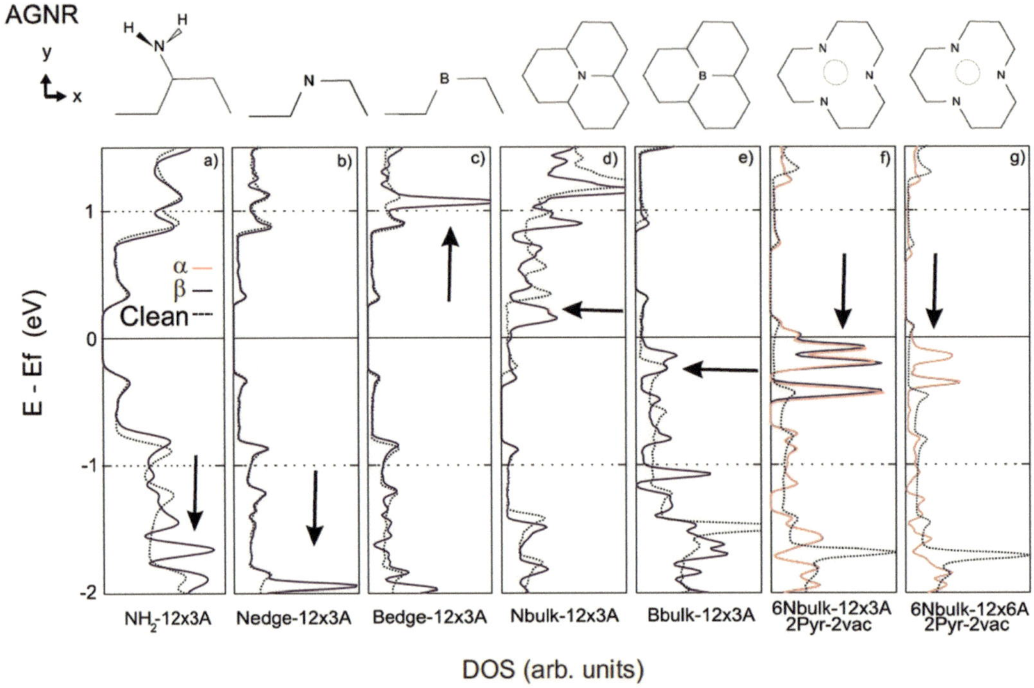

Figure 25. Spin density of α (red solid line) and β (black solid line) for functionalized AGNRs in comparison with a clean ribbon (black dotted line). 12-AGNR functionalized with a) NH_2, b) N, c) B edge substitutional atoms, N and B bulk substitutions are shown in d) and e), where a semiconductor-metal transition occurs. f) and g) Two pyridinelike substitutional doping in the bulk resulting in a metallic behavior. Arrows indicate impurity levels due to substitutional atoms. The DOS of the clean ribbon is shifted for the bulk N and B substitutions, so that the edges of the valence band of clean and doped ribbons coincide. From Ref. [166].

The most pronounced effect is the emergence of additional regions of zero transmission that proves its dependence on the pore position. The plausible reason of this effect is the quantum interference [207] between different spatial paths that electrons can follow in the scattering region. Such a sensitivity may be used for molecular passage detection via electric measurements.

5. Conclusions

In the last decade, intensive research of graphene, its precursors and derivatives revealed the great potential of nanocarbons. These materials have unique mechanical, chemical and electronic properties. Modulating the atomic structure of nanocarbons with chemically functionalized nanopores was found to be an efficient way to impart to them the desired properties and functions. The spectrum of nanocarbon applications can cover electronics, optics, and materials for general use, such as molecular filtration, sensing, detection, and recognition. In electronics, we should mention that perforation of nanocarbons with nanopores can not only trigger a metal-semiconductor transition but also a subtle magnetic phase transition, making porous nanocarbons promising materials for spin current filters and magnetoresistors. When it comes to molecular sensing and detection, one can use the exceptional sensitivity of PNC electronic structures to the presence of molecules and

http://dx.doi.org/10.5772/56247

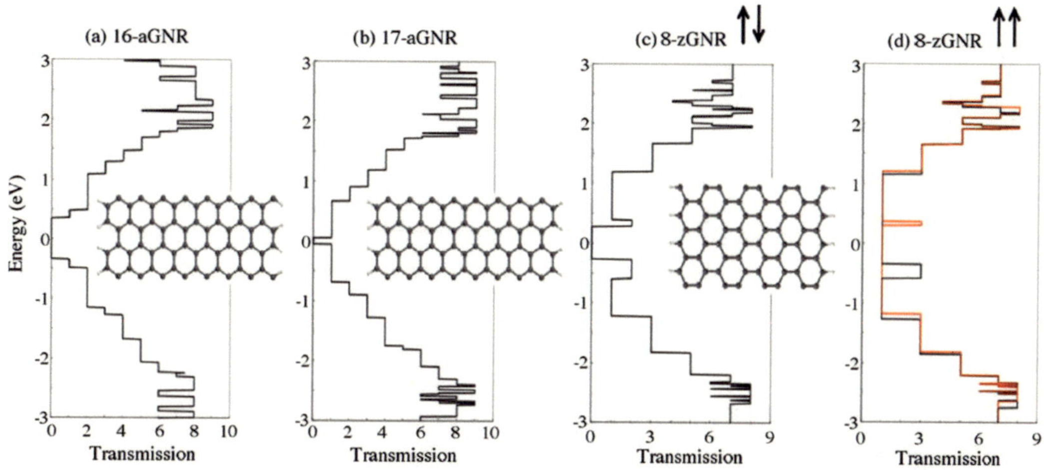

Figure 26. Quantum transport in graphene nanoribbons at zero bias. Atomic structures and electronic transmission of (a) 16-armchair GNR; (b) 17-armchair GNR, 8-zigzag GNR with (c) anti-parallel or (d) parallel spin orientations between the two magnetic edges. The spin-dependent transport is evaluated for both magnetic configurations of the 8-zGNR (c and d) but is only visible for the parallel spin orientations (ferromagnetic one). In such case, one spin orientation is labeled majority-spin (in black) while the other is labeled minority-spin (in red). Adapted from Ref. [200].

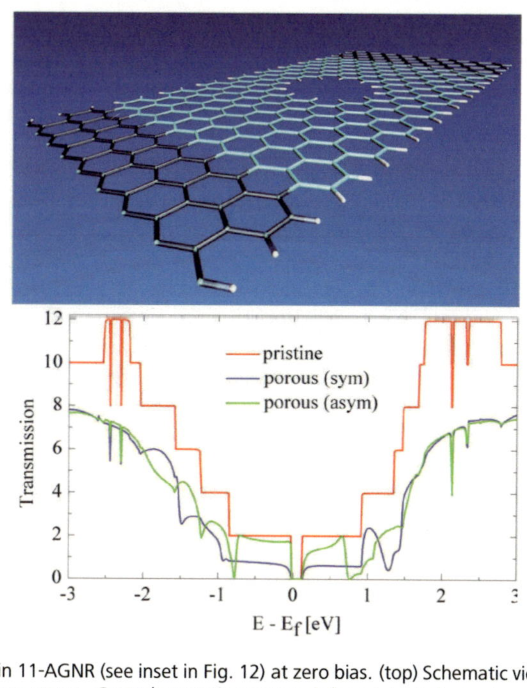

Figure 27. Quantum transport in 11-AGNR (see inset in Fig. 12) at zero bias. (top) Schematic view of two-probe porous AGNR with edges passivated by hydrogen atoms. Central scattering region, left and right electrodes are indicated. (bottom) Electron transmission spectra for the pristine 11-AGNR, 11-AGNR with centered pore, 11-AGNR with shifted pore.

passage through their pores. However, in order to benefit from material properties of porous nanocarbons more research is needed in the preparation of atomically precise porous nanocarbons.

Acknowledgments

A.B. would like to acknowledge the support of the Paaren Graduate Fellowship from the Department of Chemistry at the University of Illinois at Chicago. This work was partially supported by a grant from ACS PRF 53062. The simulations and other calculations have been realized with the NERSC, NCSA and CNM supercomputer networks.

Author details

Boyang Wang[1], Artem Baskin[2] and Petr Král[2,3,*]

* Address all correspondence to: pkral@uic.edu

1 Institute of Chemistry, Chinese Academy of Science, Beijing, R.P. China
2 Department of Chemistry, University of Illinois at Chicago, IL, USA
3 Department of Physics, University of Illinois at Chicago, IL, USA

References

[1] Novoselov K. S. Geim A. K. Morozov S. V. Jiang D. Zhang Y. Dubonos S. V. Grigorieva I. V. Firsov A. A. Electric Field Effect in Atomically Thin Carbon Films, *Science* 2004; *306*, 666–669.

[2] Lenosky T. Gonze X. Teter M. Elser V. Energetics of negatively curved graphitic carbon, *Nature* 1992; *355*, 333–335.

[3] Castro Neto A. Guinea F. Peres N. Novoselov K. Geim A. The electronic properties of graphene, *Rev. Mod. Phys.* 2009; *81*, 109–162.

[4] Abergel D. S. L. Russell A. Fal'ko V. I. Visibility of graphene flakes on a dielectric substrate, *Appl. Phys. Lett.* 2007; *91*, 063125.

[5] Cresti A. Nemec N. Biel B. Niebler G. Triozon F. Cuniberti G. Roche S. Charge transport in disordered graphene-based low dimensional materials, *Nano Research* 2008; *1*, 361–394.

[6] Freitag M. Graphene: Nanoelectronics goes flat out, *Nat Nano* 2008; *3*, 455–457.

[7] Katsnelson M. I. Novoselov K. S. Geim A. K. Chiral tunnelling and the Klein paradox in graphene, *Nat Phys* 2006; *2*, 620–625.

[8] Hwang E. H. Adam S. Das Sarma S. Carrier Transport in Two-Dimensional Graphene Layers, *Phys. Rev. Lett.* 2007; *98*, 186806.

[9] Peres N. Guinea F. Castro Neto A. Electronic properties of two-dimensional carbon, *Annals of Physics* 2006; *321*, 1559–1567.

[10] Gusynin V. P. Sharapov S. G. Unconventional Integer Quantum Hall Effect in Graphene, *Phys. Rev. Lett.* 2005; *95*, 146801.

[11] Novoselov K. S. Geim A. K. Morozov S. V. Jiang D. Katsnelson M. I. Grigorieva I. V. Dubonos S. V. Firsov A. A. Two-dimensional gas of massless Dirac fermions in graphene, *Nature* 2005; *438*, 197–200.

[12] Datta S. Strachan D. Khamis A., S.M. anf Johnson Crystallographical etching of few layer graphene, *Nano Lett.* 2008; *8*, 1912–1915.

[13] Campos-Delgado J. e. a. Bulk production of new form of sp(2) carbon: crystalline graphene nanoribbons, *Namo Lett.* 2008; *8*, 2773–2778.

[14] Yang X. e. a. Two-dimensional graphene nanoribbons, *J. Am. Chem. Soc.* 2008; *130*, 4216–4217.

[15] Li W., X.L. X.R. L., Zhang Lee S. Dai H. Chemically derived, ultrasmooth graphene nanoribbon semiconductors, *Science* 2008; *319*, 1229–1232.

[16] Chen Z. Lin Y. Rooks M. Avouris P. Graphene nano-ribbon electronics, *Phisica E* 2007; *40*, 228–232.

[17] Han M. Ozyilmaz B. Zhang Y. Kim P. Energy band-gap engineering of graphene nanoribbons, *Phys. Rev. Lett.* 2007; *98*, 206805.

[18] Yan L. Zheng Y. Zhao F. Li S. Gao X. Xu B. Weiss P. Zhao Y. Chemistry and physics of a single atomic layer: strategies and challenges for functionalization of graphene and graphene-based materials, *Chem Soc Rev.* 2012; *41*, 97–114.

[19] Kuila T. Bose S. Mishra A. K. Khanra P. Kim A. N. H. Lee J. H. Chemical functionalization of graphene and its applications, *Prog. Mater. Sci.* 2012; *57*, 1061–1105.

[20] Georgakilas V. Otyepka M. Bourlinos A. B. Chandra V. Kim N. Kemp K. C. Hobza P. Zboril R. Kim K. S. Functionalization of Graphene: Covalent and Non-Covalent Approaches, Derivatives and Applications, *Chem. Rev.* 2012; *112*, 6156–6214.

[21] Sreejith S. Ma X. Zhao Y. Graphene Oxide Wrapping on Squaraine-Loaded Mesoporous Silica Nanoparticles for Bioimaging, *J. Am. Chem. Soc.* 2012; *134*, 17346–17349.

[22] Yang J. Heo M. Lee H. J. Park S.-M. Kim J. Y. Shin H. S. Reduced Graphene Oxide (rGO)-Wrapped Fullerene (C60) Wires, *ACS Nano* 2011; *5*, 8365–8371.

[23] Wang S. Manga K. K. Zhao M. Bao Q. Loh K. P. Wrapping Graphene Sheets Around Organic Wires for Making Memory Devices, *Small* 2011; *7*, 2372–2378.

[24] Lomeda J. R. Doyle C. D. Kosynkin D. V. Hwang W.-F. Tour J. M. Diazonium Functionalization of Surfactant-Wrapped Chemically Converted Graphene Sheets, *J. Am. Chem. Soc.* 2008; *130*, 16201–16206.

[25] Dikin D. A. Stankovich S. Zimney E. J. Piner R. D. Dommett G. H. B. Evmenenko G. Nguyen S. T. Ruoff R. S. Preparation and characterization of graphene oxide paper, *Nature* 2007; *448*, 457–460.

[26] Eda G. Fanchini G. Chhowalla M. Large-area ultrathin films of reduced graphene oxide as a transparent and flexible electronic material, *Nat Nano* 2008; *3*, 270–274.

[27] Chen D. Feng H. Li J. Graphene Oxide: Preparation, Functionalization, and Electrochemical Applications, *Chem. Rev.* 2012; *112*, 6027–6053.

[28] Campos L. C. Manfrinato V. R. Sanchez-Yamagishi J. D. Kong J. Jarillo-Herrero P. Anisotropic Etching and Nanoribbon Formation in Single-Layer Graphene, *Nano Lett.* 2009; *9*, 2600–2604.

[29] Wang X. Dai H. Etching and narrowing of graphene from the edges, *Nat Chem* 2010; *2*, 661–665.

[30] Boukhvalov D. W. Katsnelson M. I. Chemical Functionalization of Graphene with Defects, *Nano Lett.* 2008; *8*, 4373–4379.

[31] Yang H. Shan C. Li F. Han D. Zhang Q. Niu L. Covalent functionalization of polydisperse chemically-converted graphene sheets with amine-terminated ionic liquid, *Chem. Commun.* 2009; *9*, 3880–3882.

[32] Quintana M. Spyrou K. Grzelczak M. Browne W. R. Rudolf P. Prato M. Functionalization of Graphene via 1,3-Dipolar Cycloaddition, *ACS Nano* 2010; *4*, 3527–3533.

[33] Cui X. Zhang C. Hao R. Hou Y. Liquid-phase exfoliation, functionalization and applications of graphene, *Nanoscale* 2011; *3*, 2118–2126.

[34] Englert J. M. Dotzer C. Yang G. Schmid M. Papp C. Gottfried J. M. Steinrück H.-P. Spiecker E. Hauke F. Hirsch A. Covalent bulk functionalization of graphene, *Nat Chem* 2011; *3*, 279–286.

[35] Bai H. Xu Y. Zhao L. Li C. Shi G. Non-covalent functionalization of graphene sheets by sulfonated polyaniline, *Chem. Commun.* 2009; *3*, 1667–1669.

[36] Cui Y. Kim S. N. Jones S. E. Wissler L. L. Naik R. R. McAlpine M. C. Chemical Functionalization of Graphene Enabled by Phage Displayed Peptides, *Nano Lett.* 2010; *10*, 4559–4565.

[37] Lin Q. Chen Y. H. Wu J. B. Kong Z. M. Effect of N-doping on band structure and transport property of zigzag graphene nanoribbons, *Act. Phys. Sin.* 2011; *60*, 097103.

[38] Sols F. Guinea F. Neto A. H. C. Coulomb Blockade in Graphene Nanoribbons, *Phys. Rev. Lett.* 2007; *99*, 166803.

[39] Hod O. Peralta J. E. Scuseria G. E. Edge effects in finite elongated graphene nanoribbons, *Phys. Rev. B* 2007; *76*, 233401.

[40] Lee E. J. H. Balasubramanian K. Weitz R. T. Burghard M. Kern K. Contact and edge effects in graphene devices, *Nat Nano* 2008; *3*, 486–490.

[41] Castro E. V. Peres N. M. R. Santos Lopes dos J. M. B. Neto A. H. C. Guinea F. Localized States at Zigzag Edges of Bilayer Graphene, *Phys. Rev. Lett.* 2008; *100*, 026802.

[42] Ritter K. A. Lyding J. W. The influence of edge structure on the electronic properties of graphene quantum dots and nanoribbons, *Nat Mater* 2009; *8*, 235–242.

[43] Liu Z. Suenaga K. Harris P. J. F. Iijima S. Open and Closed Edges of Graphene Layers, *Phys. Rev. Lett.* 2009; *102*, 015501.

[44] Kudin K. N. Zigzag Graphene Nanoribbons with Saturated Edges, *ACS Nano* 2008; *2*, 516–522.

[45] Rutter G. M. Guisinger N. P. Crain J. N. First P. N. Stroscio J. A. Edge structure of epitaxial graphene islands, *Phys. Rev. B* 2010; *81*, 245408.

[46] Liu Y. Dobrinsky A. Yakobson B. I. Graphene Edge from Armchair to Zigzag: The Origins of Nanotube Chirality, *Phys. Rev. Lett.* 2010; *105*, 235502.

[47] Jiang D.-e. Sumpter B. G. Dai S. Unique chemical reactivity of a graphene nanoribbon's zigzag edge, *J. Chem. Phys.* 2007; *126*, 134701.

[48] Yan Q. Huang B. Yu J. Zheng F. Zang J. Wu J. Gu B.-L. Liu F. Duan W. Intrinsic Current Voltage Characteristics of Graphene Nanoribbon Transistors and Effect of Edge Doping, *Nano Lett.* 2007; *7*, 1469–1473.

[49] Gunlycke D. Li J. Mintmire J. W. White C. T. Edges Bring New Dimension to Graphene Nanoribbons, *Nano Lett.* 2010; *10*, 3638–3642.

[50] Sun Z. Kohama S.-i. Zhang Z. Lomeda J. Tour J. Soluble graphene through edge-selective functionalization, *Nano Res.* 2010; *3*, 117–125.

[51] Xie L. Ling X. Fang Y. Zhang J. Liu Z. Graphene as a Substrate To Suppress Fluorescence in Resonance Raman Spectroscopy, *J. Am. Chem. Soc.* 2009; *131*, 9890–9891.

[52] Ambrosi A. Bonanni A. Pumera M. Electrochemistry of folded graphene edges, *Nanoscale* 2011; *3*, 2256–2260.

[53] Begliarbekov M. Sasaki K.-I. Sul O. Yang E.-H. Strauf S. Optical Control of Edge Chirality in Graphene, *Nano Lett.* 2011; *11*, 4874–4878.

[54] Suenaga K. Koshino M. Atom-by-atom spectroscopy at graphene edge, *Nature* 2010; *468*, 1088–1090.

[55] Gupta A. K. Russin T. J. Gutiérrez H. R. Eklund P. C. Probing Graphene Edges via Raman Scattering, *ACS Nano* 2009; *3*, 45–52.

[56] Casiraghi C. Hartschuh A. Qian H. Piscanec S. Georgi C. Fasoli A. Novoselov K. S. Basko D. M. Ferrari A. C. Raman Spectroscopy of Graphene Edges, *Nano Lett.* 2009; *9*, 1433–1441.

[57] Bell D. C. Lemme M. C. Stern L. A. Williams J. R. Marcus C. M. Precision cutting and patterning of graphene with helium ions, *Nanotechnology* 2009; *20*, 455301.

[58] Cai J. Ruffieux P. Jaafar R. Bieri M. Braun T. Blankenburg S. Muoth M. Seitsonen A. P. Saleh M. Feng X. Mullen K. Fasel R. Atomically precise bottom-up fabrication of graphene nanoribbons, *Nature* 2010; *466*, 470–473.

[59] Barone V. Hod O. Scuseria G. E. Electronic Structure and Stability of Semiconducting Graphene Nanoribbons, *Nano Lett.* 2006; *6*, 2748–2754.

[60] Son Y.-W. Cohen M. L. Louie S. G. Energy Gaps in Graphene Nanoribbons, *Phys. Rev. Lett.* 2006; *97*, 216803.

[61] Yang L. Park C.-H. Son Y.-W. Cohen M. L. Louie S. G. Quasiparticle Energies and Band Gaps in Graphene Nanoribbons, *Phys. Rev. Lett.* 2007; *99*, 186801.

[62] Safron N. S. Brewer A. S. Arnold M. S. Semiconducting Two-Dimensional Graphene Nanoconstriction Arrays, *Small* 2011; *7*, 492–498.

[63] Sint K. Wang B. Král P. Selective Ion Passage through Functionalized Graphene Nanopores, *J. Am. Chem. Soc.* 2008; *130*, 16448–16449.

[64] Kim M. Safron N. S. Han E. Arnold M. S. Gopalan P. Fabrication and Characterization of Large-Area, Semiconducting Nanoperforated Graphene Materials, *Nano Lett.* 2010; *10*, 1125–1131.

[65] Wang X. Ouyang Y. Li X. Wang H. Guo J. Dai H. Room-Temperature All-Semiconducting Sub-10-nm Graphene Nanoribbon Field-Effect Transistors, *Phys. Rev. Lett.* 2008; *100*, 206803.

[66] Sun Z. James D. K. Tour J. M. Graphene Chemistry: Synthesis and Manipulation, *J. Phys. Chem. Lett.* 2011; *2*, 2425–2432.

[67] Pei H. Li J. Lv M. Wang J. Gao J. Lu J. Li Y. Huang Q. Hu J. Fan C. A Graphene-Based Sensor Array for High-Precision and Adaptive Target Identification with Ensemble Aptamers, *J. Am. Chem. Soc.* 2012; *134*, 13843–13849.

[68] Shen H. Zhang L. Liu M. Zhang Z. Biomedical Applications of Graphene, *Theranostics* 2012; *2(3)*, 283–294.

[69] Chen J.-H. Jang C. Xiao S. Ishigami M. Fuhrer M. S. Intrinsic and extrinsic performance limits of graphene devices on SiO2, *Nat Nano* 2008; *3*, 206–209.

[70] Schedin F. Geim A. K. Morozov S. V. Hill E. W. Blake P. Katsnelson M. I. Novoselov K. S. Detection of individual gas molecules adsorbed on graphene, *Nat Mater* 2007; *6*, 652–655.

[71] Jiang D.-e. Cooper V. R. Dai S. Porous Graphene as the Ultimate Membrane for Gas Separation, *Nano Lett.* 2009; *9*, 4019–4024.

[72] Suk M. E. Aluru N. R. Water Transport through Ultrathin Graphene, *J. Phys. Chem. Lett.* 2010; *1*, 1590–1594.

[73] Garaj S. Hubbard W. Reina A. Kong J. Branton D. Golovchenko J. A. Graphene as a subnanometre trans-electrode membrane, *Nature* 2010; *467*, 190–193.

[74] Sathe C. Zou X. Leburton J.-P. Schulten K. Computational Investigation of DNA Detection Using Graphene Nanopores, *ACS Nano* 2011; *5*, 8842–8851.

[75] Koenig S. P. Wang L. Pellegrino J. Bunch J. S. Selective molecular sieving through porous graphene, *Nat Nano* 2012; *7*, 728–732.

[76] Baskin A. Kral P. Electronic structures of porous nanocarbons, *Sci. Rep.* 2011; *1*, 36.

[77] Hauser A. W. Schwerdtfeger P. Nanoporous Graphene Membranes for Efficient 3He/4He Separation, *J. Phys. Chem. Lett.* 2011; *3*, 209–213.

[78] Hauser A. W. Schrier J. Schwerdtfeger P. Helium Tunneling through Nitrogen-Functionalized Graphene Pores: Pressure- and Temperature-Driven Approaches to Isotope Separation, *J. Phys. Chem. C* 2012; *116*, 10819–10827.

[79] Paul D. R. Creating New Types of Carbon-Based Membranes, *Science* 2012; *335*, 413–414.

[80] Nair R. R. Wu H. A. Jayaram P. N. Grigorieva I. V. Geim A. K. Unimpeded Permeation of Water Through Helium-Leak-ÂĂÂŞTight Graphene-Based Membranes, *Science* 2012; *335*, 442–444.

[81] Karan S. Samitsu S. Peng X. Kurashima K. Ichinose I. Ultrafast Viscous Permeation of Organic Solvents Through Diamond-Like Carbon Nanosheets, *Science* 2012; *335*, 444–447.

[82] Gong X. Li J. Xu K. Wang J. Yang H. A Controllable Molecular Sieve for Na+ and K+ Ions, *J. Am. Chem. Soc.* 2010; *132*, 1873–1877.

[83] Hankel M. Jiao Y. Du A. Gray S. K. Smith S. C. Asymmetrically Decorated, Doped Porous Graphene As an Effective Membrane for Hydrogen Isotope Separation, *J. Phys. Chem. C* 2012; *116*, 6672–6676.

[84] Venkatesan B. M. Estrada D. Banerjee S. Jin X. Dorgan V. E. Bae M.-H. Aluru N. R. Pop E. Bashir R. Stacked Graphene-Al2O3 Nanopore Sensors for Sensitive Detection of DNA and DNA-ÂĂÂŞProtein Complexes, *ACS Nano* 2011; *6*, 441–450.

[85] Fischbein M. D. Drndic M. Electron beam nanosculpting of suspended graphene sheets, *Appl. Phys. Lett.* 2008; *93*, 113107–3.

[86] Schneider G. F. Kowalczyk S. W. Calado V. E. Pandraud G. Zandbergen H. W. Vandersypen L. M. K. Dekker C. DNA Translocation through Graphene Nanopores, *Nano Lett.* 2010; *10*, 3163–3167.

[87] Merchant C. A. Healy K. Wanunu M. Ray V. Peterman N. Bartel J. Fischbein M. D. Venta K. Luo Z. Johnson A. T. C. DrndicÂĄ M. DNA Translocation through Graphene Nanopores, *Nano Lett.* 2010; *10*, 2915–2921.

[88] Fan Z. Zhao Q. Li T. Yan J. Ren Y. Feng J. Wei T. Easy synthesis of porous graphene nanosheets and their use in supercapacitors, *Carbon* 2012; *50*, 1699–1703.

[89] Han T. H. Huang Y.-K. Tan A. T. L. Dravid V. P. Huang J. Steam Etched Porous Graphene Oxide Network for Chemical Sensing, *J. Am. Chem. Soc.* 2011; *133*, 15264–15267.

[90] Bieri M. Treier M. Cai J. Ait-Mansour K. Ruffieux P. Groning O. Groning P. Kastler M. Rieger R. Feng X. Mullen K. Fasel R. Porous graphenes: two-dimensional polymer synthesis with atomic precision, *Chem. Commun.* 2009; , 6919–6921.

[91] Jia X. Campos-Delgado J. Terrones M. Meunier V. Dresselhaus M. S. Graphene edges: a review of their fabrication and characterization, *Nanoscale* 2011; *3*, 86–95.

[92] Luo Z. Yu T. Ni Z. Lim S. Hu H. Shang J. Liu L. Shen Z. Lin J. Electronic Structures and Structural Evolution of Hydrogenated Graphene Probed by Raman Spectroscopy, *J. Phys. Chem. C* 2011; *115*, 1422–1427.

[93] Mak K. F. Sfeir M. Y. Misewich J. A. Heinz T. F. The evolution of electronic structure in few-layer graphene revealed by optical spectroscopy, *Proc. Natl. Acad. Sci USA* 2010; *107*, 14999–15004.

[94] Li Z. Q. Henriksen E. A. Jiang Z. Hao Z. Martin M. C. Kim P. Stormer H. L. Basov D. N. Band Structure Asymmetry of Bilayer Graphene Revealed by Infrared Spectroscopy, *Phys. Rev. Lett.* 2009; *102*, 037403.

[95] Zhang L. M. Li Z. Q. Basov D. N. Fogler M. M. Hao Z. Martin M. C. Determination of the electronic structure of bilayer graphene from infrared spectroscopy, *Phys. Rev. B* 2008; *78*, 235408.

[96] Phark S.-h. Borme J. Vanegas A. L. Corbetta M. Sander D. Kirschner J. Atomic structure and spectroscopy of graphene edges on Ir(111), *Phys. Rev. B* 2012; *86*, 045442.

[97] Yang X. X. Li J. W. Zhou Z. F. Wang Y. Yang L. W. Zheng W. T. Sun C. Q. Raman spectroscopic determination of the length, strength, compressibility, Debye temperature, elasticity, and force constant of the C-C bond in graphene, *Nanoscale* 2012; *4*, 502–510.

[98] Li Z. Lui C. H. Cappelluti E. Benfatto L. Mak K. F. Carr G. L. Shan J. Heinz T. F. Structure-Dependent Fano Resonances in the Infrared Spectra of Phonons in Few-Layer Graphene, *Phys. Rev. Lett.* 2012; *108*, 156801.

http://dx.doi.org/10.5772/56247

[99] Kudin K. N. Ozbas B. Schniepp H. C. Prud'homme R. K. Aksay I. A. Car R. Raman Spectra of Graphite Oxide and Functionalized Graphene Sheets, *Nano Lett.* 2007; *8*, 36–41.

[100] Niyogi S. Bekyarova E. Itkis M. E. Zhang H. Shepperd K. Hicks J. Sprinkle M. Berger C. Lau C. N. deHeer W. A. Conrad E. H. Haddon R. C. Spectroscopy of Covalently Functionalized Graphene, *Nano Lett.* 2010; *10*, 4061–4066.

[101] Daniels K. M. Daas B. K. Srivastava N. Williams C. Feenstra R. M. Sudarshan T. S. Chandrashekhar M. V. S. Evidences of electrochemical graphene functionalization and substrate dependence by Raman and scanning tunneling spectroscopies, *J. Appl. Phys.* 2012; *111*, 114306–7.

[102] Graf D. Molitor F. Ensslin K. Stampfer C. Jungen A. Hierold C. Wirtz L. Spatially Resolved Raman Spectroscopy of Single- and Few-Layer Graphene, *Nano Lett.* 2007; *7*, 238–242.

[103] Ferrari A. C. Meyer J. C. Scardaci V. Casiraghi C. Lazzeri M. Mauri F. Piscanec S. Jiang D. Novoselov K. S. Roth S. Geim A. K. Raman Spectrum of Graphene and Graphene Layers, *Phys. Rev. Lett.* 2006; *97*, 187401.

[104] Calizo I. Balandin A. A. Bao W. Miao F. Lau C. N. Temperature Dependence of the Raman Spectra of Graphene and Graphene Multilayers, *Nano Lett.* 2007; *7*, 2645–2649.

[105] Malard L. Pimenta M. Dresselhaus G. Dresselhaus M. Raman spectroscopy in graphene, *Physics Reports* 2009; *473*, 51–87.

[106] Schedin F. Lidorikis E. Lombardo A. Kravets V. G. Geim A. K. Grigorenko A. N. Novoselov K. S. Ferrari A. C. Surface-Enhanced Raman Spectroscopy of Graphene, *ACS Nano* 2010; *4*, 5617–5626.

[107] Li Z. Q. Henriksen E. A. Jiang Z. Hao Z. Martin M. C. Kim P. Stormer H. L. Basov D. N. Dirac charge dynamics in graphene by infrared spectroscopy, *Nat Phys* 2008; *4*, 532–535.

[108] Mak K. F. Lui C. H. Shan J. Heinz T. F. Observation of an Electric-Field-Induced Band Gap in Bilayer Graphene by Infrared Spectroscopy, *Phys. Rev. Lett.* 2009; *102*, 256405.

[109] Jiang Z. Henriksen E. A. Tung L. C. Wang Y.-J. Schwartz M. E. Han M. Y. Kim P. Stormer H. L. Infrared Spectroscopy of Landau Levels of Graphene, *Phys. Rev. Lett.* 2007; *98*, 197403.

[110] Eberlein T. Bangert U. Nair R. R. Jones R. Gass M. Bleloch A. L. Novoselov K. S. Geim A. Briddon P. R. Plasmon spectroscopy of free-standing graphene films, *Phys. Rev. B* 2008; *77*, 233406.

[111] Dora B. Simon F. Unusual Hyperfine Interaction of Dirac Electrons and NMR Spectroscopy in Graphene, *Phys. Rev. Lett.* 2009; *102*, 197602.

[112] Li G. Luican A. Andrei E. Y. Scanning Tunneling Spectroscopy of Graphene on Graphite, *Phys. Rev. Lett.* 2009; *102*, 176804.

[113] Ohta T. Bostwick A. McChesney J. L. Seyller T. Horn K. Rotenberg E. Interlayer Interaction and Electronic Screening in Multilayer Graphene Investigated with Angle-Resolved Photoemission Spectroscopy, *Phys. Rev. Lett.* 2007; *98*, 206802.

[114] Xu M. Fujita D. Gao J. Hanagata N. Auger Electron Spectroscopy: A Rational Method for Determining Thickness of Graphene Films, *ACS Nano* 2010; *4*, 2937–2945.

[115] Ackerman M. Clapham D. Mechanisms of disease - Ion channels - Basic science and clinical disease, *N Engl J Med* 1997; *336*, 1575–1586.

[116] Yu J. Yool A. J. Schulten K. Tajkhorshid E. Mechanism of gating and ion conductivity of a possible tetrameric pore in aquaporin-1, *Structure* 2006; *14*, 1411–1423.

[117] Jordan E. Bell R. G. Wilmer D. Koller H. Anion-Promoted Cation Motion and Conduction in Zeolites, *J. Am. Chem. Soc.* 2005; *128*, 558–567.

[118] Saufi S. Ismail A. Fabrication of carbon membranes for gas separation, *Carbon* 2004; *42*, 241–259.

[119] Duke M. C. Costa da J. C. D. Do D. D. Gray P. G. Lu G. Q. Hydrothermally Robust Molecular Sieve Silica for Wet Gas Separation, *Adv.Funct. Mater.* 2006; *16*, 1215–1220.

[120] Ghadiri M. R. Granja J. R. Buehler L. K. Artificial transmembrane ion channels from self-assembling peptide nanotubes, *Nature* 1994; *369*, 301–304.

[121] Kobuke Y. Ueda K. Sokabe M. Artificial non-peptide single ion channels, *J. Am. Chem. Soc.* 1992; *114*, 7618–7622.

[122] Hummer G. Rasaiah J. C. Noworyta J. P. Water conduction through the hydrophobic channel of a carbon nanotube, *Nature* 2001; *414*, 188–190.

[123] Skoulidas A. I. Ackerman D. M. Johnson J. K. Sholl D. S. Rapid Transport of Gases in Carbon Nanotubes, *Phys. Rev. Lett.* 2002; *89*, 185901.

[124] Holt J. K. Park H. G. Wang Y. Stadermann M. Artyukhin A. B. Grigoropoulos C. P. Noy A. Bakajin O. Fast Mass Transport Through Sub-2-Nanometer Carbon Nanotubes, *Science* 2006; *312*, 1034–1037.

[125] Majumder M. Chopra N. Andrews R. Hinds B. J. Nanoscale hydrodynamics: Enhanced flow in carbon nanotubes, *Nature* 2005; *438*, 44–44.

[126] Wang Z. Ci L. Chen L. Nayak S. Ajayan P. M. Koratkar N. Polarity-Dependent Electrochemically Controlled Transport of Water through Carbon Nanotube Membranes, *Nano Lett.* 2007; *7*, 697–702.

[127] Zhou J. Noca F. Gharib M. Flow conveying and diagnosis with carbon nanotube arrays, *Nanotechnology* 2006; *17*, 4845âĂŞ4853.

[128] Whitby M. Quirke N. Fluid flow in carbon nanotubes and nanopipes, *Nat Nano* 2007; *2*, 87–94.

[129] Liu H. Murad S. C.J. J. Ion permeation dynamics in carbon nanotubes, *J. Chem. Phys.* 2006; *125*, 084713.

[130] Corso M. Auwörter W. Muntwiler M. Tamai A. Greber T. Osterwalder J. Boron Nitride Nanomesh, *Science* 2004; *303*, 217–220.

[131] Geim A. K. Novoselov K. S. The rise of graphene, *Nat Mater* 2007; *6*, 183–191.

[132] Berner S. Corso M. Widmer R. Groening O. Laskowski R. Blaha P. Schwarz K. Goriachko A. Over H. Gsell S. Schreck M. Sachdev H. Greber T. Osterwalder J. Boron Nitride Nanomesh: Functionality from a Corrugated Monolayer, *Angew. Chem. Int. Ed.* 2007; *46*, 5115–5119.

[133] Laskowski R. Blaha P. Gallauner T. Schwarz K. Single-Layer Model of the Hexagonal Boron Nitride Nanomesh on the Rh(111) Surface, *Phys. Rev. Lett.* 2007; *98*, 106802.

[134] Ohta T. Bostwick A. Seyller T. Horn K. Rotenberg E. Controlling the Electronic Structure of Bilayer Graphene, *Science* 2006; *313*, 951–954.

[135] Bunch J. S. Verbridge S. S. Alden J. S. Zande van der A. M. Parpia J. M. Craighead H. G. McEuen P. L. Impermeable Atomic Membranes from Graphene Sheets, *Nano Lett.* 2008; *8*, 2458–2462.

[136] Liu Y. Dong X. Chen P. Biological and chemical sensors based on graphene materials, *Chem. Soc. Rev.* 2012; *41*, 2283–2307.

[137] Jagerszki G. Takacs A. Bitter I. Gyurcsanyi R. E. Solid-State Ion Channels for Potentiometric Sensing, *Angew. Chem. Int. Ed.* 2011; *50*, 1656–1659.

[138] Titov A. V. Wang B. Sint K. Král P. Controllable Synthetic Molecular Channels: Biomimetic Ammonia Switch, *J. Phys. Chem. B* 2009; *114*, 1174–1179.

[139] Stampfer C. Guttinger J. Molitor F. Graf D. Ihn T. Ensslin K. Tunable Coulomb blockade in nanostructured graphene, *Appl. Phys. Lett.* 2008; *92*, 012102–3.

[140] Stankovich S. Dikin D. A. Dommett G. H. B. Kohlhaas K. M. Zimney E. J. Stach E. A. Piner R. D. Nguyen S. T. Ruoff R. S. Graphene-based composite materials, *Nature* 2006; *442*, 282–286.

[141] Gilje S. Han S. Wang M. Wang K. L. Kaner R. B. A Chemical Route to Graphene for Device Applications, *Nano Lett.* 2007; *7*, 3394–3398.

[142] The NVT ensemble with periodic boundary conditions is used, where the electrostatic coupling is calculated by the PME method. The Langevin dynamics is applied to the systems with the damping coefficient of 0.01 ps-1, in order to minimize the unphysical loss of momenta to the reservoirs. The time step is always 1 fs.

[143] Darden T. York D. Pedersen L. Particle Mesh Ewald-an N.Log(N) method for Ewald sums in large systems, *J. Chem. Phys.* 1993; *98*, 10089–10092.

[144] Wang B. Král P. Coulombic Dragging of Molecules on Surfaces Induced by Separately Flowing Liquids, *J. Am. Chem. Soc.* 2006; *128*, 15984–15985.

[145] Phillips J. Braun R. Wang W. Gumbart J. Tajkhorshid E. Villa E. Chipot C. Skeel R. Kale L. K. S. Scalable molecular dynamics with NAMD, *J. Comput. Chem.* 2005; *26*, 1781–1802.

[146] MacKerell A. D. Bashford D. Bellott Dunbrack R. L. Evanseck J. D. Field M. J. Fischer S. Gao J. Guo H. Ha S. Joseph-McCarthy D. Kuchnir L. Kuczera K. Lau F. T. K. Mattos C. Michnick S. Ngo T. Nguyen D. T. Prodhom B. Reiher W. E. Roux B. Schlenkrich M. Smith J. C. Stote R. Straub J. Watanabe M. Wiórkiewicz-Kuczera J. Yin D. Karplus M. All-Atom Empirical Potential for Molecular Modeling and Dynamics Studies of Proteins, *J. Phys. Chem. B* 1998; *102*, 3586–3616.

[147] The calculations are done with 3-21g basis set. Charges are obtained by the Natural Population analysis.

[148] Frisch M. et al. *Gaussian 03, Revision C.02* 2004; *Gaussian, Inc., Wallingford, C.T.*

[149] Auffinger P. Bielecki L. Westhof E. Symmetric K^+ and Mg^{2+} ion-binding sites in the 5S rRNA loop E inferred from molecular dynamics simulations, *J. Mol. Biol.* 2004; *335(2)*, 555–571.

[150] Wang B. Král P. Cover Picture: Optimal Atomistic Modifications of Material Surfaces: Design of Selective Nesting Sites for Biomolecules (Small 4/2007), *Small* 2007; *3*, 580–584.

[151] Merlet C. Rotenberg B. Madden P. A. Taberna P.-L. Simon P. Gogotsi Y. Salanne M. On the molecular origin of supercapacitance in nanoporous caaarbon electrodes, *Nat Mater* 2012; *11*, 306–310.

[152] Du H. Li J. Zhang J. Su G. Li X. Zhao Y. Separation of Hydrogen and Nitrogen Gases with Porous Graphene Membrane, *J. Phys. Chem. C* 2011; *115*, 23261–23266.

[153] Shan M. Xue Q. Jing N. Ling C. Zhang T. Yan Z. Zheng J. Influence of chemical functionalization on the CO_2/N_2 separation performance of porous graphene membranes, *Nanoscale* 2012; *4*, 5477–5482.

[154] Hauser A. W. Schwerdtfeger P. Methane-selective nanoporous graphene membranes for gas purification, *Phys. Chem. Chem. Phys.* 2012; *14*, 13292–13298.

[155] Blankenburg S. Bieri M. Fasel R. Müllen K. Pignedoli C. A. Passerone D. Porous Graphene as an Atmospheric Nanofilter, *Small* 2010; *6*, 2266–2271.

[156] Wang E. N. Karnik R. Water desalination: Graphene cleans up water, *Nat Nano* 2012; *7*, 552–554.

[157] Cohen-Tanugi D. Grossman J. C. Water Desalination across Nanoporous Graphene, *Nano Lett.* 2012; *12*, 3602–3608.

[158] Bae S. Kim H. Lee Y. Xu X. Park J.-S. Zheng Y. Balakrishnan J. Lei T. Ri Kim H. Song Y. I. Kim Y.-J. Kim K. S. Ozyilmaz B. Ahn J.-H. Hong B. H. Iijima S. Roll-to-roll production of 30-inch graphene films for transparent electrodes, *Nat Nano* 2010; *5*, 574–578.

[159] Venkatesan B. M. Bashir R. Nanopore sensors for nucleic acid analysis, *Nat Nano* 2011; *6*, 615–624.

[160] Wells D. B. Belkin M. Comer J. Aksimentiev A. Assessing Graphene Nanopores for Sequencing DNA, *Nano Lett.* 2012; *12*, 4117–4123.

[161] Postma H. W. C. Rapid Sequencing of Individual DNA Molecules in Graphene Nanogaps, *Nano Lett.* 2010; *10*, 420–425.

[162] Nelson T. Zhang B. Prezhdo O. V. Detection of Nucleic Acids with Graphene Nanopores: Ab Initio Characterization of a Novel Sequencing Device, *Nano Lett.* 2010; *10*, 3237–3242.

[163] Saha K. K. Drndic M. Nikolic B. K. DNA Base-Specific Modulation of Microampere Transverse Edge Currents through a Metallic Graphene Nanoribbon with a Nanopore, *Nano Lett.* 2011; *12*, 50–55.

[164] Palacios J. J. Fernández-Rossier J. Brey L. Vacancy-induced magnetism in graphene and graphene ribbons, *Phys. Rev. B* 2008; *77*, 195428.

[165] Balog R. Jorgensen B. Nilsson L. Andersen M. Rienks E. Bianchi M. Fanetti M. Laegsgaard E. Baraldi A. Lizzit S. Sljivancanin Z. Besenbacher F. Hammer B. Pedersen T. G. Hofmann P. Hornekaer L. Bandgap opening in graphene induced by patterned hydrogen adsorption, *Nat Mater* 2010; *9*, 315–319.

[166] Cervantes-Sodi F. Csányi G. Piscanec S. Ferrari A. C. Edge-functionalized and substitutionally doped graphene nanoribbons: Electronic and spin properties, *Phys. Rev. B* 2008; *77*, 165427.

[167] Furst J. A. Pedersen J. G. Flindt C. Mortensen N. A. Brandbyge M. Pedersen T. G. Jauho A.-P. Electronic properties of graphene antidot lattices, *New Journal of Physics* 2009; *11*, 095020.

[168] Pedersen T. G. Flindt C. Pedersen J. Mortensen N. A. Jauho A.-P. Pedersen K. Graphene Antidot Lattices: Designed Defects and Spin Qubits, *Phys. Rev. Lett.* 2008; *100*, 136804.

[169] Vanevic M. Stojanovic V. M. Kindermann M. Character of electronic states in graphene antidot lattices: Flat bands and spatial localization, *Phys. Rev. B* 2009; *80*, 045410.

[170] Gao Y.-D. Kumazaki H. Terai J. Chida K. Hosoya H. Topological factors governing the HOMO-LUMO band gap of the density of states of periodic hydrocarbon polymer networks, *Journal of Mathematical Chemistry* 1993; *12*, 279–308.

[171] Fujita M. Wakabayashi K. Nakada K. Kusakabe K. Peculiar localized states at zigzag graphite edges, *J. Phys. Soc. Jpn.* 1996; *65*, 1920.

[172] Shima N. Aoki H. Electronic structure of super-honeycomb systems: A peculiar realization of semimetal/semiconductor classes and ferromagnetism, *Phys. Rev. Lett.* 1993; *71*, 4389–4392.

[173] Sato T. Imade M. Yamabe T. Electronic structure of porous nanotube, *Synthetic Metals* 1999; *103*, 2519–2520.

[174] Hatanaka M. Band structures of porous graphehes, *Chem. Phys. Lett.* 2010; *488*, 187.

[175] Ivanciuc O. Bytautas L. Klein D. J. Mean-field resonating-valencebond theory for unpaired π-electrons in benzenoid carbon species, *J. Chem. Phys.* 2002; *116*, 4735.

[176] Nakada K. Fujita M. Dresselhaus G. Dresselhaus M. S. Edge state in graphene ribbons: Nanometer size effect and edge shape dependence, *Phys. Rev. B* 1996; *54*, 17954–17961.

[177] Miyamoto Y. Nakada K. Fujita M. First-principles study of edge states of H-terminated graphitic ribbons, *Phys. Rev. B* 1999; *59*, 9858–9861.

[178] Ezawa M. Peculiar width dependence of the electronic properties of carbon nanoribbons, *Phys. Rev. B* 2006; *73*, 045432.

[179] Pisani L. Chan J. A. Montanari B. Harrison N. M. Electronic structure and magnetic properties of graphitic ribbons, *Phys. Rev. B* 2007; *75*, 064418.

[180] Yu S. Wen Q. Zheng W. Jiang Q. Electronic properties of graphene nanoribbons with armchair-shaped edges, *Molecular Simulation* 2008; *34*, 1085–1090.

[181] Kharche N. Zhou Y. O'ÂĂÂŹBrien K. P. Kar S. Nayak S. K. Effect of Layer Stacking on the Electronic Structure of Graphene Nanoribbons, *ACS Nano* 2011; *5*, 6096–6101.

[182] Sanchez-Portal D. Ordejon P. Artacho E. Soler J. M. Density-functional method for very large systems with LCAO basis sets, *Int. J. Quantum Chem.* 1997; *65*, 453–461.

[183] Perdew J. P. Zunger A. Self-interaction correction to density-functional approximations for many-electron systems, *Phys. Rev. B* 1981; *23*, 5048–5079.

[184] Lu Y. H. et al. Effects of edge passivation by hydrogen on electronic structure of armchair graphene nanoribbons and band gap engineering, *Appl. Phys. Lett.* 2009; *94*, 122111.

[185] Boukhvalov D. W. Katsnelson M. I. Lichtenstein A. I. Hydrogen on graphene: Electronic structure, total energy, structural distortions and magnetism from first-principles calculations, *Phys. Rev. B* 2008; *77*, 035427.

[186] Weiss D. Roukes M. L. Menschig A. Grambow P. Klitzing von K. Weimann G. Electron pinball and commensurate orbits in a periodic array of scatterers, *Phys. Rev. Lett.* 1991; *66*, 2790–2793.

[187] Flindt C. Mortensen N. A. Jauho A.-P. Quantum Computing via Defect States in Two-Dimensional Antidot Lattices, *Nano Lett.* 2005; *5*, 2515–2518.

[188] Pedersen J. Flindt C. Asger Mortensen N. Jauho A.-P. Spin qubits in antidot lattices, *Phys. Rev. B* 2008; *77*, 045325–.

[189] Lieb E. H. Two theorems on the Hubbard model, *Phys. Rev. Lett.* 1989; *62*, 1201–1204.

[190] Inui M. Trugman S. A. Abrahams E. Unusual properties of midband states in systems with off-diagonal disorder, *Phys. Rev. B* 1994; *49*, 3190–3196.

[191] Fazekas P., *Lecture Notes on Electron Correlation and Magnetism* (World Scientific, New York, 1999).

[192] Tylianakis E. Dimitrakakis G. K. Melchor S. Dobado J. A. Froudakis G. E. Porous nanotube network: a novel 3-D nanostructured material with enhanced hydrogen storage capacity, *Chem. Commun.* 2011; *47*, 2303–2305.

[193] Zhang J. Ong K. P. Wu P. The Influence of Out-of-Plane Deformation on the Band Gap of Graphene Nanoribbons, *J. Phys. Chem. C* 2010; *114*, 12749–12753.

[194] Lu Y. Guo J. Band gap of strained graphene nanoribbons, *Nano Research* 2010; *3*, 189–199.

[195] Banhart F. Kotakoski J. Krasheninnikov A. V. Structural Defects in Graphene, *ACS Nano* 2010; *5*, 26–41.

[196] Mele E. J. Král P. Electric Polarization of Heteropolar Nanotubes as a Geometric Phase, *Phys. Rev. Lett.* 2002; *88*, 056803.

[197] Xue Y. Ratner M. A. Microscopic study of electrical transport through individual molecules with metallic contacts. I. Band lineup, voltage drop, and high-field transport, *Phys. Rev. B* 2003; *68*, 115406.

[198] Bauschlicher J., Charles W. Lawson J. W. Current-voltage curves for molecular junctions: Effect of substitutients, *Phys. Rev. B* 2007; *75*, 115406.

[199] Sahin H. Senger R. T. First-principles calculations of spin-dependent conductance of graphene flakes, *Phys. Rev. B* 2008; *78*, 205423.

[200] Dubois S. M.-M. Zanolli Z. Declerck X. Charlier J.-C. Electronic properties and quantum transport in Graphene-based nanostructures, *Eur. Phys. J. B* 2009; *72*, 1–24.

[201] Topsakal M. Bagci V. M. K. Ciraci S. Current-voltage (I-V) characteristics of armchair graphene nanoribbons under uniaxial strain, *Phys. Rev. B* 2010; *81*, 205437.

[202] Terrones M. Botello-Mendez A. R. Campos-Delgado J. Lopez-Urias F. Vega-Cantu Y. I. Rodriguez-Macias F. J. Elias A. L. Munoz-Sandoval E. Cano-Márquez A. G. Charlier J.-C. Terrones H. Graphene and graphite nanoribbons: Morphology, properties, synthesis, defects and applications, *Nano Today* 2010; *5*, 351–372.

[203] Kim W. Y. Kim K. S. Prediction of very large values of magnetoresistance in a graphene nanoribbon device, *Nat Nano* 2008; *3*, 408–412.

[204] Biel B. Blase X. Triozon F. Roche S. Anomalous Doping Effects on Charge Transport in Graphene Nanoribbons, *Phys. Rev. Lett.* 2009; *102*, 096803.

[205] Datta S. *Electronic Transport in Mesoscopic Systems* (Cambridge University Press, Cambridge, England, 1995).

[206] Brandbyge M. Mozos J.-L. Ordejón P. Taylor J. Stokbro K. Density-functional method for nonequilibrium electron transport, *Phys. Rev. B* 2002; *65*, 165401.

[207] Hansen T. Solomon G. C. Andrews D. Q. Ratner M. A. Interfering pathways in benzene: An analytical treatment, *J. Chem. Phys.* 2009; *131*, 194704–12.

Chapter 7

Chemically Functionalized Graphene and Their Applications in Electrochemical Energy Conversion and Storage

Yuhai Hu and Xueliang Sun

Additional information is available at the end of the chapter

http://dx.doi.org/10.5772/55666

1. Introduction

Electrochemical energy conversion and storage (EECS) is one of the important strategies to address the strong demand on clean energy supply. Rechargeable batteries, fuel cells (FC) and supercapacitors are the typical electrochemical devices. Now, the challenge for the EECS is to make these devices bear enough capacities so as to fulfill the demand of the emerging markets, particularly, transportation applications and portable smart electronics. In this regard, the limitation of graphite, which has been a key component in the existing electrochemical devices, is seen. Much effort is being made toward exploring new morphologies for carbon, which are expected to provide some novel properties that would overcome the drawbacks of graphite. Among these, success in exfoliation of graphene from graphite has broken the ground.

Graphene is a one-atom-thick planar sheet consisting of sp^2 carbon atoms that are densely packed in a honeycomb crystal lattice [1], as shown in Fig. 1. Graphene has many unique properties such as high surface area, high electronic conductivity, high Young's modulus, high thermal conductivity and high optical transmittance [2,3]. Thus, graphene is considered a versatile building material for fabrication of electrochemical devices. Ever since the mechanical exfoliation of single-layer graphene from graphite succeeded in 2004, graphene has been receiving extensive research interest in EECS. The introduction of graphene brings some innovative properties to the electrochemical devices. Many review papers have focused on this topic [2,4,5].

Pristine graphene, which is composed of only sp^2 carbon atoms, is a zero-gap semiconductor, and its Fermi level exactly crosses the Dirac point. For practical application, an energy gap is essential and thus, it is necessary and crucial to develop new methods to precisely control the carrier type and concentration in graphene for further development of graphene-based devices. Normally, graphene has two oxidation states: graphene oxide (GO) and reduced graphene oxide (RGO). GO is water soluble with low electronic conductivity, while RGO reveals good conductivity but poor solubility in water. The excellent solubility of GO in aqueous solution is primarily due to its rich oxygen-containing and hydrophilic groups, such as hydroxyl, epoxide, carboxyl and carboxylic. Upon reduction, most of the oxygen-containing groups, in particular the hydroxyl, epoxide and carboxyl, will be completely removed. Thus, GO will then be converted to π-conjugation-rich graphene, i.e., RGO. The π-conjugation in graphene sheets (GNS) can restore the conductivity of graphene but reduces its solubility in water and other organic solvents. The decline in solubility may inevitably reduce processability of graphene and therefore limit its applications. All these render functionalization of graphene highly desirable.

Functionalization of graphene is one of the key topics in graphene research. Generally, there are two main categories of functionalization: chemical and nonchemical. Chemical functionalization is realized through the formation of new covalent bonds between the atoms native to RGO/GO and the guest functional groups; in contrast, nonchemical functionalization is mainly based on π interaction between guest molecules and RGO/GO, i.e., mainly a physical interaction. Both types of functionalization can induce some property changes for graphene, but the chemical routes are more effective. So far, various chemical routes have proposed, which successfully incorporate a large number of different atoms/organic groups into graphene, including heterogeneous atoms doping, diazonium coupling, amidation, silanization, esterization, substitution, cycloaddition, etc. The research also demonstrates that the size of graphene (particularly the thickness of the ribbons) strongly affects the reactivity of graphene and the application of the functionalized graphene [6,7]. The thinner the ribbons, the more reactive they are. Moreover, it is also confirmed that carbon atoms on the ribbon edge are more reactive than those in the middle of the ribbon [6,7]. The chemical functionalization greatly alters the electric conductivity of graphene and introduces some defect sites, significantly affecting the application of the functionalized graphene in EECS applications. So far, heterogeneous atoms doping evidently improves the electric conductivity of graphene and hence, the resulting functionalized graphene is widely investigated [8,9]; on the other hand, incorporation of organic groups into graphene reduces the electric conductivity, preventing its electrochemical applications. Nevertheless, such functionalized graphene finds more extensive application in polymer science and technology, a topic that is not the focus of this chapter.

In this chapter, we review the chemical routes for functionalizing graphene and summarize the advances in the applications of the functionalized graphene in EECS, including FCs, lithium ion batteries (LIBs) and supercapacitors. Noticeably, chemical functionalization of graphene is a big field that involves organic chemistry/polymer chemistry. Various kinds of organic molecules have been chemically anchored onto graphene. It is impossible for us to list all kinds of molecules in this short review; instead, we address this in term of what kind of typical organic reaction is adopted for the functionalization. The organic reactions in this chapter are easy to understand and follow. For each reaction, the most original and representative studies are selected for review. Moreover, a large number of references are involved in

http://dx.doi.org/10.5772/55666

this topic, which can not be listed totally due to the page limit. Hence, only the references that are strictly about functionalized graphene and their applications are provided here. Finally, research in this field is advancing quickly, thus, some new results might not be addressed.

Figure 1. Schematic diagrams of graphene and graphene-based 0D, 1D and 3D structured carbon (Nature Mater. 2007;6(3) 183-191.)

2. Chemical functionalization of graphene

Generally, the functionalization is realized in two ways. The first is to dope heterogeneous atoms onto the basal plane (form covalent bond directly with C atoms of graphene), mainly tuning the electronic structure and introducing defect sites; the second is to establish covalent bond between the functional group native to GO and the guest functional group. Various kinds of functional groups have been chemically anchored onto graphene, giving rise to a diversity of graphene based nanocomposites that were widely investigated in various applications. In this part, we will provide a comprehensive picture for this subject in term of how the covalent bonds are formed.

2.1. Characterization of the functionalized graphene

Chemical functionalization is a chemical (or electrochemical) process that proceeds at an atomic/molecular level. Raman, Infra red spectroscopy (IR), X-ray photoemission spectrometry (XPS), High resolution transmission electric microscopy (HRTEM), Atomic force microscopy (AFM), and Cyclic voltammetry (CV) are the typical instruments that are widely employed for understanding the properties resulting from the functionalization. Raman and IR are sensitive to the formation and change of chemical bonds, helping to determine whether chemical functionalization is successful. Moreover, they also allow semi-quantity analysis. XPS is the powerful instrument for functionalized graphene characterization in terms of both qualification and quantization. For example, XPS can provide decisive information on how many N atoms are doped into graphene and the types of the doped N atoms. HRTEM can tell how the morphology and structure changes following chemical functionalization. AFM can show the thickness and the morphology of graphene. CV measurement is a direct electrochemical method for understanding the electrochemical properties resulting from the functionalization, a topic of the focus in this chapter.

Guest functional group	Chemical process	References
N atoms	CVD	[13-16]
	N-plasma or N^+-ion irradiation	[17,18]
	Arc discharge	[19,20]
	Electrothermal reactions	[22]
	Chemical synthesis	[23-25]
B atoms	CVD	[20,26]
S atoms	Chemical synthesis	[27]
	Pyrolysis	[28,67]
F atoms	Arc discharge	[29]

Table 1. A summary of the chemical routes for functionalization of graphene by heterogeneous atoms doping

2.2. Heterogeneous atoms doping

Heterogeneous atoms doping is one of the most important approaches to chemically functionalize graphene, and such an approach mainly alters the electric properties of graphene. Understanding the electric properties of the doped graphene is a hot topic in physics and semiconductors, for which the studies usually take a combination of experimental measurement and computer simulation. As a typical example, Novoselov and Geim, the two Nobel Prize laureates in 2010 because of graphene, did some pioneer research on this topic. In 2008, they chose the NO_2 system providing both open-shell single molecules and closed-shell dimers N_2O_4 to study the doping due to adsorbates by combining ab initio theory with transport measurements [10]. A general relation between the doping strength and whether adsorbates are open- or closed-shell systems is demonstrated with the NO_2 system: The single, open shell NO_2 molecule is a strong acceptor (attracts electrons from graphene), whereas its closed shell dimer N_2O_4 causes only weak doping. This effect is pronounced by graphene's peculiar density of states (DOS), which provides an ideal situation for model studies of doping effects in semiconductors.

Strictly speaking, GO, which contains a variety of heterogeneous atoms or groups such as -O, -COOH, and -OH, is a typical functionalized graphene. But due to the low electronic conductivity, their application in EECS is limited and hence, there are few studies focusing on GO in this field. Nevertheless, these functional groups play important roles in the chemical functionalization of gaphene. Considering these, in this chapter, we will take GO as graphene. Doping atoms, such as N, B, P and S and chlorides, into graphene is highly desired because the doping effectively converts graphene from gapless structure to semiconductor. The resultant functionalized graphenes are either p-type or n-type, depending on the electronic structure of the guest atoms. Moreover, the doping also induced more defect sites that facilitate deposition of other electrochemically functional components (e.g., growth of Pt on graphene) and stabilize them. Theoretically, all these atoms can be covalently bonded to graphene. However, in practice, doping N is much easier than doping other elements, and has received more extensive attention. Success in B, S and F doping is reported recently. To make this part more understandable, Table 1 summarizes the most typical chemical approaches adopted in the existing studies.

2.2.1. N-doped graphene

N doping gives rise to n-type (negative) graphene. The well-bonded nitrogen atoms improve the electronic conductivity and offer more active sites (defects). Nitrogen doping has proved to be an effective method to improve both the microstructure and the electrochemical properties of graphene. So far, many chemical routes have been developed for synthesis of N-doped graphene (N-graphene), including chemical vapour deposition (CVD), N-plasma treatment, arc discharge, electrothermal reaction, eletrochemical reaction, chemical synthesis, etc. Now, CVD and chemical synthesis are more widely adopted for large-scale production.

2.2.1.1. CVD method

Chemical incorporation of nitrogen atoms into graphite by CVD method was first reported by Johansson, et al., in 1990s. Now, CVD has become one of the important methods for synthesizing carbon based nanomaterials including the graphene and N-graphene. A typical CVD process involves three key components, i.e., catalyst, reactants and high temperature. The mechanism can be expressed as following [11-13]: (I) at high temperature (e.g., >800 ºC) the catalyst (transition metals) is liquidized, acting as the catalytic sites for absorption and dissociation of the gas reactants including N-containing reactant (mainly NH_3), (II) the catalyst becomes saturated with the atoms/fragments from the dissociation of the reactants and hence; (III) solid graphitic carbon (graphene layers) grows from the saturated catalyst by means of precipitation, with the adsorbed N atoms precipitating into the graphitic lattice, giving rise to N-doped carbon. The CVD synthesis of N-graphene succeeded in 2009 [13]. In principle, a Cu film grown on a Si substrate was employed as the catalyst. The substrate was placed in a quartz tube with a flow of hydrogen and argon. When the center of the furnace reached 800 °C, CH_4 and NH_3 were introduced into the flow as the carbon source and nitrogen source respectively, and then the substrate was rapidly moved to the high temperature region. After 10 min of growth, the sample was cooled to room temperature under H_2 ambient. Most of the products are few-layer graphene, and single-layer graphene can be occasionally found. The doped N atoms are localized in three different sites, i.e., the well-known graphitic N, pyridinic N and pyrrolic N, respectively. Their configurations are schematically described in Fig. 2. The typical TEM images are shown in Fig. 3. Atomic percentage of N in the sample is about 8.9 %. Following this pioneer work, many modified CVD methods have been developed, of which, annealing graphene in NH_3 stream is relatively easier to achieve [14-16].

2.2.1.2. Nitrogen plasma or N^+-ion irradiation

Nitrogen plasma is another important method for the N doping in the earlier period of N-graphene research. Typically, graphene was first synthesized on Si substrate using CH_4/N_2 plasma at 800 W on microwave plasma enhanced CVD (MPECVD) system [17]. Further N_2 plasma treatment of the as-synthesized graphene was carried out using in situ electron cyclotron resonance (ECR) plasma at a low-pressure and room-temperature. For N doping, a working pressure of ~0.025 Pa and a microwave power of 150 W were applied for 5 min. N content varies from approximately 0.7 at% to nearly 6.3 at% (following ECR plasma treatment). N^+-ion irradiation synthesis involves N^+-ions bombardment of graphene and subsequently annealing of the bombarded graphene in NH_3 [18]. To achieve this goal, N^+-ion irradiation (30 KeV) was first carried out on the mechanically exfoliated single-layer graphene grown on a 300 nm SiO_2/Si substrate at room temperature in a vacuum chamber. The bombardment introduces defects into the plane of the graphene. The defect-possessing graphene was then annealed in NH_3 atmosphere at 1100 °C for 30s, resulting in restoration of some defect sites and N doped into the plane. N atoms exist mainly as graphitic N.

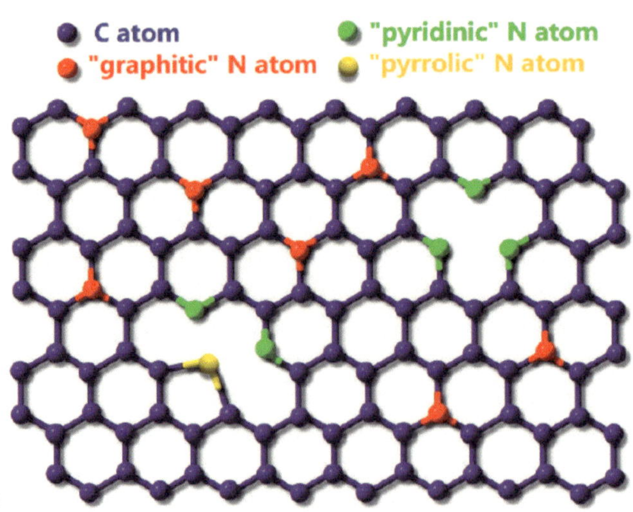

Figure 2. Schematic diagram of the three types of doped N atoms in N-graphene (Nano Lett. 2008;9(5) 1752-1758.)

http://dx.doi.org/10.5772/55666

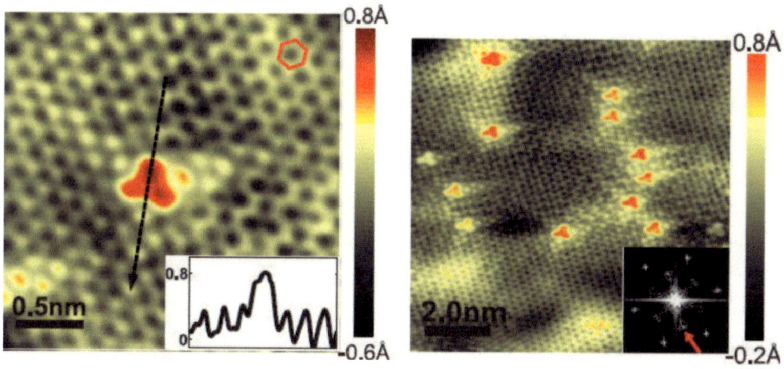

Figure 3. TEM images of N-doped graphene on copper foils: Left, with a single graphitic N dopant; right, with 14 graphitic N dopants and strong intervalley scattering tails. (Science 2011;333, 999-1003.)

2.2.1.3. Arc discharge

The most typical study focusing on fabrication of N-graphene by arc discharge was reported by Li, et al., a method that allows large scale synthesis of N-doped multi-layered (2-6 layers) sheets [19]. In detail, direct current (DC) arc-discharge was carried out in a water-cooled stainless steel chamber. Two electrodes were both Φ8 mm pure graphite rod. The current was held at 120 A. As the rods were brought close together, discharge occurred resulting in the formation of plasma. The anode was vaporized in He and NH_3 mixing atmosphere of 760 Torr. As the anode was consumed, the rods were kept at a constant distance of about 1 mm by rotating the cathode. When the discharge ended, the soot generated was collected under ambient conditions. The condition that is favorable for obtaining N-doped multi-layer graphene is the high proportion of NH_3 (above 50%, by volume). However, no information on how many N atoms can be incorporated into the plane was provided. Rao et al., [20] modified this method using H_2+pyridine or H_2+ammonia as starting materials, making use of the fact that in the presence of hydrogen, GNS do not readily roll into nanotubes. The result N-graphene only contains about 1 at% of N. It should be mentioned that there is a major drawback in using arc discharge for carbon materials synthesis (including graphene). This process gives rise to a large amount of unwanted products; in other word, the selectivity is lower. As a result, the process requires complicated and well controlled purification steps [21].

2.2.1.4. Electrothermal reactions

Electrothermal reaction is relatively more complicated and of higher cost. It was developed in Dai's lab [22]. This method involves high-power electrical joule heating of GNS in ammonia gas. To achieve this goal, they first fabricated a field emission like device using graphene nanoribbons. The graphene nanoribbon device was then e-annealed in a ~1 torr NH_3/Ar environment with carefully designed sequences and control experiments. In NH_3, they applied similar e-annealing sequences as in vacuum. After e-annealing, the chamber was pumped down to base pressure to fully remove physisorbed NH_3 molecules. It is proposed that chemical reactions between graphene nanoribbons and NH_3 lead to nitrogen functionalization, most likely take place at the more reactive edge carbon atoms.

2.2.1.5. Chemical synthesis

Irrespective of the progress made on N-graphene synthesis, continuous pursuing in this area is still highly demanded. Future applications of N-graphene rely on easy-to-operate methods, which can deliver products in a reasonably large quantity and at obviously reduced cost. On this aspect, chemical synthesis has drawn much attention. The first attempt for this is the synthesis using embedded carbon and nitrogen sources in metal [23]. First, boron and nickel layers were sequentially deposited on SiO_2/Si substrate by electron beam evaporation, where a trace amount of nitrogen species were incorporated into the boron layer spontaneously. Then, the sandwiched Ni(C)/B(N)/SiO_2/Si substrate was vacuum annealed at a desired temperature (800-1100 °C) at a slow rate of 15-20 °C min^{-1}. The samples were maintained at a desired temperature for 0-60 min with a pressure of 10^{-3}-10^{-4} Pa, and then cooled down to room temperature at a rate of 2-50 °C min^{-1}. Typically, the N/C atomic ratios, estimated by XPS, range from 0.3 to 2.9 at%. Later, a novel method was developed for one-pot direct synthesis by the hydrothermal reaction of lithium nitride with tetrachloromethane under mild conditions (in a stainless steel autoclave in nitrogen with temperatures below 350 ºC), allowing fabrication in gram scale [24]. The nitrogen content varies in the range of 4.5-16.4 at% with the aid of cyanuric chloride. This method also allows adjusting the ratio of nitrogen species (graphitic N, pyrrolic N and pyridinic N). Except the high cost of the starting material (Li_3N), this method is indeed suitable for massive N-graphene synthesis. Recently, N-graphene with the nitrogen level as high as 10.13 at% was synthesized via a simple hydrothermal reaction of graphene oxide (GO) and urea [25]. N-doping and reduction of GO were achieved simultaneously under the hydrothermal reaction. In the fabrication, the nitrogen-enriched urea plays a pivotal role in forming the N-graphene nanosheets with a high nitrogen level. During the hydrothermal process, the N-doped urea could release NH_3 in a sustained manner, accompanied by the released NH_3 reacting with the oxygen functional groups of the GO and then the nitrogen atoms doped into graphene skeleton, leading to the formation of N-graphene. The nitrogen level and species could be conveniently controlled by tuning the experimental parameters, including the mass ratio between urea and GO and the hydrothermal temperature.

2.2.2. B-doped graphene

B doping gives rise to P-type (positive) graphene. As compared to N doping, B doping is harder to achieve. So far, to our knowledge, there are only a couple of research groups reporting successful synthesis of B-graphene. Rao et al., are the first group that succeeded [20]. Typically, they used two methods for the synthesis. For the first, B-graphene was prepared by performing the arc discharge of graphite electrodes in the presence of hydrogen, helium, and diborane (B_2H_6). B_2H_6 vapor was generated by the addition of BF_3-diethyl etherate to sodium borohydride in tetraglyme. B_2H_6 vapor was carried to the arc chamber by flowing hydrogen and subsequently He through the B_2H_6 generator. For the second, B-graphene was prepared by carrying out the arc discharge using boron-packed graphite electrodes (3 at% B) in the presence of H_2 and He. XPS results indicated that B content could be 3.1 at%. Wang, et al., [26] explored a more novel method. B-graphene was prepared by reacting CCl4, K and BBr3 at 160 °C for 20 h in a sealed teflon-lined stainless steel autoclave, i.e., a typical solvent thermal process. The content of boron was determined to be 1.1 at% based on the XPS intensity measurement.

2.2.3. *S-doped graphene*

For S doping, the S atoms exist either as thiophene S (One S atom bonded to two C atoms of graphene) and SO_x. Sulfur doped graphene (S-graphene) was reported successful only recently [27]. The doping proceeds through thermal treatment of GO and benzyl disulfide (BDS) in Ar. In detail, GO and BDS were first ultrasonically dispersed in ethanol for about 30 min. The resulting suspension was spread onto an evaporating dish and dried, forming a uniform solid mixture. The mixtures were placed into a quartz tube in Ar atmosphere and were annealed at 600-1050 °C. The contents and bonding configurations of sulfur in the resultant S-graphene can be adjusted by adjusting the mass ratio of GO and BDS as well as the annealing temperatures. This approach gives rise to a highest S content of ~1.53 wt%. Another typical method involves pyrolysis of sulphur-containing liquid precursor [28]. Sulfur powder was dissolved in hexane under ultrasound to form a transparent liquid, which was used as the growth precursor. Cu foil was placed in a quartz tube, and the whole system was pumped down to 10^{-2} Torr. The substrate was heated up with a flow of hydrogen and argon mixture. When the temperature reached 950 °C, the H_2-Ar flow was shut off and S-hexane mixture vapor was introduced into the reaction chamber. After 2.5 min of growth, the sample was rapidly cooled down to 800 °C, and then the whole system was cooled down to room temperature within the hydrogen and Ar atmosphere. The doping is about 0.6 at% as estimated by XPS.

2.2.4. *Fluorine doped graphene*

Fluorine-doped GNS (F-graphene) were synthesized by arc discharge [29]. The arc-discharge process for preparing GNS was carried out in a water-cooled chamber and graphite rods with purity of 99.99% were used as the cathode and anode. After the pressure in the chamber reached 1 Pa, the chamber was filled by H_2 and He. During the discharge, the current was maintained at 140 A. For preparing F-graphene, a hollow graphite rod filled with powdery graphite fluoride (fluorine content is 60 wt%) was used as the anode. The as-obtained powders were collected only in the inner and top wall of the chamber in order to exclude relatively heavy products (such as unexfoliated graphite or graphite fluoride) dropped to the bottom of the chamber during the arc discharge process. The functionalized GNS contain about 10 wt% F and are highly hydrophobic.

2.3. Chemical functionalization of graphene with organic groups

As compared to element doping, the functionalization with more complicated organic groups is an easier procedure. The functionalization usually takes advantage of some typical organic reactions, through which a wide spectrum of functional groups has been chemically anchored onto graphene. Besides the evidently different electrical properties from undoped graphene, such kind of functionalization also brings about some other intriguing properties, such as thermal stability, mechanical strength, high processability, etc, and makes a platform for fabricating 3D materials. There are a huge number of papers published. In this section, we will review the related studies in this area in detail. Considering the major diversity of the organic groups, the review is run in term of how the covalent bonds are formed between graphene/GO and the functional groups, including diazonium coupling, amidation, silanization, esterization, substitution, cycloaddition, etc. The organic molecules that chemically bonded to graphene are listed in Table 2.

2.3.1. Electrochemical reaction

Electrochemical reaction is a powerful method to functionalize graphene with inert precursors, e.g., H_2. In 2009, Zhang's group reported direct electrochemical reduction of single-layer graphene oxide (GO) and subsequent functionalization with glucose oxidase [30]. This approach involves several key steps (as shown in Fig. 4), i.e., adsorption of graphene on the surface of a glassy carbon electrode (GCE), formation and reduction of GCE-APTES-GO (APTES stands for 3-aminopropyltriethoxysilane), electrografting of N-Succinimidyl Acrylate (NSA) on the surface of GCE-APTES-rGO, and anchoring glucose oxidase (GOx) on the surface of GCE-APTES-rGO-pNSA. Most recently, Daniels, et al., reported electrochemical graphene functionalization and its substrate dependence [31]. For this purpose, nanocrystalline graphene layers grown on non-polar faces of SiC were used as control samples for the functionalized graphene on Si-face. Atomic hydrogen was generated using a home-built electrochemical setup, with current applied through a 10% H_2SO_4 acid solution. A 99.6% Pt wire and exposed graphene (approximately a 4 mm diameter circular area) were used as the anode and cathode, respectively. With this setup, under applied voltage, H^+ cations are attracted to the negatively charged graphene cathode electrostatically where they can be reduced by addition of an electron, leading to current flow. Oxidation occurs at the Pt anode during this process. A voltage <1.2 V was used to prevent the splitting of H_2O which causes the formation of H_2 bubbles.

2.3.2. Diazonium coupling

The reduction of diazonium salts has been widely employed for grafting aryl groups to the surface of sp^2-hybridized carbon materials including glassy carbon, HOPG, and carbon nanotubes. The sp^2 atoms of graphene are prone to react with diazonium salts. Recently, this type of chemistry has also been applied to chemically functionalize graphene in several research groups, who also investigated the generated unique properties [32-36]. Haddon, et al., are one of the several groups early reporting success in this study [32]. The surface modification with nitrophenyl groups was achieved through the spontaneous reaction of the diazonium salt (4-nitrophenyl diazonium tetrafluoroborate, denoted as 4-NPD) with the graphene layer, as illustrated in Fig. 5. Intrinsically, the reaction is a result of spontaneous electron transfer from the graphene layer and its substrate to the diazonium salt. The functionalization changes the electronic structure and transport properties of the epitaxial graphene from near-metallic to semiconducting. Following this work, later, the same group [33] further proved that aryl-radical functionalization of epitaxial graphene not only changes the

http://dx.doi.org/10.5772/55666

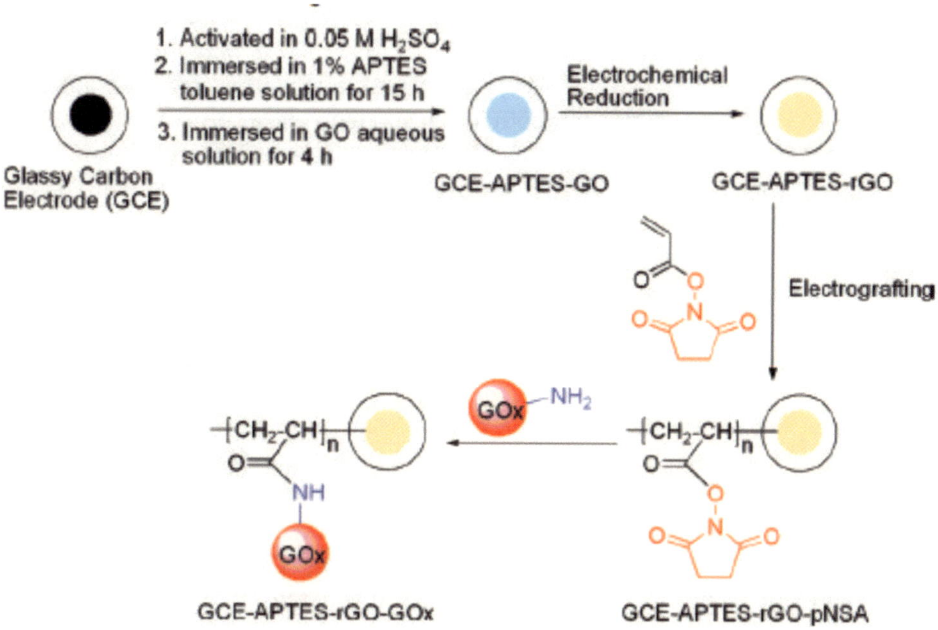

Figure 4. A schematic diagram for the direct electrochemical reduction of single-layer graphene oxide (GO) and subsequent functionalization with glucose oxidase (J. Phys. Chem. C 2009, 113, 14071-14073.)

electronic properties but leads to disordered magnetism in the sheet, which consists of a mixture of ferromagnetic (ferromagnetic), superparamagnetic, and antiferromagnetic regions.

The diazonium coupling reaction of graphene is strongly dependent on graphene layers. Strano, et al., found that single GNS are almost 10 times more reactive than bi- or multilayers of graphene [6]. More interestingly, the reactivity of edges is at least two times higher than the reactivity of the bulk single GNS. This suggestion is supported by the research by Lim [7] and Sinitskii, et al [34].

Besides the aryl groups, through the diazonium reduction chemistry, more complicated organic groups, e.g., polymers, were also reported to be chemically anchored onto graphene, a strategy for fabricating graphene based polymer nanocomposites. Nutt, et al., reported grafting of polystyrene chain on a single-layer GNS [35]. Grafting density and polystyrene chain lengths are controlled by modulating the concentrations of diazonium compound and the monomer of target polymer during the grafting reaction of the initiator and the succeeding atomic transfer radical polymerization. Polystyrene chains grafted on the surface of single-layer GNS exhibited remarkably confined relaxation behavior. An increase in the glass transition temperature (T_g) of up to 18 °C is observed for high grafting density, low molecular weight polymer-grafted graphene samples. The low grafting density, high molecular weight sample shows an increase in T_g of ~9 °C, which is attributed to superior heat conduction efficiency. The measured thermal conductivity for the polystyrene composite film with 2.0 wt % single-layer GNS increases by a factor of 2.6 compared to that of the pure polystyrene.

For practical uses, an easy-access approach was developed to functionalize individually dispersed, highly soluble, and conductive GNS by diazonium reduction chemistry [36]. Typically, requisite amounts of GO and NMP were placed in a flask fitted with a condenser. The mixture was treated within an ultrasonic bath for 1 h and then placed on a magnetic stirrer with an oil bath. After the mixture was bubbled with nitrogen, a given amount of azide compounds (e.g., Az-OH, Az-COOH, Az-NH_2, Az-Br, Az-C16, Az-PEG, Az-PS) was added. The reaction mixture was then heated and maintained around 160 °C in a nitrogen atmosphere under constant stirring. After being cooled to room temperature, the mixture was separated by repeated centrifugation and washed, affording the final product of functionalized graphenes.

2.3.3. Chemical functionalization of GO

Compared to graphene, there are many heterogeneous atoms or atom groups on GO, such as OH, O, O-O, COOH, etc, which are covalently bonded to the C atoms (sp^3) of graphene. These groups are relatively more reactive than C atoms of GNS and therefore, are considered the predominating sites to which the functional groups are bonded. In this section, we will summarize the research aiming at chemically functionalizing GO. Noticeably, no matter how complicated a functional group is, the functionalization generally takes place through typical organic reactions, such as silanization, amidation, esterization, substitution, cycloaddition, etc. The review of this part also follows this line. Such an approach may not be able to comprise all the papers but is effective enough for the readers to understand the chemistry and therefore to explore new methods for the functionalization of graphene.

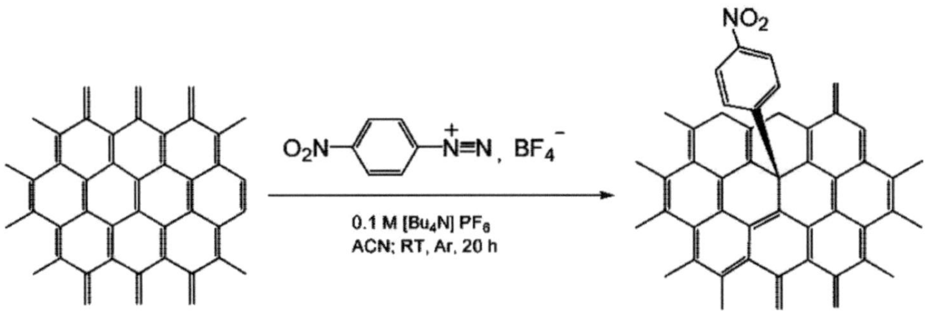

Figure 5. Schematic diagram for the functionalization of graphene using diazonium reduction reaction (J. Amer. Chem. Soc. 2009;131(4) 1336-1337.)

2.3.3.1. Silanization

R-O-Si bond in silane is very reactive toward protic group, i.e., OH. This chemistry has been widely applied to anchor silane groups onto carbon surface. Introduction of silane into GO has also succeeded. In 2010, Hou, et al., reported chemical functionalization of GNS with N-(trimethoxysilylpropyl) ethylenediamine triacetic acid (EDTA) [37]. As described in Fig. 6, the reaction is assumed to proceed through two steps: (1) hydrolysis of the trialkoxy groups of

http://dx.doi.org/10.5772/55666

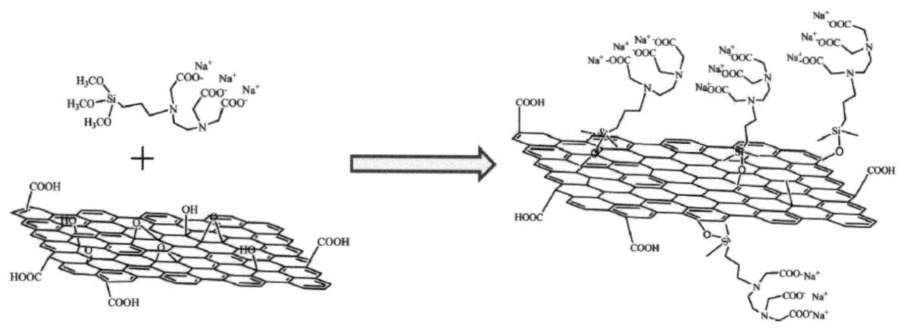

Figure 6. Schematic diagram for the functionalization of graphene using silanization reaction (J. Phys. Chem. C 2009;114 (35), 14915-14921.)

silane generates -Si-OH groups; and (2) the reaction between Si-OH and C-OH of graphene links EDTA to the graphene surface through Si-O-C bond. One CH_4 is released. The silanized GO sheets were further reduced with hydrazine to produce graphene derivatives (EDTA-graphene). The EDTA-graphene showed improved dispersion in water. A relatively simpler method was developed with quaterthiophene molecules (T4-Si) functionalizing using microwaves [38]. GO dispersed in dimethylformamide (DMF) were introduced in a microwave oven reactor and irradiated at 80 °C (100 W) for 40 min. The chemical tethering to GO hinders T4-T4 electronic interactions and aggregation.

2.3.3.2. Amidation

On GO, amidation proceeds mainly through the reaction between the -COOH group native to GO and the $-NH_2$ group of the functional molecule, or vice versa. Amidation is a most effective way to chemically functionalize graphene, and is more widely employed than other methods. Various $-NH_2$ terminated functional groups have been anchored onto graphene [39-44]. The most typical example is Xu's study in 2009 [39]. They synthesized porphyrin-graphene nanohybrid through 5-4 (aminophenyl)-10, 15, 20-triphenyl porphyrin (TPP) functionalization of GO. The synthesis procedure is shown in Fig. 7. TPP-NH_2, as the reaction precursor, was first synthesized. GO was refluxed in $SOCl_2$ in the presence of DMF at 70 ºC for 24 h under Ar atmosphere. In the presence of triethylamine (Et_3N), the above product was allowed to react with TPP-NH_2 in DMF at 130 ºC for 72 h under Ar. The product was isolated by filtration and washed thoroughly. Attachment of TPP-NH_2 significantly improves the solubility and dispersion stability of the graphene-based material in organic solvents. In this donor-acceptor nanohybrid, the fluorescence of photoexcited TPP-NH_2 is effectively quenched by a possible electron-transfer process.

2.3.3.3. Esterization

The presence of -COOH groups native to graphene renders functionalization of GO with CH_2OH-terminated functional groups feasible, i.e., esterization. There are a number of studies addressing this process [45,46]. As a typical molecule that possesses CH_2OH-terminated group, poly(3-hexylthiophene) (P3HT) has been employed to chemically modify GO [45]. End-functionalized regioregular P3HT with methylene hydroxy groups was first synthesized. In a typical experiment for the synthesis of the P3HT-grafted graphene, dried GO sample was refluxed in thionyl chloride for 24 h, followed by the removal of excess thionyl chloride under vacuum. CH_2OH-terminated P3HT in THF was then added through a syringe to the thionyl chloride treated GO under stirring, followed by the addition of triethylamine in nitrogen atmosphere. After sonication, the reaction mixture was vigorously stirred, leading to a dark suspension. The solid in the suspension was removed by centrifuging, and the solvent in the clear solution thus prepared was partially removed by evaporation. It was further purified by precipitating in methanol, filtering, and solvent-washing thoroughly to remove the excess triethylamine.

This reaction also allows fabrication of more complicated graphene based nanocomposites. Pham, et al., explored a facile strategy for covalent functionalization of GO with polyglycerol and used the resultant composites as templates for anchoring magnetic nanoparticles [46]. Pristine graphite was firstly oxidized to obtain GO with hydroxyl functional groups. Then, the covalent grafting of polyglycerol onto the surface of GO was carried out based on *in situ* ring-opening polymerization of glycidol. For the construction of novel hybrid nanostructure, Fe-core/Au-shell nanoparticles were functionalized using 4-mercaptophenylboronic acid through the well-developed Au-S chemistry. Subsequently, magnetic nanoparticles were anchored on the surface of polyglycerol-grafted graphene nanosheets via boroester bonds. The synthesized novel hybrid nanostructures could be stably dispersed in water over 3 months.

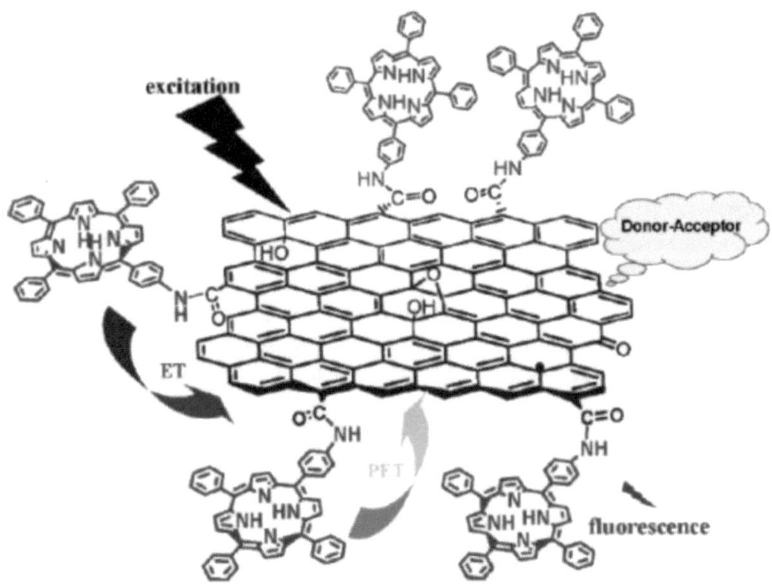

Figure 7. Chemical functionalization of graphene using amidation reaction (Adv. Maters. 2009;21(4) 1275-1279.)

http://dx.doi.org/10.5772/55666

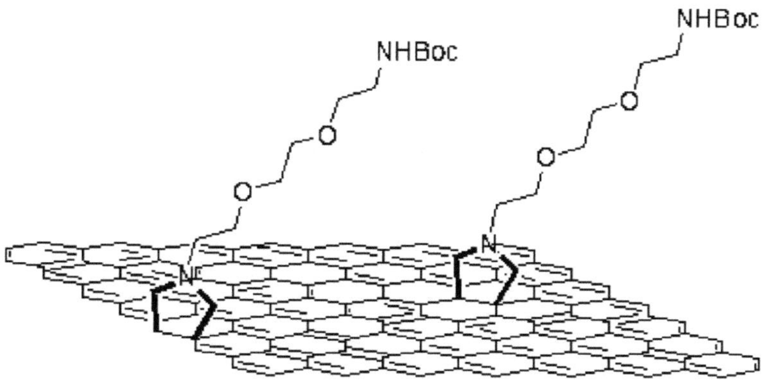

Figure 8. Chemical functionalization of graphene using cycloaddition reaction (ACS Nano 2010;4(6) 3527-3533.)

2.3.3.4. Substitution

Chloride atoms, either bonded to graphene or to functional group, can be substituted by OH group, a reaction that is an alternative route for chemical functionalization. Shen, et al.,[47] report a scalable, fast, and easy method for preparation of organophilic chemically functionalized GNS. The basic strategy involved the complete exfoliation of graphite oxide into GO sheets, followed by reacting with 1-bromobutane. The resulting organic dispersions are homogeneous, exhibit long-term stability, and are made up of graphene sheets a few hundred nanometers large. Pramoda, et al., describes a new route to chemically bonded polymer-graphene nanocomposites and the subsequent enhancement in thermal and mechanical properties [48]. At first, the graphite oxide is functionalized with octadecylamine (ODA). The ODA functionalized graphite oxides are reacted with methacryloyl chloride to incorporate polymerizable -C=C- functionality at the nanographene platelet surfaces, which were subsequently employed in *in situ* polymerization of methylmethacrylate to obtain covalently bonded poly(methyl methacrylate) (PMMA)-graphene nanocomposites. The obtained nanocomposites show significant enhancement in thermal and mechanical properties compared with neat PMMA. Even with 0.5 wt % graphene nanosheets, the T_g increased from 119 °C for neat PMMA to 131 °C for PMMA-graphene nanocomposite, and the respective storage modulus increased from 1.29 to 2 GPa.

2.3.3.5. Cycloaddition

Arynes are recognized as useful reactive intermediates during nucleophilic aromatic substitution in organic synthesis and are frequently used in a variety of reactions. To date, quite a few attempts at aryne cycloaddition have been performed for carbon nanomaterials, demonstrating that fullerenes and derivatives can be successfully functionalized. Recently, chemical modification of graphene by aryne cycloaddition has been reported by several groups. Zhong, et al., developed a novel and convenient approach to chemically functionalize graphene by this reaction with benzene precursors [49]. Commercial GNS from the arc-discharge method were used. 2-(Trimethylsilyl) phenyl triate was employed as the functional group. A typical procedure for preparing aryne-modified graphene sheets is given in Fig. 8. On the basis of weight loss, it is estimated that the degree of functionalization is approximately over 1 functional group per 17 carbon atoms. The resultant graphene shows remarkable solubility and thermal stability. The functionalized graphene is stable at temperature high to 500 ºC. With the similar reaction, chemical functionalizations using the 1,3-dipolar cycloaddition of azomethine ylides and cyclopropanated malonate also succeeded[50,51].

Chemical process	Organic group	References
Diazonium coupling	Nitrophenyl	[6,7, 32-34]
	Polystyrene chain	[35]
	$C_{16}H_{34}$, Polyethylene glycol	[36]
Silanization	EDTA	[37]
	Quaterthiophene	[38]
Amidation	Porphyrin	[39]
	Zinc phthalocyanine	[41]
	Oligothiophene	[40]
	Poly(N-vinylcarbazole)	[42]
	2-amino-4,6-didodecylamino-1,3,5-Triazine	[44]
	Chitosan	[43]
Etherization	poly(3-hexylthiophene) (P3HT)	[45]
	Polyglycerol	[46]
Substitution	1-bromobutane	[47]
	Methacryloyl chloride	[48]
Cycloaddition	2-(Trimethylsilyl) phenyl triate	[49]
	Azomethine ylides	[50]
	Cyclopropanated malonate	[51]

Table 2. A summary of the chemical functionalization of graphene with organic groups

3. Chemically functionalized graphene for electrochemical energy conversion and storage

Although a wide spectrum of functional groups (from single atoms to polymers) has been chemically grown onto graphene, it is a fact that for EECS, only element doped graphene has been widely investigated on EECS devices including FCs [16, 17, 29, 52-61], LIBs [14, 62-67]

and supercapacitors [27, 68, 69]. The related studies are tabulated in Table 3. This is because the covalent bonding with most of the organic groups frequently damages the inherent properties of graphene, especially the electronic conductivity significantly, resulting in decline in the electron transfer and hence decay of electrochemical performances. In this section, we will comprehensively outline all the characteristic research in this area, emphasizing on FCs, LIBs and then supercapacitors.

3.1. Fuel cells

A major impediment to the commercialization of proton exchange membrane fuel cells (PEMFC) is the high cost and the stability of Pt-based electrocatalysts (Pt/C). Much effort is focusing on seeking either new support materials to improve the distribution of Pt and hence to lower its loading or non-noble metal electrocatalysts to decrease the cost. Nitrogen doped carbon material is recognized as a good support for Pt based catalyst. The doped nitrogen atoms not only provide the anchoring sites for the metal particles, but also act as chemically active sites for catalytic reactions. Moreover, recent intense research has found that nitrogen-doped carbon materials (especially, vertically aligned nitrogen-containing carbon nanotubes, nitrogen doped ordered mesoporous graphitic carbon and silk-derived carbon (0.8% nitrogen in the carbon network) could act as effective non-noble metal-free electrocatalysts. The real active sites of nitrogen-doped carbon materials in such applications remain unclear, but the doped nitrogen atoms are believed to (such as graphite-like, pyridine-like, pyrrole-like, and quaternary nitrogen atoms, see Fig. 2) play a crucial role for oxygen reduction reaction (hereafter denoted as ORR). Usually, the greater the extent of graphitization of the carbon material, the longer the durability is. The existing data prove that N-graphene exhibit high electrocatalytic activity and long-term operation stability for the ORR. Besides N doping, S-graphene is also investigated for FCs application. The related research is reviewed in two separate parts here.

3.1.1. Chemically functionalized graphene as a catalyst support

Wu, et al., reported identification of the nitrogen species on N-graphene layers and the performance of Pt/NG (Pt/N-graphene) composite catalyst in direct methanol fuel cells (DMFC) [16]. The N-graphene was synthesized through thermal-treatment of GO in ammonia stream at various temperatures, offering various distributions of nitrogen species. In electrochemical tests, Pt/NG, Pt/graphene and Pt/CB (carbon black, Vulcan XC 72) composites were investigated in anodic half-cell reaction in DMFC. A Pt/NG-800 composite, for which the graphene was treated in NH_3 flow at 800 ºC, exhibits outstanding electrocatalytic activities for methanol oxidation. The mass specific oxidation current for Pt/NG-300, Pt/G-300 and Pt/G-800 is 40, 41 and 34 mA mg^{-1}, respectively. In sharp contrast, the methanol oxidation current of Pt/NG-800 (135 mA mg^{-1}) is three times higher than the other three composites and the commercial Pt/CB (27 mA mg^{-1}). The higher methanol oxidation activity from Pt/NG-800 is assumed to be associated with (1) the high dispersion state of the Pt nanoparticles facilitated by the pyridinic nitrogen doped in the carbon network; and (2) higher conductivity as a result of reduced defect sites by the higher temperature treatment. Ramaprabhu compared the activities of Pt nanoparticles supported on graphene nanoplatelets and N-graphene nanoplatelets, respectively, as catalysts for ORR in PEMFC [52]. Nitrogen plasma treatment created pyrrolic nitrogen defects, which act as good anchoring sites for the deposition of platinum nanoparticles. Pt was dispersed over these support materials using the conventional chemical reduction technique. A maximum power density of 440 mW cm^{-2} was obtained with the Pt loaded N-graphene, and 390 mW cm^{-2} for the Pt loaded graphene nanoplatelets, respectively. The improved performance of FCs with N-graphene as catalyst supports was attributed to the increased electrical conductivity and the enhanced carbon-catalyst binding.

Enhanced electrochemical sensitivity of bimetal PtRh electrodes coated with N-graphene (PtRh@N-graphene) was reported by Wang, et al [53]. Deposition of N-graphene onto the surface of a metal electrode (PtRh alloy mesh) was performed by dipping the mesh into the dispersion and withdrawing. The suspended GNS was thus deposited onto the surface of the PtRh alloy mesh. The coating effectively improves the electron transfer kinetics. Nanoflower-like N-graphene with pure sp^2 hybridized carbon and designed nitrogen types was synthesized by a low temperature solvothermal process [54]. The product consists of many N-graphene sheets with about 6-10 graphitic layers. The sheets are rich of pyridine-like and quaternary nitrogen atoms, and possess high graphitized structure and ultra-micropores. As such, electrochemical test showed that N-graphene has a much higher durability as a Pt support for FCs than commercial carbon black (Vulcan XC-72).

A $MnCo_2O_4$/N-graphene hybrid material was developed as highly efficient ORR electrocatalyst in alkaline conditions [55]. The covalent coupling between $MnCo_2O_4$ nanoparticles and N-graphene sheets offers much higher activity and stronger durability than the physical mixture of both components. At a same mass loading, the $MnCo_2O_4$/N-graphene hybrid can outperform Pt/C in ORR current density at potential <0.75 V vs RHE. The stability is also superior to that of the Pt/C. Semi-quantitatively, an optimum range of $MnCo_2O_4$ content between 65 and 80 wt % was found to give similarly high ORR performance from carbon fiber paper measurements. Out of this range, too low a $MnCo_2O_4$ content could lead to fewer ORR active sites in the hybrids, while too high a $MnCo_2O_4$ content could result in aggregation of nanoparticles and even free growth, which were less active than the nanoparticles directly grown on GNS.

3.1.2. Chemically functionalized graphene as a catalyst

3.1.2.1. Nitrogen-doped graphene

N-graphene as metal-free electrocatalyst for FCs is drawing rapidly increasing attention. Plausible electrocatalytic activities were observed. Dai, et al., are the first group reporting the use of N-graphene as metal-free catalysts for ORR [56]. N-graphene was synthesized by CVD of methane in the presence of ammonia. For comparison, electrocatalytic selectivities of the N-graphene electrode were tested against the electro-oxidation of various commonly used fuel molecules, including hydrogen gas, glucose and methanol. The N-graphene electrode demonstrated a stronger and more stable amperometric response from the ORR than the Pt/C catalyst does. The current remained unchanged after the addition of hydrogen gas, glucose, and methanol. They ascribed the high selectivity of the N-graphene electrode and remarkably

http://dx.doi.org/10.5772/55666

good tolerance to crossover effect to the much lower ORR potential than that required for oxidation of the fuel molecules. Moreover, the N-graphene electrode was insensitive to CO, but the Pt/C electrode was rapidly poisoned under the same conditions. Continuous potential cycling test indicated that for the N-graphene electrode, no obvious decrease in current was observed after 200000 continuous cycles between -1.0 and 0 V in air-saturated 0.1 M KOH. Shao, et al., synthesized N-graphene by exposing graphene to nitrogen plasma and carried out detailed studies on its electrochemical activities [17]. The N-graphene exhibits much higher electrocatalytic activity toward ORR and H_2O_2 reduction than graphene does, and much higher durability and selectivity than the Pt/C. The ORR kinetic currents for graphene, N-graphene and Pt/C (ETek, 20 wt%) at -0.03 V (Hg/HgO) are ~1, 65 and 255 μA, respectively. The ORR over-potential is greatly decreased on the N-graphene in comparison with graphene. After accelerated degradation test, the N-graphene exhibits a slight increase in ORR activity; however, Pt/C degraded by ~85%. Moreover, when performing in O_2-saturated 0.1 M KOH, no changes in surface chemistry of the N-graphene were observed. The ORR on N-graphene is not influenced by the fuel molecules (e.g., methanol). Geng, et al., reported high ORR activity and durability of N-graphene synthesized by the thermal treatment of graphene with ammonia at different temperatures [57]. 2.8 at% nitrogen was introduced for the N-graphene (900 ºC), which also exhibits a highest ORR activity in alkaline solution. The onset potentials for ORR (E_{ORR}) of graphene, N-graphene (800), N-graphene (900), and N-graphene (1000) were 0.046, 0.184, 0.308, and 0.204 V, respectively. N-graphene (900) catalyst was assumed to effectively promote the desired $4e^-$ ORR activity in alkaline solution. Only quaternary type nitrogen species plays the most important role for the ORR. In comparison to the commercial Pt/C catalyst (4.85 μgPt cm^{-2}), the electrochemical performance of the N-graphene (900) catalyst is also better, as shown in Fig. 9. Besides the higher onset potential (0.308 V), oxygen-reduction half-wave potential is 43 mV more positive on the N-graphene catalysts. Almost no change in the voltammetric charge was found after 5000 cycles of the potential sweep for the N-graphene, however, the current density for Pt/C dropped dramatically.

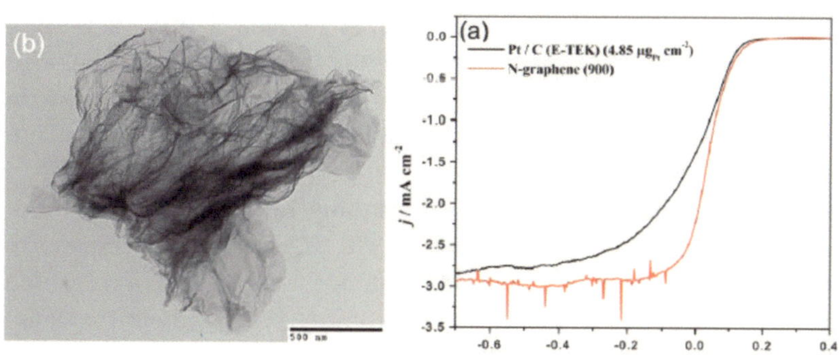

Figure 9. TEM image of nitrogen doped graphene and its electrochemical performance as a metal free catalyst for ORR. (Energy & Environ. Sci. 2011;4(3) 760-764.)

ORR on N-graphene was also tested in acid media [58]. The mass activity of the N-graphene catalyst (1234 mA g^{-1}_{catal}) is about 35 times higher than that of the Vulcan carbon (35 mA g^{-1}_{catal}).

This activity is about 50% lower than that of Pt/C catalysts (2791 mA g^{-1}_{catal}). High electric conductivity, high surface area, large amount of edge sites and pyridinic N site in reduced graphene sheets are assumed to make major contribution to the high ORR activity. Lee, synthesized N-doped carbon nanosheets from pyrolysis of collagen at 800 °C, which possesses a surface area of 695 m^2 g^{-1} and nitrogen content of ~1% [59]. With the contribution of N-doped structures, the carbon nanosheets showed a specific capacitance of 102 F g^{-1} at 25 mV s^{-1}, and 80% capacitance retention in 0.5 M H_2SO_4 at 1000 mV s^{-1}. In comparison with a commercial electrocatalyst, 20% Pt on Vulcan XC-72, the carbon nanosheets display a positive shift in the onset potential and superior electrocatalytic activity toward the ORR. The onset potential of the ORR for the N-doped carbon nanosheets is 0.95 V (the commercial Pt electrocatalyst is 0.83 V). The significant catalytic activity toward the ORR of the N-doped carbon nanosheets is assumed to mainly come from the contribution of pyridinic-N and quaternary-N, which change the adsorption of the oxygen molecule on carbon materials from end-on to side-on type, thus weakening the O-O bond to facilitate the ORR. On the other hand, Luo, et al., was focused on exploring the contribution of different doped N atoms to the ORR reaction of the N-graphene [60]. For this purpose, single layer graphene doped with pure pyridinic N was synthesized by CVD of hydrogen and ethylene on Cu foils in the presence of ammonia. By adjusting the flow rate of ammonia, the atomic ratio of N and C can be modulated from 0 to 16%. The pyridinic N efficiently changes the valence band structure of graphene, including the increase in the density of π states near the Fermi level and the reduction of work function. Interestingly, on the base of the 2e reduction mechanism of ORR on the resultant CN_x graphene revealed by rotating disk electrode voltammetry, they suggested that the pyridinic N may not be an effective promoter for the ORR activity of carbon materials as previously expected. Obviously, this question remains open for further investigation.

3.1.2.2. S-doped graphene

S-graphene as catalysts for PEMFC was first reported by Mullen's group [61]. The S-graphene was synthesized by annealing graphene in H_2S at temperatures from 500 to 1000 °C. S exists as thiophene like S and oxidized S. A highest S content is obtained in the sample doped at 900 ºC (1.7%). The electrocatalytic activity was investigated via RRDE linear sweep voltammetry at a scan rate of 100 mV s^{-1} in O_2 and Ar-saturated 0.1 M KOH solution as well as O_2-saturated 0.1 M KOH solution with 3M methanol. The S-graphene sheets exhibit good electrocatalytic activity, long durability and high selectivity when employed as metal-free catalysts for ORR, depending on the annealing temperature. As the annealing temperature increases from 500 to 900 °C, the electron transfer number of S-graphene slightly decreases from 3.5 to 3.2, a result of the reduction of the amount of sulfur (from 1.7% to 1.2%), because the S-C bonds play a key role to affect the catalytic process of S-graphene for the ORR. Moreover, the S-graphene and commercial Pt/C were further compared by separately introducing O_2 and fuel molecules (e.g., methanol) into the electrolyte to examine their possible selectivity and crossover effects via chronoamperometric measurements. The S-graphene also outperforms the commercial Pt/C. Similarly, it was reported that S-graphene through thermal treatment of graphene oxide (GO) and benzyl disulfide (BDS) in Ar showed higher electrocatalytic activity than Pt/C catalysts do [29].

http://dx.doi.org/10.5772/55666

3.2. Lithium ion batteries

The research on the application of chemically functionalized graphene to LIBs is also developing rapidly. For LIBs, the functionalized graphene is used both as anode materials and as doping composite. In both cases, improved lithium storage capacities were observed.

3.2.1. N-doped graphene as anode materials

In 2010, Reddy, et al., published the first paper about N-graphene for LIB application [62]. Few layers of graphene and N-doped graphene films on Cu substrate were synthesized by CVD technique using hexane and acetonitrile precursors, respectively. For the half cell measurements, an electrochemical test cell was assembled using the graphene/N-dgraphene as working electrode, lithium metal foil as the counter/reference electrode and 1 M solution of $LiPF_6$ in 1:1 (v/v) mixture of ethylene carbonate (EC) and dimethyl carbonate (DMC) as electrolyte. Through introduction of surface defects and pyridinic N atoms into the graphene structure, intercalation of Li-ion in N-graphene electrode was obviously enhanced. Reversible discharge capacity of N-graphene remains beyond 0.6 mAh cm^{-1} after 50 cycles, while that of pristine graphene is ~0.48 mAh cm^{-1}. Cho, et al., [63] deposited N-graphitic layers on silicon nanowires by CVD for LIB anode. Graphite-like and pyridine-like structures were selectively chosen for 3 and 10% N doping, respectively. Increasing the thickness of the undoped graphitic layers from 20 to 50 nm led to an increase in the charge capacity from 800 to 1040 mAh g^{-1} after 45 cycles. Graphite-like 3% N-doping in the 50 nm-thick shell increases the charge capacity by 21% (i.e., to 1260 mAh g^{-1}), while pyridine-like 10% N-doping in the 20 nm-thick shell increases it by 36% (i.e., to 1090 mAh g^{-1}). This suggests that both pyridine- and graphite-like structures can be effective for lithium intercalation.

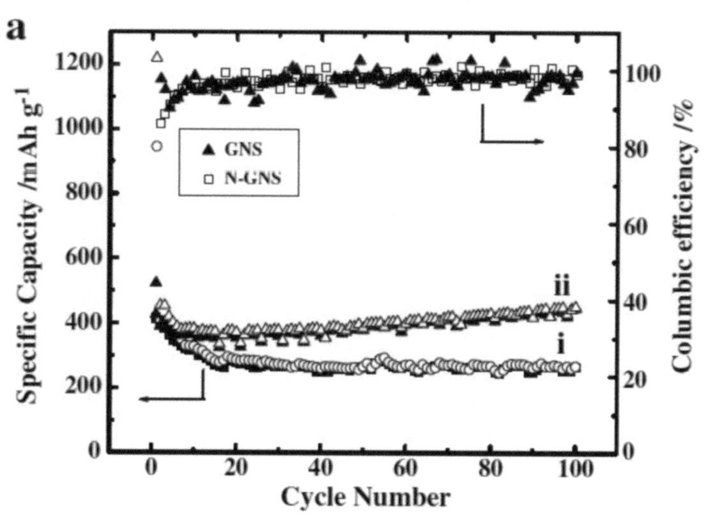

Figure 10. Electrochemical performances of nitrogen doped graphene as anode for LIBs (Electrochem. Comm. 2011;13(8) 822-825.)

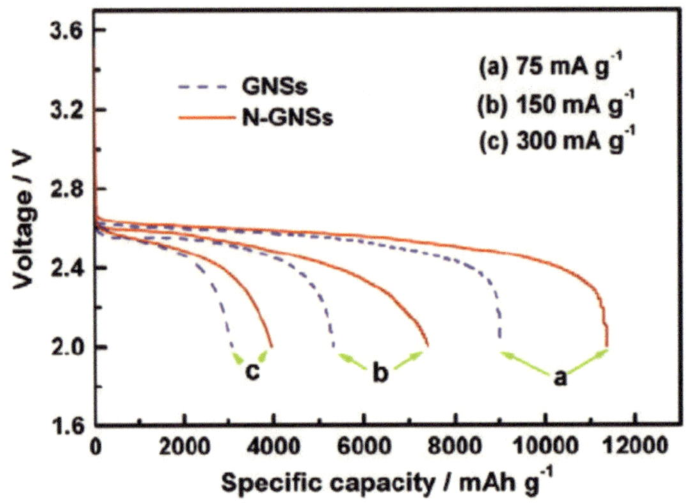

Figure 11. Electrochemical performances of nitrogen doped graphene as cathode for lithium oxygen batteries (Electrochem. Comm. 2011;18(1) 12-15.)

Powder based N-graphene with excellent lithium storage was also reported by Cui and Li respectively [14, 64]. In Cui's research, the N-graphene nanosheets were prepared by heat treatment of graphite oxide under an ammonia atmosphere at 800 °C for 2 h. The doping level is about 2% nitrogen, where the N binding configuration of the graphene includes 57.4% pyridinic, 35.0% pyrrolic and 7.6% graphitic N atoms. The N-graphene nanosheets exhibited a high reversible capacity (900 mAh g^{-1} at 42 mA g^{-1}), excellent rate performance (250 mAh g^{-1} at a current density of 2.1 A g^{-1}), and significantly enhanced cycling stability. The doped graphene in Li's study was obtained by high-temperature annealing graphene in NH_3. Interestingly, the specific capacity of the doped graphene evidently increases with charge/discharge cycles. As shown in Fig. 10, after 100 charge-discharge cycles, the specific capacity was ~480 mAh g^{-1} for the N-graphene, while only 290 mAh g^{-1} for the undoped graphene. The obtained significant improvement is attributed to the incorporated nitrogen to graphene planes with a result of more structural defects during cycling, as indicated by Raman analysis.

N-graphene also indicates excellent performances when used as cathode for Li air batteries. Sun's research group fabricated nonaqueous lithium-oxygen battery with N-graphene nanosheets as cathode materials, which delivered a discharge capacity of 11660 mAh g^{-1}, as shown in Fig. 11 [65]. This capacity is about 40% higher than that with the pristine GNS. The excellent electrochemical performance of N-graphene is attributed to the defects and functional groups as active sites introduced by nitrogen doping.

3.2.2. N-doped graphene as an electrode composite

Wang, et al., synthesized N-graphene-SnO_2 sandwich papers as anode for LIBs through a complicated procedure [66]. The materials show excellent electrochemical performances. The

N-doped G-SnO_2 paper delivered an initial reversible capacity of 918 mAh g^{-1}, which is higher than those obtained for both pure SnO_2 nanoparticles (881 mAh g^{-1}) and graphene paper (548 mAh g^{-1}). Moreover, the N-doped G-SnO_2 papers also exhibited enhanced rate capacities. The specific capacity was maintained at values as high as 683 mAh g^{-1} and 619 mAh g^{-1} after the current density was increased to 1000 mA g^{-1} and 2000 mA g^{-1}, respectively. Even at a current density of 5000 mA g^{-1}, the material still delivered a capacity of 504 mAh g^{-1}.

3.2.3. S-doped graphene as cathode for lithium oxygen batteries

Recently, Sun's research group reported the use of S-graphene as cathode for lithium oxygen batteries [67]. The S-doped graphene was obtained through pyrolysis of p-toluenesulfonic acid. The initial discharge capacity of S-graphene electrode is about 4300 mAh g^{-1}, which is lower than that of the pristine graphene electrode (~8700 mAh g^{-1}). However, the initial charge capacity of the S-graphene electrode is about 4100 mAh g^{-1}, but it is only ~170 mAh g^{-1} for the undoped graphene electrode. Moreover, the discharge capacities in the second cycle of the S-graphene and undoped graphene electrodes are 3500 and 220 mAh g^{-1}, respectively. Importantly, S-doping in graphene results in morphological change of reaction products from nanoparticles to nanorod-like.

Guest functional group	Electrochemical devices	References
N atoms	Fuel cells	[16,17,52-60]
	Lithium ion batteries	[14,62-66]
	Supercapacitors	[27,68,69]
B atoms	Fuel cells	[70]
S atoms	Fuel cells	[29,61]
	Lithium ion batteries	[67]

Table 3. A summary of the functionalized graphene applied in EECS.

3.3. Supercapacitors

Although various carbon nanomaterials including activated carbon, carbon nanotubes and graphene have demonstrated high-performance for supercapacitors, the capacitances they can hold are still much lower than the demanded. It is a challenge to research and development of qualified materials for high-performance supercapacitors. Application of N-graphene in supercapacitors slightly falls behind FCs and LIBs [27,68,69]. The first report concerning N-graphene for supercapacitor appeared in May 2011 by Hyung, et al., [68]. The N-graphene was produced by a simple plasma treatment of graphene. The N-graphene supercapacitors gave rise to a highest capacitance of ~280 $F/g_{electrode}$, which is about 4 times larger than those of undoped graphene based counterparts. Moreover, the supercacapacitors also demonstrated excellent cycle life (>200000), high power capability (a power density up to ~8 ×10^5 W kg^{-1} and an energy density up to ~48 Wh kg^{-1}), and compatibility with flexible substrates. In the same study, using local N-configuration mappings during plasma treatment along with side binding energy calculated by density functional theory, they also articulated that the origin of the improved capacitance is a certain N-configuration at basal planes. N-graphene nanosheets with the nitrogen level as high as 10.13 at% were synthesized from GO and urea in a hydrothermal process [27]. The N-graphene has a large surface area of 593 m^2 g^{-1}, and exhibits excellent capacitive behaviors (326 F g^{-1}, 0.2 A g^{-1}), superior cycling stability (maintaining initial capacity even) and high coulombic efficiency (99.58%) after 2000 cycles. An energy density of 25.02 Wh kg^{-1} could be achieved at power density of 7980 W kg^{-1} by a two-electrode symmetric capacitor test. Both the N-content and the N-type are very significant for the capacitive behaviors. In more detail, the pyridinic-N and pyrrolic-N play main roles for improving pseudo-capacitance by the redox reaction, while quaternary-N could enhance the conductivity of the materials which is favorable to the transport of electrons during the charge/discharge process.

4. Conclusion

Chemical functionalization of graphene is one of the key subjects in graphene science and technology, and has been extensively studied. The funtionalization proceeds mainly through chemical atom doping, diazonium coupling, silanization, amidation, etherisation, substitution and cycloaddition, etc. Various innovative properties have been brought about and therefore the functionalized graphene exhibits potential application in a wide spectrum of fields. In electrochemical energy storage and conversion, graphene doped with N, S and B has shown plausible performances and has been widely investigated. Nitrogen doped graphene was used either as a support for Pt or as a metal-free catalyst for fuel cells. Its application in lithium ion batteries and supercapacitors also offers higher lithium storage capacities and rate capability and higher capacitance. S doped graphene also results in improved oxygen reduction reaction and lithium storage capacities for lithium oxygen batteries. This subject is still under rapid development. B doped graphene also improved oxygen reduction reaction. However, functionalized graphene with complicated organic molecules has yet been applied to this field due to the lower electronic conductivity. An atomic understanding of interaction between functional group and graphene, the change in the property of the functionalized graphene during performance, etc., should be pursued in the future studies.

Author details

Yuhai Hu and Xueliang Sun*

*Address all correspondence to: xsun9@uwo.ca

Department of Mechanical and Materials Engineering, The University of Western Ontario, London, ON, Canada

References

[1] Novoselov K, Geim A, Morozov S, Jiang D, Zhang Y, Dubonos S, Grigorieva I, Firsov A. Electric field effect in atomically thin carbon films. Science 2004;306(22) 666-669.

[2] Allen M, Tung V, Kaner B. Honeycomb carbon: a review of graphene. Chemical Reviews 2010;110(1) 132-145.

[3] Dutta S, Pati S. Novel properties of graphene nanoribbons: a review. Journal of Materials Chemistry 2010;20(8) 8207-8223.

[4] Geim A. Graphene: Status and prospects. Science 2009:324(5934) 1530-1534.

[5] Brownson D, Banks C. Graphene electrochemistry: an overview of potential applications. Analyst 2010;135(11) 2768-2778.

[6] Sharma R, Baik J, Perera C. Anomalously large reactivity of single graphene layers and edges toward electron transfer chemistries. Nano Letters 2010;10(2) 398-405.

[7] Lim H, Lee J, Shin H. Spatially resolved spontaneous reactivity of diazonium salt on edge and basal plane of graphene without surfactant and its doping effect. Langmuir 2010;26(14) 12278-12284.

[8] Sun Y, Wu Q, Shi G. Graphene based new energy materials. Energy & Environmental Science 2011;4(4) 1113-1132.

[9] Wang H, Maiyalagan T, Wang X. Review on recent progress in nitrogen-doped graphene: synthesis, characterization and its potential application. ACS Catalysis 2012;2(5) 781-794.

[10] Wehling T, Novoselov K, Morozov S. Molecular doping of graphene. Nano Letters 2008;8(1) 173-177.

[11] Dato A, Radmilovic V, Lee Z. Substrate-free-gas phase synthesis of graphene sheets. Nano Letter 2008;8(7) 2012-2016.

[12] Wang X, Liu Y, Zhu D. Controllable growth, structure, and low field emission of well-aligned CN_x nanotubes. Journal of Physical Chemistry B 2002;106(9) 2186-2190.

[13] Wei D, Liu Y, Wang Y. Synthesis of N-doped graphene by chemical vapor deposition and its electrical properties. Nano Letters 2009;9(5) 1752-1758.

[14] Li X, Geng D, Zhang Y. Superior cycle stability of nitrogen-doped graphene nanosheets as anode for lithium ion batteries. Electrochemistry Communications 2011;13(8) 822-825.

[15] Geng D, Yang S, Zhang Y. Nitrogen doping effects on the structure of graphene. Applied Surface Science 2011;257(21) 9193-9198.

[16] Zhang L, Liang X, Song W. Identification of the nitrogen species on N-doped graphene layers and Pt/NG composite catalyst for direct methanol fuel cell. Physical Chemistry Chemical Physics 2010;12(38) 12055-12059.

[17] Shao Y, Zhang S, Engelhard M. Nitrogen doped graphene and its electrochemical applications. Journal of Materials Chemistry 2010;20(35) 7491-7496.

[18] Guo B, Liu A, Chen E. Controllable N-doping of graphene. Nano Letters 2010;10(12) 4975-4980.

[19] Li N, Wang Z, Zhao K. Large scale synthesis of N-doped multi-layered graphene sheets by simple arc-discharge method. Carbon 2010;48(1) 255-259.

[20] Panchokarla L, Subrahmanyam K, Saha S. Synthesis, structure and properties of boron and nitrogen doped graphene. Advanced Materials 2009;21(46) 4726-+.

[21] Kim H, Kim H. Preparation of carbon nanotubes by DC arc discharge process under reduced pressure in an air atmosphere. Materials Science and Engineering B 2006;133 (1-3) 241-244.

[22] Wang X, Li X, Zhang L. N-doping of graphene through electrothermal reactions with ammonia. Science 2009;324(5928) 768-771.

[23] Zhang C, Fu L, Liu N. Synthesis of nitrogen-doped graphene using embedded carbon and nitrogen sources. Advanced Materials 2011;23(8) 1020-1024.

[24] Deng D, Pan X, Yu L. Toward N-doped graphene via solvothermal synthesis. Chemistry of Materials 2011;23(5) 1188-1193.

[25] Sun L, Tian C, Tan T. Nitrogen-doped graphene with high nitrogen level via a one-step hydrothermal reaction of graphene oxide with urea for superior capacitive energy storage. RSC Advances 2012;2 4498-4506.

[26] Lin T, Huang F, Jiang J. A facile preparation route for boron-doped graphene, and its CdTe solar cell application. Energy & Environmental Science 2012;4(3) 862-865.

[27] Yang Z, Yao Z, Li G. Sulfur-doped graphene as an efficient metal-free cathode catalyst for oxygen reduction. ACS Nano 2012;6(1) 205-211.

[28] Gao H, Liu Z, Song L. Synthesis of S-doped graphene by liquid precursor. Nanotechnology 2012;23(27) Article number 275605.

[29] Shen B, Chen J, Xue Q. Synthesis of fluorine-doped multi-layered graphene sheets by arc-discharge. RSC Advances 2012;2 6761-6764.

[30] Wang Z, Zhou X, Zhang J. Direct electrochemical reduction of single-layer graphene oxide and subsequent functionalization with glucose oxidase. Journal of Physical Chemistry C 2009;113(32) 14071-14075.

[31] Daniels K, Daas K, Srivastava. Evidences of electrochemical graphene functionalization and substrate dependence by Raman and scanning tunneling spectroscopies. Journal of Applied Physics 2012;111(11) 114306.

[32] Bekyarova E, Itkis M, Ramesh P. Chemical modification of expitaxial graphene: spontaneous grafting of aryl groups. Journal of the American Chemical Society 2008;131(4) 1336-+.

[33] Hong J, Niyogi S, Bekyarrova, Effect of nitrophenyl functionalization on the magnetic properties of epitaxial graphene. Small 2011;7(9) 1175-1180.

[34] Sinitskii A, Dimiev A, Corley D. Kinetics of diazonium functionalization of chemically converted graphene nanoribbons. ACS Nano 2010;4(4) 1949-1954.

[35] Fang M, Zhang Z, Li J. Constructing hierarchically structured interphases for strong and tough epoxy nanocomposites by amine-rich graphene surface. Journal of Materials Chemistry 2010;20(43) 9635-9643.

[36] He H, Gao C. General approach to individually dispersed, highly soluble, and conductive graphene nanosheets functionalized by nitrene chemistry. Chemistry of Materials 2010;22(17) 5054-5064.

[37] Hou S, Kasner M, Su S. Highly sensitive and selective dopamine biosensor fabricated with silianized graphene. Journal of Physical Chemistry C 2010;114(35) 14915-14921.

[38] Melucci M, Treossi E, Ortolani L. Facile covalent functionalization of graphene oxide using microwaves: bottom-up development of functional graphitic materials. Journal of Materials Chemistry 2010;20(41) 9052-9060.

[39] Xu Y, Liu Z, Zhang X. A graphene hybrid material covalently functionalized with porphyrin: synthesis and optical limiting property. Advanced Materials 2009;21(12) 1275-1279.

[40] Liu Y, Zhou J, Zhang X. Synthesis, characterization and optical limiting property of covalently oligothiophene-functionalized graphene material. Carbon 2009;47(13) 3113-3121.

[41] Zhu J, Li Y, Chen Y. Graphene oxide covalently functionalized with zinc phthalocyanine for broadband optical limiting. Carbon 2011;49(6) 1900-1905.

[42] Zhang B, Chen Y, Zhuang X. Poly(N-vinylcarbazole) chemically modified graphene oxide. Journal of Polymer Science Part A-Polymer Chemistry 2010;48(12) 2642-2649.

[43] Bao H, Pan Y, Ping Y. Chitosan-functionalized graphene oxide as a nanocarrier for drug and gene delivery. Small 2011;7(11) 1569-1578.

[44] Tang X, Li W, Yu Z. Enhanced thermal stability in graphene oxide covalently functionalized with 2-amino-4,6-didodecylamino-1,3,5-triazine. Carbon 2011;49(4) 1258-1265.

[45] Yu D, Yan Y, Durstock M. Soluble P3HT-grafted graphene for efficient bilayer-heterojunction photovoltaic devices. ACS Nano 2010;4(10) 5633-5640.

[46] Pham T, Kumar N, Jeong Y. Covalent functionalization of graphene oxide with polyglycerol and their use as templates for anchoring magnetic nanoparticles. Synthetic Metals 2010;160(17-18) 2028-2036.

[47] Shen J, Li N, Shi M. Covalent synthesis of organophilic chemically functionalized graphene sheets. Journal of Colloid and Interface Science 2010;348(2) 377-383.

[48] Pramoda K, Hussain H, Koh H. Covalent bonded polymer graphene nanocomposites. Journal of Polymer Science Part A-Polymer Chemistry 2010;48(19) 4262-4267.

[49] Zhang X, Jin J, Li S. Aryne cycloaddition: highly efficient chemical modification of graphene. Chemical Communications 2010;46(39) 7340-7342.

[50] Quintana M, Spyrou K, Grzelczak M. Functionalization of graphene via 1,3-dipolar cycloaddition. ACS Nano 2010;4(6) 3527-3533.

[51] Economopoulos S, Rotas G, Miyata Y. Exfoliation and chemical modification using microwave irradiation affording highly functionalized graphene. ACS Nano 2010;4(12) 7499-7507.

[52] Jafri R, Imran R, Ramaprabhu. Nitrogen doped graphene as catalyst support for oxygen reduction reaction in proton exchange membrane fuel cell. Journal of Materials Chemistry 2010;20(34) 7114-7117.

[53] Wang D, Gentle I, Lu G. Enhanced electrochemical sensitivity of PtRh electrodes coated with nitrogen-doped graphene. Electrochemistry Communication 2010;12(10) 1423-1437.

[54] Geng D, Hu Y, Li Y. One-pot solvothermal synthesis of doped graphene with the designed nitrogen type used as a Pt support for fuel cells. Electrochemistry Communication 2012;22 65-68.

[55] Liang Y, Wang H, Zhou J. Covalent hybrid of spinel manganese-cobalt oxide and graphene as advanced oxygen reduction electrocatalysts. Journal of the American Chemical Society 2012;134(7) 3517-3523.

[56] Qu L, Liu Y, Baek J. Nitrogen doped graphene as efficient metal-free electrocatalysts for oxygen reduction in fuel cell. ACS Nano 2010;4(3) 1321-1326.

[57] Geng D, Chen Y, Chen Y. High oxygen reduction activity and durability of nitrogen doped graphene. Energy & Environmental Science 2011;4(3) 760-764.

[58] Lee K, Lee K, Lee J. Electrochemical oxygen reduction on nitrogen doped graphene sheets in acid media. Electrochemistry Communications 2010;12(8) 1052-1055.

[59] Lee Y, Lee Y, Chang K. Synthesis of N-doped carbon nanosheets from collagen for electrochemical energy storage/conversion. Electrochemistry Communications 2011;13(1) 50-53.

[60] Luo Z, Lim S, Tian Z. Pyridinic N doped graphene: synthesis, electronic structure, and electrocatalytic property. Journal of Materials Chemistry 2011;21(22) 8038-8044.

[61] Yang S, Zhi L, Tang K. Efficient synthesis of heteroatom (N or S)-doped graphene based on ultrathin graphene oxide-porous silica sheets for oxygen reduction reactions. Advanced Functional Materials 2012;22(17)3634-3640.

[62] Reddy A, Srivastava A, Gowda S. Synthesis of nitrogen doped graphene films for lithium ion batteries. ACS Nano 2010;4(11) 6337-6342.

[63] Cho Y, Kim H, Im H. Nitrogen-doped graphitic layers deposited on silicon nanowires for efficient lithium ion battery anodes. Journal of Physical Chemistry C 2011;115(19) 9451-9457.

[64] Wang H, Zhang C, Liu Z. Nitrogen-doped graphene nanosheets with excellent lithium storage properties. Journal of Materials Chemistry 2011;21(14) 5430-5434.

[65] Li Y, Wang J, Li X. Nitrogen-doped graphene nanosheets as cathode materials with excellent electrocatalytic activity for high capacity lithium-oxygen batteries. Electrochemistry Communications 2012;18 12-15.

[66] Wang X, Cao X, Bourgeois L. N-doped graphene-SnO_2 sandwich paper for high-performance lithium-ion batteries. Advanced Functional Materials 2012;22(13) 2682-2690.

[67] Li Y, Wang J, Li X. Discharge product morphology and increased charge performance of lithium-oxygen batteries with graphene nanosheet electrodes: the effect of sulphur doping. Journal of Materials Chemistry 2012;22(38) 20170-20174.

[68] Jeong H, Lee J, Shin W. Nitrogen-doped graphene for high-performance ultracapacitors and the importance of nitrogen-doped sites at basal planes. Nano Letters 2011;11(6) 2472-2477.

[69] Wen Z, Wang X, Mao S. Crumpled nitrogen-doped graphene nanosheets with ultrahigh pore volume for high-performance supercapacitor. Advanced Materials 2012;24(41) 5610-5616.

[70] Shen Z, Gao H, Bao W. Synthesis of boron doped graphene for oxygen reduction reaction in fuel cells. Journal of Materials Chemistry 2012;22(2) 390-395.

Chapter 8

Graphene Oxide Based Surface Plasmon Resonance Biosensors

Nan-Fu Chiu, Teng-Yi Huang and Hsin-Chih Lai

Additional information is available at the end of the chapter

http://dx.doi.org/10.5772/56221

1. Introduction

Graphene oxide (GO), an amorphous insulatormaterial, has consists of a hexagonal ring based carbon network having both sp^2-hybridized carbon atoms and sp^3-hybridizedcarbons bearing hydroxyl andepoxide functional groups on either side of the sheet, whereas the sheet edges are mostly decorated by carboxyl and carbonyl groups [1-6]. These unique properties hold great promise for potential applications in many technological aspects such as nanoelectronics [7-10], nanophotonics [11-16], and bio-sensors [9, 17-21], and nanocomposites [22, 23].

This chapter is intended to demonstrate a facile and effective method to construct GO sheets onto the self-assembled monolayer (SAM) at Au films based surface plasmon resonance (SPR) technique for proteins immunization and deoxyribonucleic acid (DNA) detection. The amine functionalized grapheme oxide has demonstrated various applications. Given the high number of functional groups in the GO, such as -OH, -COOH, and epoxides, compared to those in Au electrode. The oxygen functional groups on graphene oxide surfaces, including carboxylic acid groups at the edge and epoxy/hydroxyl groups on the basal plane can be utilized to change the surface functionality of graphene oxide.

We introduced a novel and simple methods for fabricating high-sensitivity, high-resolution GO based SPR biosensors that provide high accuracy and precision over relevant ranges of analyte measurement. We used SPR technique todetect the binding phenomenonbetween proteins and GO films. SPR is a surface-sensitive optical technique that very suitable for monitoring of bio-molecular interactions occurring in very close vicinity to sensor surfaces. It allows real-time and label-free detection of analysis by exploiting the interfacial refractive index changes associated with any affinity binding interaction between a biomolecule immobilized on a sensor surface and its biospecific partner in solution. Rapid and real-time analysis without labeling, highly specific detection with extremely low detection limits creating the unique potential for the application of SPR in biosensor. Owing to these powerful advantages of GO-based SPR technique, many applications of GO-based biosensors in such areas as molecular engineering, biomedical diagnostics, drug discovery, environmental monitoring and food analysis.In this chapter, we report progresses in the following areas:

- Characterization of functionalized graphene oxide
- Characterization optical of graphene oxide sheets
- Graphene oxide based SPR biosensorstechnological
- Graphene oxide layer effect
- Biomolecular detection

This chapter addresses the preparation, characterization and potential biosensing applications of graphene oxide-based SPR, which have beenparticularly relevant in our research group at bio-plasmonic. Special attention will be dedicated to GO precursor as a building block for the preparation of these nanocomposites. Interesting and promising applications for these materials are also discussed.

2. Characterization of functionalized graphene oxide

The oxidized counterpart of graphene, functionalized graphene sheets, which is also called graphene oxide (GO) sheets as shown at Fig. 1(a), usually has abundant functional groups which are advantageous for biosensor applications [9, 17-21]. It has been used as a platform for the detection of proteins [24] and DNA [25] by utilizing its good water dispensability, and versatile surface modification. There have many method to evidencing the presence of –COOH, –OH, and C=O groups at the edge of the GO sheet, like infrared spectroscopic, extensive nuclear magnetic resonance, and electron diffraction.The basal plane of GO is covered with mostly epoxide and –COOH groups as shown at Fig. 1(b). Semi-quantitative study of the EDC/NHS activation of acid terminal groups at modified graphene oxide surfaces. Thus, we use GO as a medium for the covalent binding reaction with proteins. It is expected that forming the novel GO sheets will taking advantages of the striking properties of GO [26-28]. Therefore, we can obtain hybrid material manipulation and new electrochemical properties.

2.1. Characterization optical of graphene oxidesheets

The GO is synthesized by the oxidation of graphene, and its unique atomic and electronic structure has been revealed toshow small sp^2 carbon clusters that are isolated by the sp^3 matrix [29, 30]. GOsheets bearing carboxyl groups (-COOH) located atthe sheet edges andhydroxyl (-OH) located at the sheet surface, these functional groups makes GO sheets have strongly hydrophilic, which allows GOto readily swell and disperse in water[31].Unlike other conventional materials, the valence band and conduction band in graphene are smooth-sided cones that meet at a point, called the Dirac point, and graphene has no energy band-gap and readily

http://dx.doi.org/10.5772/56221

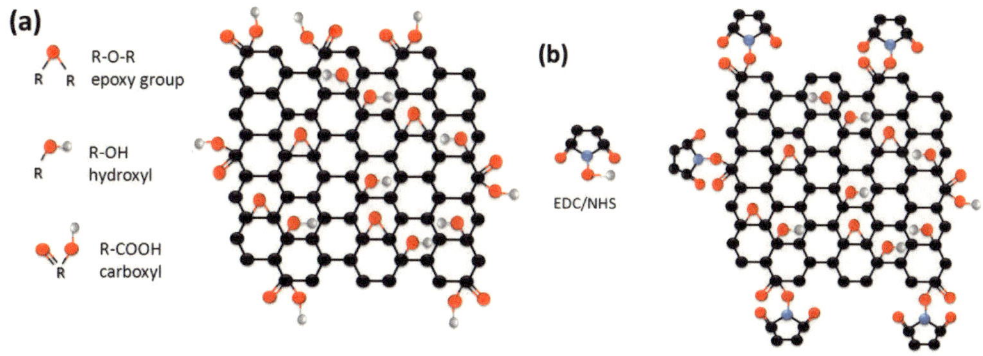

Figure 1. (a) graphene oxide structure, (b) chemical modificationamine-reactive EDC/NHS esters can be created on any carboxyl-containing molecule.

absorbs allphotons at any wavelength[32].Due to the adsorption of oxygen inthe form of epoxy, carboxyl, hydroxyl, andether groups on graphemeoxide surface not only opens up the band gap, but also provides anoption to tune the electronic, optical, andmechanical properties by means of controlled oxidation [15].

The nonlinear optical properties andultrafast relaxation dynamics shows that the unique atomic and electronic structure of GO leads to the emergence of variable absorption processes under different input intensities, and the different absorption processes are distributed between the sp^2 clusters and the sp^3 domains of GO [33]. Consequently, the nonlinear optical properties of GO are determined by the combined action of the sp^2 and the sp^3domains. Owingto theunique and large two-dimensional (2D) π-electron conjugation systems of graphene, much like that in zero-dimensional (0D) fullerene and one-dimensional (1D) carbon nanotubes (CNTs), we expect that GO would exhibit good nonlinear optical properties. Finally, Zhibo Liu et al. [34] had demonstrated that graphene oxide has a large two photon absorption coefficient.

Thus, we used ITO as substrate, attaching with graphene oxide by APTES, forming GO sheets (0.275 mg/ml), and get multilayer GO sheets through layer by layer (LBL) technique [35]. In Fig. 2., we observed that the relative transmittance fall down when GO layers rises. The result here was corresponded to Liu's [34] theory. The GO sheets that we used hasmany advantages for optical sensing of biomolecules detection which compared with conventional immune interactions.These advantages arise from the unique physical and chemical structure of GO.

2.2. Graphene oxide based SPR biosensors technological

The surface plasmon resonance (SPR) behavior of free electrons or plasma at the interface of a metal-dielectric material has been widely studied [36–43]. Prism coupling between Kretschmann and Otto configurations has been used extensively to study the optical properties of metallic thin films, including refractive index (n), extinction coefficient (k), thickness (d), and roughness [44, 45]. Conventionally, SPR biosensors are used in biochemistry and biology to detect molecular concentration, thickness, and specific chemistry analytes. In biochemistry,

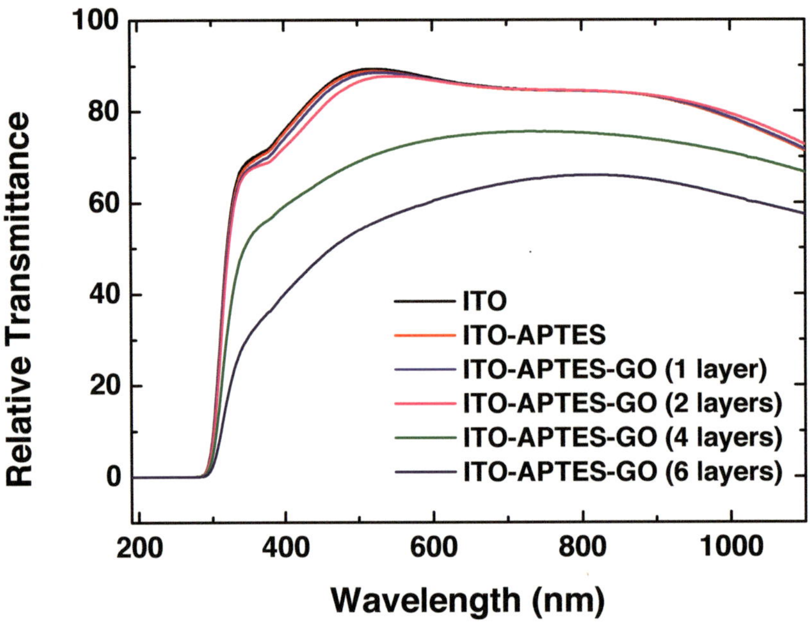

Figure 2. Relative Transmittance (%T) of ITO film, ITO-APTES film, ITO-APTES-GO (0.275 mg/ml) 1 layer, 2 layers, 4 layers, and 6 layers, respectively.

analyte concentration is determined from the SPR angle shift by a biosensor operating in the angular interrogation mode. The shift or difference between the initial and final values of the SPR angles provides a quantitative measurement of the analyte concentration.

The SPR is another interesting field of research for rising applications of GO. SPR occurs when radiant energy is coupled or transferred to electrons in a metal film. The wavelength of light at which coupled occurs depends on the characteristics of the metal that is illuminated and the optical properties of the surrounding environment. When there is a match or resonance between the energy and wavenumber of the light photons and the electrons at the metal surface, a transfer of energy occurs.

The electrons charge on a metal boundary can perform coherent fluctuations which are called surface plasmon oscillations. Their existence has been demonstrated in plasma losses by fast electrons at thin films experiments.The volume and surface plasma losses ofmaterial can be identified in reflection experiments by varying the primary electron beamenergy and observing the intensity variation of the energy loss peaks at a fixed scattering angle. The frequency $\omega$of these longitudinal oscillations is tied to its wave vector k_x by a dispersion relation $e(k_x)$. These charge fluctuations, which can be localized in the Zdirection within the Thomas-Fermi screening length of about 1Å, are accompanied by a mixed transversal and longitudinal electromagnetic field which disappears at $|Z| \rightarrow \infty$, Figure 3(a), and has its maximum in the

http://dx.doi.org/10.5772/56221

surface Z=0, typical for surface waves. This explains their sensitivity to surface properties. Maxwell's equations yield the retarded dispersion relation for the plane surface of a semi-infinite metal $(\varepsilon_m = \varepsilon_m^{'} + i\varepsilon_m^{''})$, adjacent to a medium dielectric , Fig. 3(b), has an interface metal $(\varepsilon_m = \varepsilon_d^{'} + i\varepsilon_d^{''})$/dielectric (ε_m), on which electromagnetic fields propagates in the x direction.The coupling of light into a metal surface produces a plasmon. The plasmon, in turn, generates an electro-magnetic field that typically extends on the order of about 100 nm above and below the metal surface and oscillates with optical frequencies [36, 46-48].

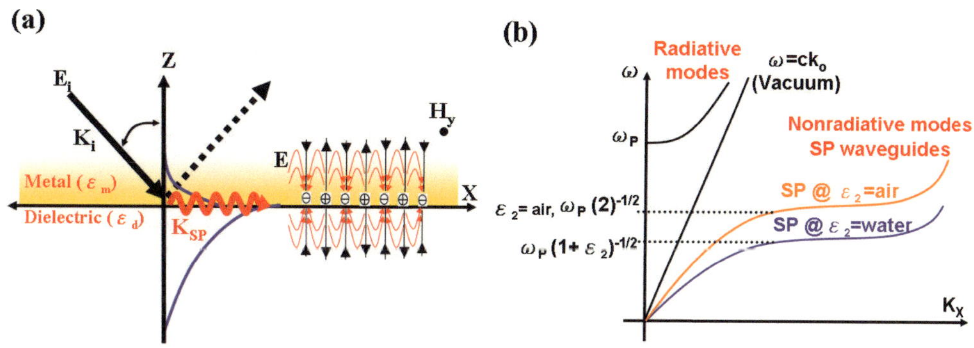

Figure 3. Schematic presentation of surface plasmons is constituted of resonantly oscillating surface charges at a metal/dielectric interface and of the electromagnetic surface wave that originates from these surface charges. (a) Electric field lines of a surface plasmon on a smooth surface and the electromagnetic field of SPs propagating on a surface in the x direction. (b) The dispersion relation of non-radiative SP modes and radiative modes.

The most common approach to excitation of surface plasmons is by means of a prism coupler and the attenuated total reflection method (ATR). There are two configurations of the ATR method-Otto geometry [36, 37] and Kretschmann geometry [36, 38]. If light wave is reflected at a metal surface covered with a dielectric medium (ε_0>1) e.g., with a quartz prism, its momentum becomes (ε_d) instead of $(\hbar\omega / c)\sqrt{\varepsilon_p}$ and its projection on the surface as shown in Fig. 4(a). The plasmonic dispersion relation for SPs propagating on the interface $(\hbar\omega / c)$ can thus be satisfied between the lines c, beyond which limit total reflection at an interface $\mathbf{E_d} / \mathbf{E_m}$ takes place.

The grapheme oxide can provide new perspectives in SPR ways: (1) as a surface which directly supports surface plasmons at visible range, (2) as a tunable propagation and excitation platform whose optical properties can be tuned by an external electromagnetic field, and (3) as an oxygen-functional groups coating for the existing plasmonic devices. These oxygen-containing functional groups on graphene oxide not only opens up the band gap, but also provides an option to tune the electronic, optical, and mechanical properties by means of controlled oxidation [15, 16, 49-51].

The formation scheme of gold surface multilayer modification system for site-oriented immobilization of GO sheets, GO is assembled on a gold film by a layer-by-layer technique using a self-assembled linker monolayer. The GO modified electrode is represented as Au/

linker/GO. Protein contains both amine group (–NH2) and carboxylic acid group (–COOH). It is evident that GO possess a number of carboxylic acid group (–COOH) that can easily bind with the free –NH2 terminals of the protein to result in a strong amide covalent linkage. Figure 4(b) shown schematic representation of the sensing configuration for the GO based SPR immunoassay.

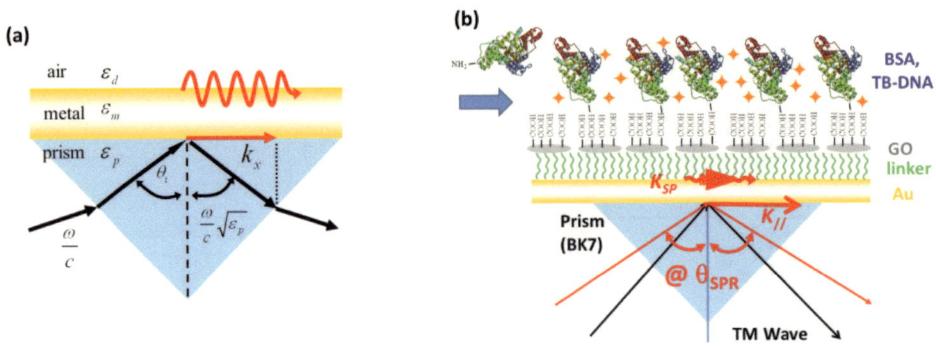

Figure 4. Excitation of surface plasmon in the (a) Kretschmann geometry and (b) functionalized graphene oxide based SPR biosensing mechanisms.

When a light wave propagating in the prism is made incident on the metal film a part of the light is reflected back into the prism and a part propagates in the metal in the form of an inhomogeneous electromagnetic wave [36-38]. This inhomogeneous wave decays exponentially in the direction perpendicular to the prism-metal interface and is therefore referred as to an evanescent wave. If the metal film is sufficiently thin (less than ~50nm), the evanescent wave penetrates through the metal film and couples with a surface plasmon at the outer boundary of the film [36]. In order for the coupling between the evanescent wave and the surface plasmon to occur, the propagation constant of the evanescent wave and that of the surface plasmon have to be equal (1) and (2):

$$\mathbf{E_p / E_m} \tag{1}$$

$$k_x = \sqrt{\varepsilon_p}\frac{\omega}{c}\sin\theta_i \tag{2}$$

The dispersion relation of free photos in a dielectric $k_{sp} = \frac{\omega}{c}\sqrt{\frac{\varepsilon_d \varepsilon_m}{\varepsilon_d + \varepsilon_m}}$, and in a coupling prism $c / \sin\theta_i$, compared to the dispersion relation for non-radiative surface plasmons at the metal/dielectric interface as shown in Fig. 5.

http://dx.doi.org/10.5772/56221

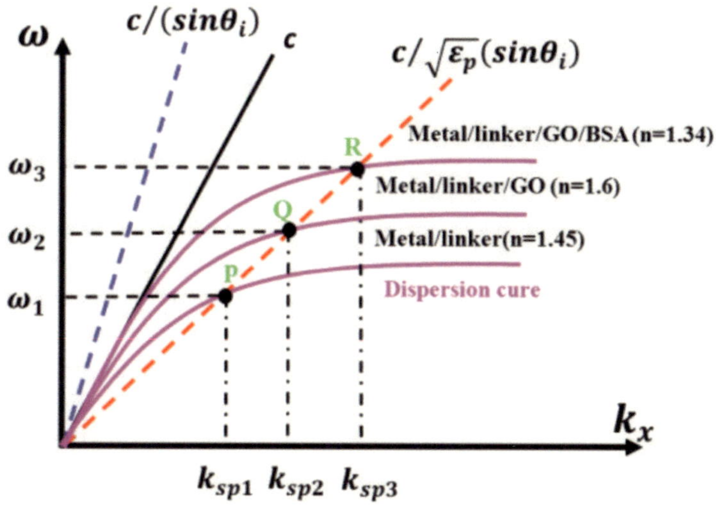

Figure 5. The ATR method: Dispersion relation of SPs for a prism/metal/air system $c / \sqrt{\varepsilon_p} sin\theta_i$, c is light line in vacuum, $\varepsilon_d = 1$ is light line in the medium $c / \sqrt{\varepsilon_p}$. Since the light line ε_p lies to the right of the dispersion relation up to a certain $c / \sqrt{\varepsilon_p}$, light can excite SPs of frequencies k_x below the crossing point P, Q and R on the metal and dielectric interface.

To excite the SPR, the layer needs to be thicker than this, but still thin enough for the field to sample the dielectric half-space. For the appropriate metal film thickness the SPR can be excited and the characteristic dip in the reflectance angle scan appears. This phenomenon is illustrated in Fig. 6, which shows reflectance dips produced by the excitation of surface plasmons at the different angle.

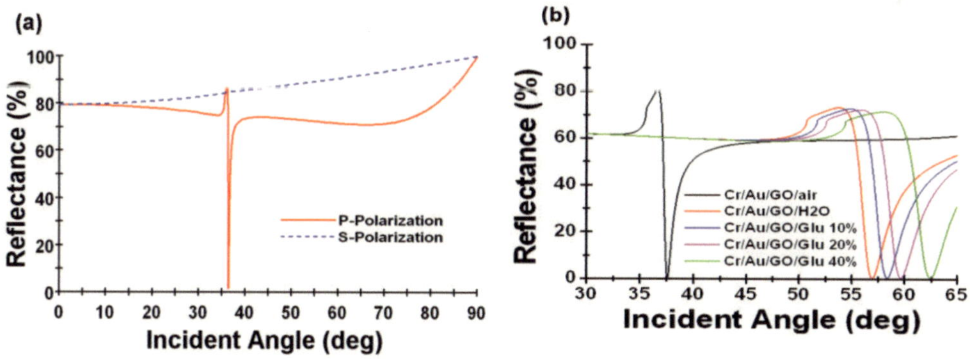

Figure 6. SPR curves with prism coupling in the Kretschmann configuration. (a) It demonstrates that the p-polarization illumination has higher sensitivity than s-polarization. (b) TM reflectance as a function of angle of incidence using the rigorous Fresnal reflection theory. Configuration: SF10 glass/Cr/Au/GO and five different dielectric layers air, Water, glucose 10%, glucose 20% and glucose 40%, respectively.

2.3. Graphene oxide layer effect

It is a novel Au-SAM-GO sheets and taking advantages of the striking properties of both grapheme oxide and Au film, fundamental understanding in hybrid material manipulation and new biosensors properties can be obtained.The sensitivity of proteins immunization detection was determined by GO sheets and compared with that of a conventional Au films. We used bovine serum albumin (BSA) to binding with GO sheets and conventional Au films, respectively. As the meanwhile, we also real-time detect the binding phenomenon by SPR technique. The results shows that GO sheets for SPR biosensor offers a potentially powerful assay, with a highly sensitive analysis, which can reach to 100 of pg/ml of BSA. However, the conventional Au films can just reach to 10μg/ml of BSA. Therefore, the results described herein that GO sheets are a promising approach towards highly sensitive for proteins immunization detection.The sensitivity of TB detection in the DNA-based assay for the amplification of the Insertion Sequence 6110 (IS6110) samples was determined by a GO/Au thin film andcompared with that of a conventional Au/Cr chips. The results show that a GO/Au SPR offers a potentially powerful assay, with a highly sensitive analysis, that may be applicable as an important tool for bio-marker detection.

The GO solution was purchased from GRAPHENE SUPERMARKET. 1-Ethyl-3-(3-dimethylaminopropyl) carbodiimide hydrochloride(EDC), bovine serum albumin(BSA) and 8-Mercaptooctanoic acid($C_8H_{16}O_2S$; MOA)were purchased from Sigma-Aldrich (USA). Cystaminedihydrochloride (Cys, 97+%), 1-Octadecanethiol ($C_{18}H_{37}SH$; ODT) and N-hydroxysuccinimide (NHS)were purchased from Alfa Aesar (USA). 10X Phosphate-buffered saline(PBS) solution was purchased from UniRegion Bio-Tech (Taiwan).

a. Preparation of conventional Au film

Standard glass microscope slides (18 × 18 mm; High Precision Cover Glass, NO. 1.5H) purchased from Superior Marienfeld (Germany), were used as base substrates.To remove any contaminants, the glass substrates were subjected to ultrasonic cleaningby acetone, IPA, anddeionized water for 5 min before deposition, respectively. The slides were finally dried with nitrogen gas. The thin film grown on the glass substrates was shown in Fig.7. Cr and Au films were deposited by an electron beam evaporator at a vacuum level of about 3 × 10^{-6} Torr.

Au	47 nm
Cr	2 nm
Glass (n=1.52)	170 μm

Figure 7. Schematic diagram of the thin film structure (not drawn to scale) in Au/Cr

http://dx.doi.org/10.5772/56221

b. Surface functionalization of conventional Au film (Au-MOA)

Prior tosurface functionalization, the gold-coated glass slide was washed with acetone, IPA, anddeionized water for 3 min, respectively. After that,the gold-coated glass was rinsed with deionized water with sonication. The assay developed was designed as a bio-affinity immobilization assay. Finally, the gold-coated slide was immersed into 1mM MOA solution at room temperature for 24 h to form a self-assembled monolayer on the surface as shown in Fig. 8(a), we called conventional Au film.

c. Surface functionalization of GO sheets (Au-ODT-GO)

Prior tosurface functionalization, the gold-coated glass slide was washed with acetone, IPA, and deionized water for 3 min, respectively. After that,the gold-coated glass was rinsed with deionized water with sonication. The assay developed was designed as a bioaffinity immobilization assay. Finally, the gold-coated slide was immersed into an10mM 1-Octadecanethiol (ODT) solution at room temperature for 24 h to form a self-assembled monolayer on the surface. During the process, organosilane SAMs was formed on Au, we called Au-ODT film. After the Au-ODT substrate was rinsed with ethanol and dried with $N_{2,}$ the substrate was immersed into a GO solution for 5 h in order to adsorbing GO on the Au-ODT surface, forming GO sheets as shown in Fig. 8(b).

d. Preparation of GO sheets

It is well known that $C_{18}H_{37}SH$ (ODT) can form a stable SAM on Au substrate through the formation of Au-S bond. As shown in Fig. 8(c), Au films modified with SAM of ODT were prepared by immersing the films into ethanol solution of ODT (10 mM) for 24 h at room temperature. The SAM films were then thoroughly rinsed with ethanol and dried with pure N_2 [52], forming GO sheets.

For adsorption of the GO onto the SAM films, the SAM films were immersed into GO solution, 0.275 mg/ml *vs.* 2 mg/ml for 5 h. The concentration of GO was adjusted to achieve films with different surface coverage of GO. The films (denoted as GO sheets) were thoroughly rinsed with deionized water to remove GO which unstable adsorbed onto the Au film surface and then dried with pure N_2 before use.

e. Optical characterization of GO sheets

Figure 9 shows digital microscopy image of bare Au film, Au-ODT film and different concentration of GO sheets. In Fig. 9(a) and 9(b), we cannot observe anything at the surface of the Au film. However, in Fig. 9(c) and 9(d), we observed that GO sheet (2 mg/ml) has higher surface density compared with lower concentration of GO sheet (0.275 mg/ml).

Fig. 10 shows relative transmittance spectra of bare Au film, Au-ODT film and different concentration of GO sheets. The relative transmittance at 670 nm decreased following the formation of Au-ODT film, the relative transmittance of GO sheets also lower than Au-ODT film. Fig. 10(a) shows relative transmittance spectra of bare Au film, Au-ODT film and different concentration of GO sheets. The value of transmittance decreased following the formation of Au-ODT film, the relative transmittance of GO sheets also lower than Au-ODT film. It shows that higher concentration of GO sheet (2 mg/ml) have more opportunities to covalent binding with the following biomolecules and improved the detection limit.

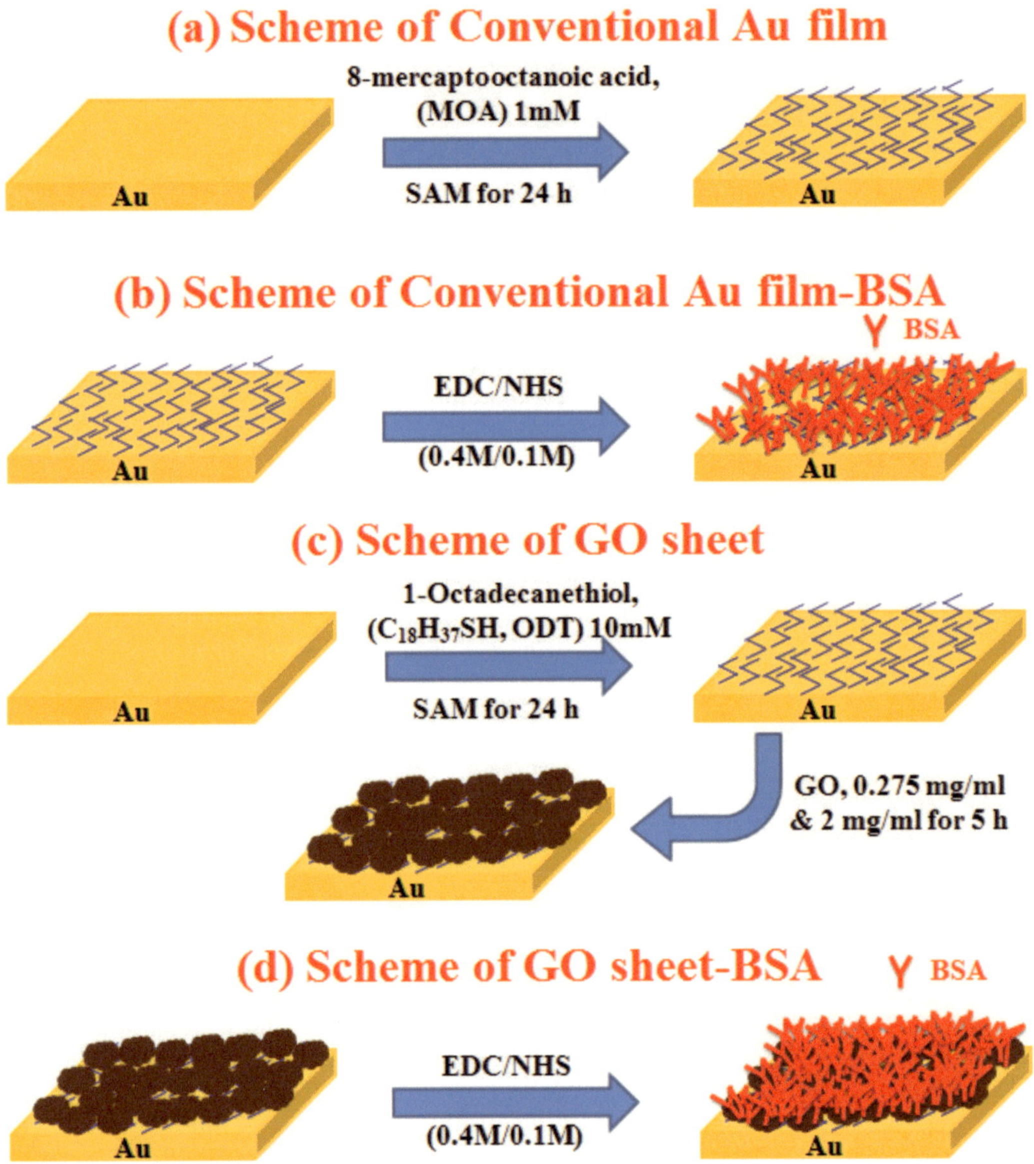

Figure 8. Scheme of (a) conventional Au film, (b) conventional Au film-BSA, (c) GO sheets (0.275 mg/ml & 2 mg/ml) and (d) the GO sheets with BSA for the preparation of GO sheets-BSA conjugate through the activated functional groups at the surface of GO sheets.

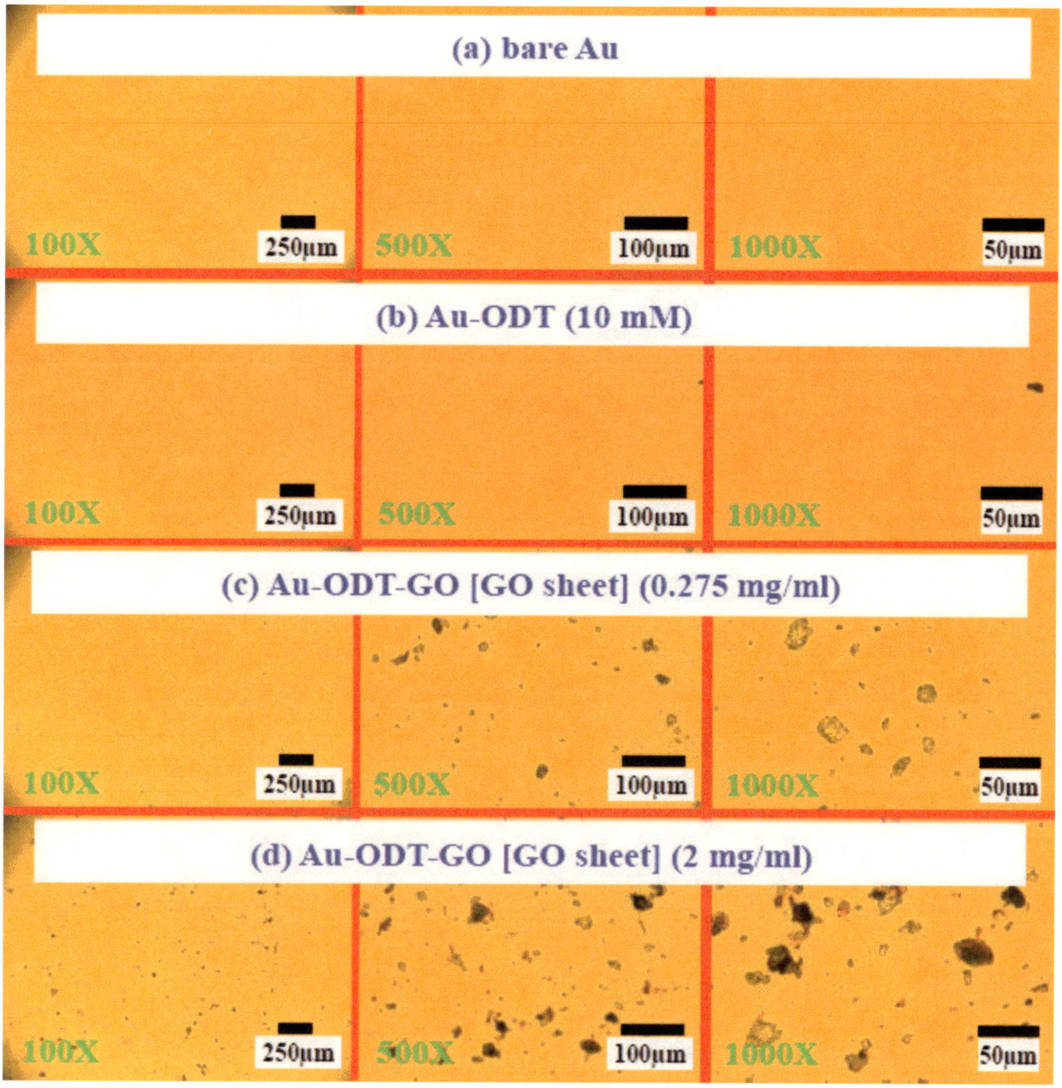

Figure 9. Digital microscopy image of (a) bare Au film, (b) Au-ODT (10 mM) film, (c) GO sheet (0.275 mg/ml) and (d) GO sheet (2 mg/ml).

As shown in Fig. 10(b), we used different films to observe the normalized reflectance intensity and SPR angle. The results show that when the concentration of GO rise, both intensity and SPR angle were risen, too.It demonstrated that GO solutions were successfullyadsorbed on Au-ODT films surface.

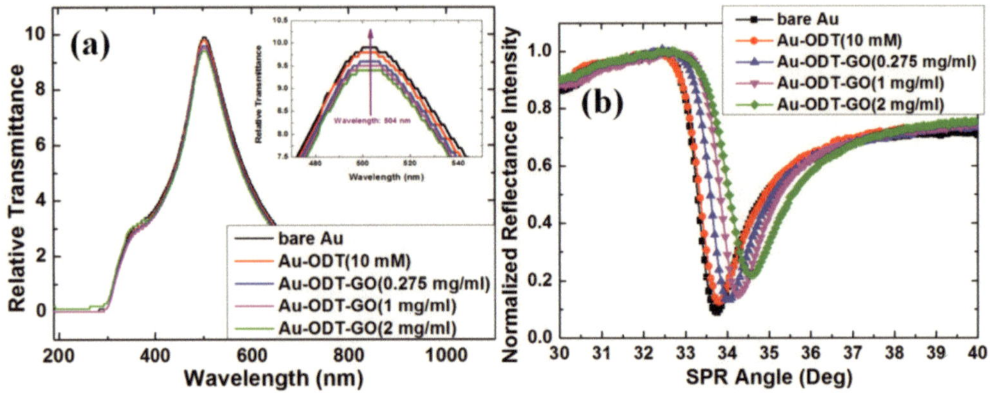

Figure 10. (a) Relative transmittance spectra of bare Au film, Au-ODT film and GO sheets (0.275 mg/ml, 1 mg/ml, and 2 mg/ml) from 190 nm to 1100 nm(Inset: The relative transmittance at 504 nm); (b) Normalized reflectance intensity of bare Au film, Au-ODT film and GO sheets (0.275 mg/ml, 1 mg/ml and 2 mg/ml).

f. Instruments and measurement

The transmission spectra were recorded using a Hitachi U-2900 Ultraviolet-vis (UV-vis) spectrophotometer (Hitachi High-Tech worldwide, Japan) with wavelength in range of 190-1100 nm. The surface plasmon resonance (SPR) measurements were carried out using BI-3000 (Biosensing Instrument, Tempe, AZ) with Kretschmann prism-coupling. These sample devices were prepared to measure the intensities of reflected light at a fixed angle. We are able to provide a real-time detection of immune interaction. The change in reflective intensity wasrecorded as the difference in the lowest point of the SPR reflectivity curve.

2.4. Biomoleculardetection

2.4.1.Detection of proteins immunization using a graphene oxide sheets

Here, we report the GO sheets with BSA for the preparation of GO sheets-BSA conjugate through the activated functional groups at the surface of GO sheets, and real-time observing the binding phenomenon between GO sheets and BSA by SPR technique as shown in Fig. 8(d). When we analysis various type of proteins, serum albumins are the major soluble protein constituents of the circulatory system and have many physiological functions, moreover, it can also serving as transporters for a variety of compounds. Especially, bovine serum albumin (BSA) has been one of the most extensively studied proteins in this group of proteins, because of its structural homology with human serum albumin (HSA) [53].

The whole preparation process contains two steps: firstly, carboxylic acid groups on GO sheets are activated by N-(3-Dimethylaminopropyl)-N'-ethylcarbodiimide hydrochloride (EDC). Stable active ester is formed in the presence of N-hydroxysuccinimide (NHS). Secondly, the activation surface of GO sheets reacted with the amine groups of BSA, forming a strongly covalent bond. This preparation process will cause no denaturing of the protein species and guarantee the uniform attachment of BSA on GO sheets. In addition, the present method is promising to be used for bridging other biological systems to GO sheets-based materials.

a. Principle of proteinbinding with GO sheets

It is know that there were abounding epoxy groups at the edge of the GO surface, like carboxylic acid groups and hydroxyl groups [26, 27]. BSA proteins used in this study contains 60 amino moieties in lysine residues and 26 arginine moieties in guanidine side chains and 17 disulfide bonds with one free thiol in cysteine residues [54]. One of the universal methods for connecting proteins to GO sheets surface is covalent binding. Activating agents as EDC/NHS was usually used. After that, the amine groups (-NH_2) of BSA and the carboxylic acid groups (-COOH) of GO sheets on their surface will forming a strongly covalent bond.

b. GO sheets for protein biosensing by SPR technique

Figure 11 shows the SPR angle of conventional Au film and GO sheets (0.275 mg/ml *vs.* 2 mg/ml) towards different concentration of BSA. Compared with conventional Au film, GO sheets had higher sensitivity and lower detection limit.

The high BSA loading is attributed to the large surface area of GO and oxygen rich functional groups on its surface. Thus, due to the higher surface density, GO sheet (2 mg/ml) is much sensitive than GO sheet (0.275 mg/ml). The BSA detection limit of conventional Au film is 10 μg/ml of BSA. However, the GO sheet (2 mg/ml) can reach to 100 pg/ml of BSA.

Figure 12 shows SPR Angle of conventional Au film and GO sheets (0.275 mg/ml, 1 mg/ml and2 mg/ml) towards different concentration of BSA. Compared with conventional Au film, GO sheets have higher sensitivity and lower detection limit.The detection limit of conventional Au film is 1μg/ml.However, the GO sheet (2 mg/ml) can reach to 100 pg/ml of BSA.The high BSA loading is attributed to the large surface area of GO and functional groups on its surface. Furthermore, owing to the high surface density, the sensitivity of GO sheet (2 mg/ml) was superior to GO sheet (0.275 mg/ml).

c. Kinetic Analysis

Figure13 shows kinetic analysis for conventional Au film and GO sheets. Compared with conventional Au film, GO sheets have higher association constant (K_A) and lower dissociation constant (K_D), it demonstrated thatGO sheets weremuch more sensitive than conventional Au films [55]. The value of K_A also related to the concentration of GO sheets.

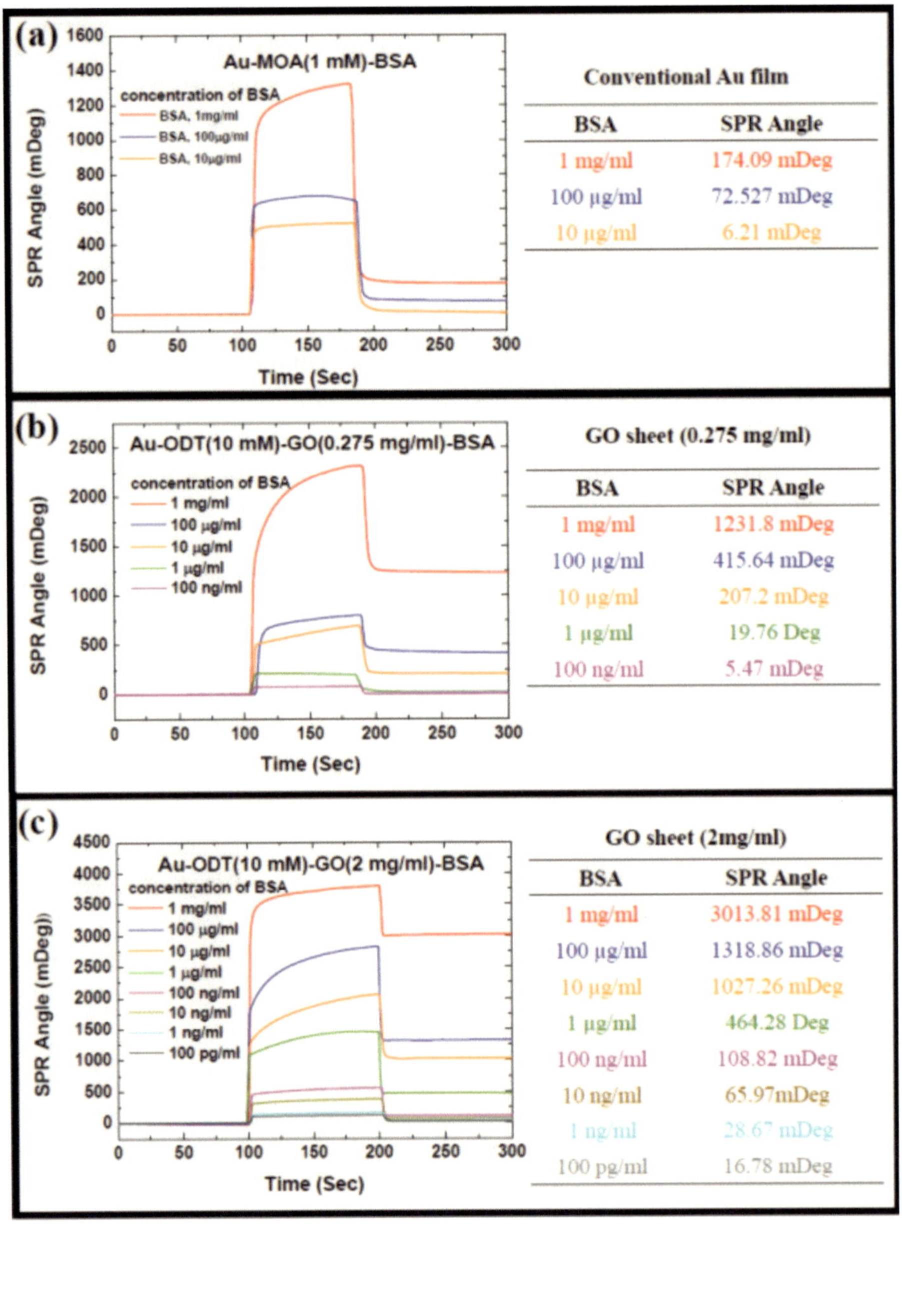

Figure 11. SPR Angle of (a)conventional Au film, (b) GO sheet (0.275 mg/ml) and (c) GO sheet (2 mg/ml) with various concentration of BSA. It shows high sensitivity of GO sheets SPR biosensor.

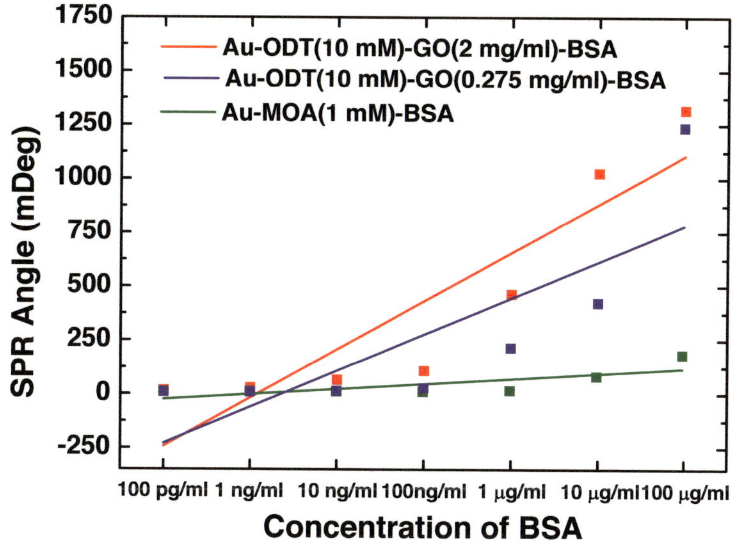

Figure 12. SPR Angle of conventional Au film (Au-MOA) and GO sheets with different concentration of BSA.

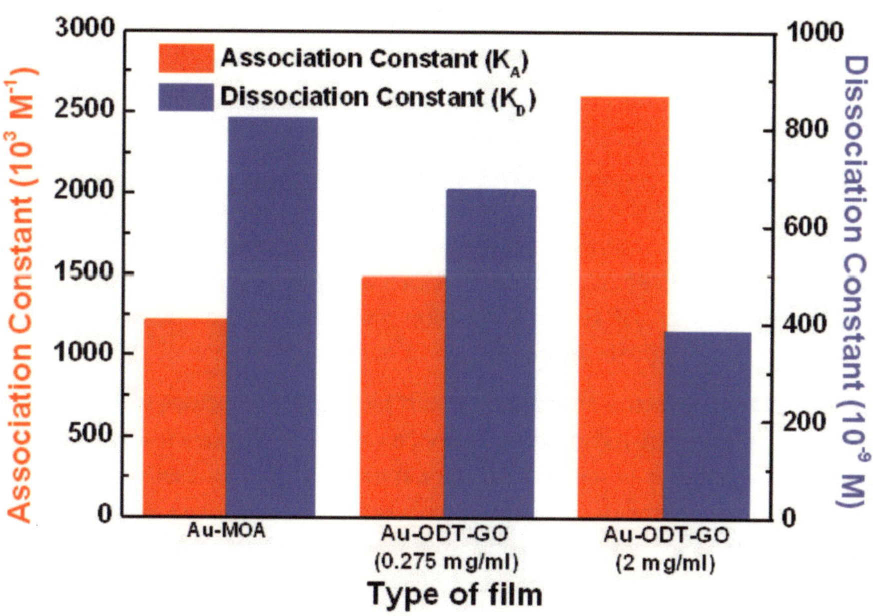

Figure 13. Kinetic analysis of conventional Au film and GO sheets.

2.4.2. Detection of Deoxyribonucleic Acid (DNA) using a graphene oxide sheets

Human tuberculosis bacillus (TB) is a highly contagious bacterial infection that is passed from person to person through droplets in the air. Infection is caused by the bacteria multiplying inside the body, causing tissue and organ damage. The infection of TB-virus often is more difficult to diagnose since the patient does not have the normal signs and symptoms associated with pulmonary TB. Therefore, development of a rapid diagnosis of the viral agent will be important for prevention of TB-viral infection and spreading.According to the estimation of WHO [56], in 2010, there were 8.8 million (range, 8.5–9.2 million)incident cases of TB, 1.1 million (range, 0.9–1.2 million)deaths from TB among HIV-negative people and an additional 0.35 million (range, 0.32–0.39 million) deaths from HIV-associated TB.Between 1990 and2010, prevalence rates were halved, mortality rates fell by almost 80% and TB incidence rates fell by 3.4% per year.

Development the rapid and efficient diagnosis of infectious pulmonary TB is one of the crucial issues in the global fight to control this disease [57, 58]. TB is one of the most important chronic infectious diseases that cause millions of deaths annually.Traditional TB diagnostic techniques include smear microscopy and Mycobacterial culture and nucleic acid diagnosis. However these techniques are not suitablefor mass screening for some disadvantages: smear microscopyis low sensitivity (105 bacteria/ml), Mycobacterial culture need 30day bacteria culture, and nucleic acid diagnosis is too expensive [53, 57-60]. In addition, the PCR-based [56, 57] assays are among the most promising and attractive methods compared to the routine methods. However, the long reaction time of traditional PCRamplification and the high costof thermal cycler are still major issues for such a screening application.Therefore, it is important to develop a rapid and accurate diagnostic device.

In 2000, Notormi et al. [57] reported a technique for the amplification of nucleic acid has been described, named loop-mediated isothermal amplification (LAMP). The LAMP is advantages as compared to the conventional PCR-based methods.In the LAMP methodare labeling the fluorescent primers method, the so called "Fluo-LAMP" method. It needs using UV irradiation requirements for Loop-amp fluorescent detection reagent. Due to the advantage of Fluo-LAMP reaction are efficient amplication of the target DNA and proceeding at constant temperature, we want integrated SPR-LAMP (surface plasmon resonance) technique to develop a without fluorescent primers and high-sensitivitydetection devices.

We demonstrate the interfacing of graphene with biological molecules to build a novel SPR-LAMP device and a TB DNA-hybridization device with excellent sensitivity. It is evident that graphene possess a number of reactive functional groups that can easily bind with the free-NH2 terminals of the enzyme or protein to result in a strong amide covalent linkage [61-65]. As a result, in this work, we employ graphene as an electrode material for glucose biosensing. SPR biosensors based on immobilize biomolecules are suitablefor a highly specific, sensitive, and rapid analysis of various biological species in vivo and in vitro.

The synthesis and investigation of the graphene oxide (GO) material properties of these stiff materials have been undertaken, although little has been done to explore GO and reduction

http://dx.doi.org/10.5772/56221

graphene oxide (rGO) in the analysis ofbiomolecules. Herein us studies the ability of water-soluble GO and rGO as a biochip for the sensitiveand high selective detection of TB-DNA.A new SPR-LAMP for TBdevice based on a GO material, which allowed immobilizing gold (Au) film onto the TB DNA antigens detection, was constructed. Figure 14 shows a schematic of a SPR-LAMP chips using graphene sheets with assay DNA format. Most important of all, we combined with the surface plasmon resonance to real-time monitoring the thickness of single layer GO.Surface roughness and specificity of the SPR sensor systems were also evaluated.This dispersion also facilitated the intimate mixing of the graphene oxide sheets with matrix polymers, providing a novel synthesis route to make GO-DNA nanocomposites. In the development of GO-DNA based biosensors, self-assembled monolayers (SAMs) or ordered structures of thiolated DNA on a gold surface can be achieved via use of spacer molecules between DNA and the thiol moiety, since DNA is known to form a disordered formless globular structure.

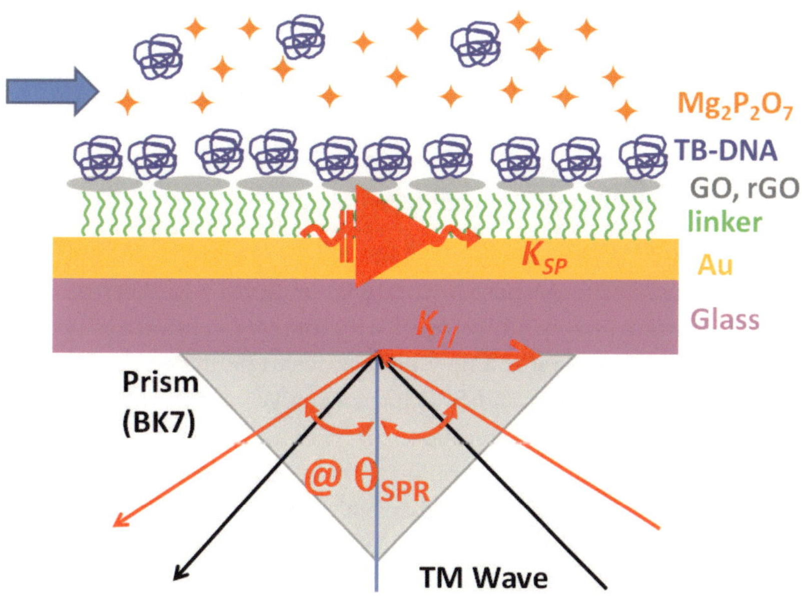

Figure 14. Schematic illustration of the SPR-LAMP detection chips for angle change of the TB DNA-GO binding. The experiments were compared with GO-linked immunosorbent assay, a conventional immunological method. GO-based SPR showed more advantages in providing label-free and real-time detection. Additionally, the high sensitivity and specificity for the detection of TBinfection showed its potential for future development of biosensor for TB DNA diagnosis.

10 X buffer	5ul
10X DIG labeling Mix (0.35mM)	x
dNTP(2mM)	5ul
F2(1uM)	5ul
B2(1uM)	5ul
Enzyme Mix (3.5U/ul)	0.75ul
ddH2O	24.25ul
LAMP DNA product	5 ul
Total	50ul

Table 1. Detection and confirmation of TB-LAMP reaction assay

a. Preparation of primers for TB-LAMP reaction

In the experiments, LAMP reactions are performed at a fixed temperature(60-65 °C) heatdenaturation and rapid cooling were applied to aDNA sample containing the target sequence (IS6110)and the four specific primers [57, 58]. We use sensitive LAMP-based assay for the detection of *Mycobacterium tuberculosis complex* (MTBC), based on the amplification of the *Insertion sequence 6110* (IS6110). IS 6110 is specific for the members, and first report on a LAMP assay for detection of MTBC in which IS6110 is used as a target sequence [57, 58]. The IS6110-specific LAMP primers were designed according to the general criteria described by Notomi et al. [57] and Aryan et al. [58]. Total volume of sample concentrations of LAMP DNA product assay was 50 μl, as shown in table 1.

b. GO sheets forTB-DNAbiosensing by SPR technique

For the SPR-LAMP analyses, we prepared threedevicesin 10X dilution of LAMP DNA productas shownin Fig.15. The purpose of combination SPR-LAMP technology is that it has excellent ability for surface analysis, high sensitivity for oxygen refractive coefficient, and the ability for real-time detection, so it is a new kind of electrode materials with potential applications in electrochemical sensing and SPR biosensing.We can get the best structure for –COOH groups of GO which have great biocompatible and binding properties. The surface functionalization of GO is covalent functionalization with TB DNA. After NaOH injected, the baseline didn't decrease. It shows the strongly covalent bonding between Au-Cys-GO surface and TB DNA. However, GO is compatible with DNA molecule. Table 2.shows the performance comparisons of the resolution for the three devices.

This measurement results shown, the SPR angle shift of TB DNA was measured for evidence of the adsorption of TB-DNA on GO. We demonstrated the GO sheets for the detection of TB DNA.

Comparison of the detection limits at the SPR signal and agarose gel electrophoresis of the LAMP product assays. We analyzed three different concentrations of LAMP products for 1X, 25X and 100X dilution. In the Au-Cys-Go device, the SPR real-time resonance angle was measured. The experimental results shown in Fig. 16(a) illustrate the performance of Au-Cys-

http://dx.doi.org/10.5772/56221

Go devices in TB DNA solutions of dilution percentages(deionized water), 1X, 25X and 100X for 1102.499, 85.335 and 23.564 mDeg of SPR angle shift. Following the validation of LAMP reaction, the amplified products elongated to a length of several kbp and when generated they showed complex cauliflower-like structures. We demonstrated that the positive sample reveals many bands of different sizes afteragarose gel electrophoresis (Fig. 16b). Comparison of the detection limits, the SPR-LAMP has high-performance and simple, rapid an advantageous position.

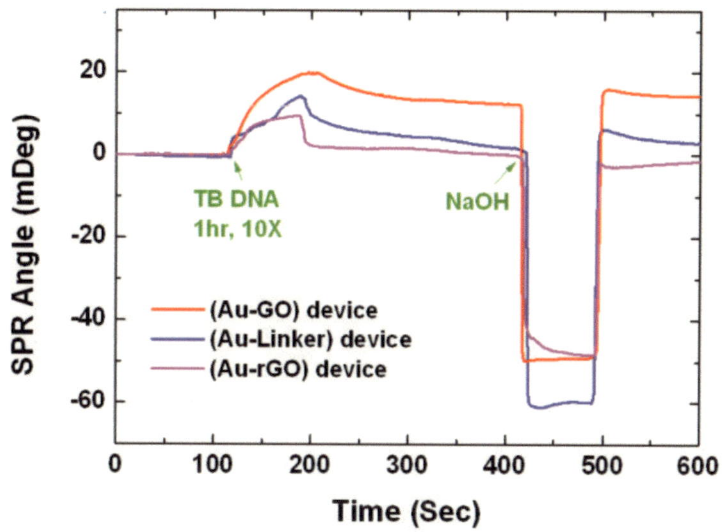

Figure 15. SPR real-time monitoring of the response of the three devices in LAMP product of TB DNA solutions.

Type of chips	SPR shift angle for 10X TB DNA
Au-GO	15.646 mDeg
Au-linker	5.418 mDeg
Au-rGO	2.312 mDeg

Table 2. Resonance angle shift at different SPR devices

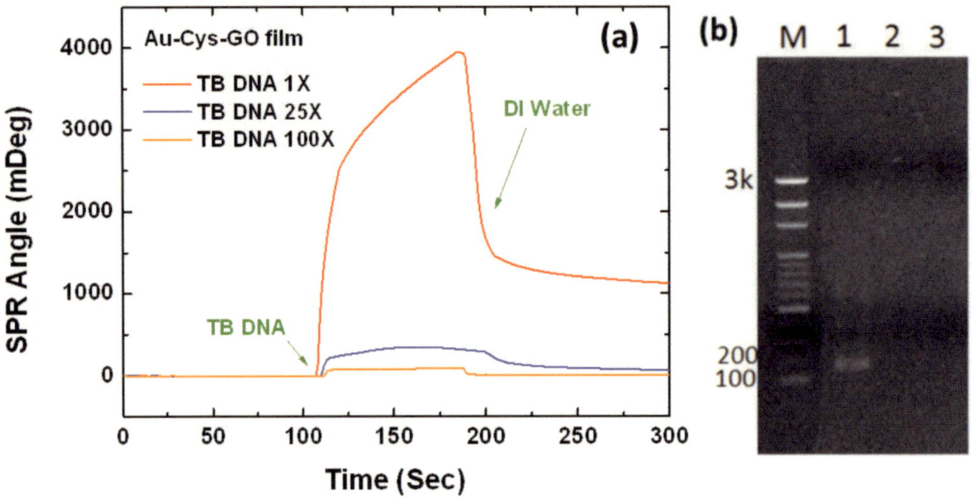

Figure 16. Comparison of the detection limits at the (a) SPR signal and (b) agarose gel electrophoresis of the LAMP product assays. Restriction enzyme analysis of the MTBC-LAMP products. Aliquots of 2.5ul LAMP products were electrophoresed in 2% agarose gels (1×TBE) and then stained with SYBR Green I dye for verification by fluorescent imager. M: 100 bp DNA ladder; 1: TB LAMP reaction (positive control) for 1X; 2: B LAMP reaction (positive control) for 25X; 2: B LAMP reaction (positive control) for 100X.

3. Conclusions

A SPR biosensor of high sensitivity and good detection limit with the employment of a GO functionalized Au-ODT composite (GO sheets) has been successfully demonstrated in this work. The linear range was dramatically improved. We observed a detection limit of pg/ml using the proposed GO sheets, whereas the conventional Au film can just reach to μg/ml. Although this work only reported the detection of BSA in vitro, it is likely that the method based on the GO sheets will ultimately apply in vivo and on-line detection. Therefore, the results described herein that GO sheets are a promising approach towards highly sensitive for diseases with immunization detection. The covalent functionalization of –COOH group utilizes the strong interactions between the graphene oxide and DNA molecules. We demonstrated the GO sheets for the detection of TB DNA. However, the techniques have a high sensitivity. For the practical measurement and mass production in the future, we designed a low-cost, simple, rapid and convenient SPR-LAMP for TB chip. This work can reduce human operation errors and make measurement more easily. Moreover, the SPR-LAMP sensing technique is useful for preventive medicine and personalized medicine. Therefore, another mission to attempt here is to commercialize SPR-LAMP chip and detection platform. The optimization of GO capture layer has been done by using protein BSA and TB DNA with measurable results.

http://dx.doi.org/10.5772/56221

Acknowledgements

This work was supported in part by the National Science Council of the Republic of China, Taiwan under Grant NSC 100-2325-B-182-007 and Grant NSC 99-2218-E-003-002-MY3, and by the Emerging Research and Development Projects of National Taiwan Normal University under Grant NTNU100-D-01.

Author details

Nan-Fu Chiu[1*], Teng-Yi Huang[1] and Hsin-Chih Lai[2]

*Address all correspondence to: nfchiu@ntnu.edu.tw

1 Institute of Electro-Optical Science and Technology, National Taiwan Normal University, Ting-Chou Road, Taipei, Taiwan

2 Department of Medical Biotechnology and Laboratory Science, Chang Gung University, Gueishan Shiang, Taoyuan, Taiwan

References

[1] Singh, V, Joung, D, Zhai, L, Das, S, Khondaker, S. I, & Seal, S. Graphene based materials: Past, present and future," Progress in Materials Science, (2011). , 56, 1178-1271.

[2] Zhu, Y, Murali, S, Cai, W, Li, X, Suk, J. W, Potts, J. R, & Ruoff, R. S. Graphene and Graphene Oxide: Synthesis, Properties, and Applications," Advanced Materials, (2010). , 22, 3906-3924.

[3] Mao, S, Pu, H, & Chen, J. Graphene oxide and its reduction: modeling and experimental progress," RSC Adv.,(2012). , 2, 2643-2662.

[4] Marcano, D. C, Kosynkin, D. V, Berlin, J. M, Sinitskii, A, Sun, Z, Slesarev, A, Alemany, L. B, Lu, W, & Tour, J. M. Improved Synthesis of Graphene Oxide," ACS Nano, (2010). , 4, 4806-4814.

[5] Pei, S, & Cheng, H. -M. The reduction of graphene oxide," Carbon, (2012). , 50, 3210-3228.

[6] Chen, C, Yang, Q. -H, Yang, Y, Lv, W, Wen, Y, Hou, P. -X, Wang, M, & Cheng, H. -M. Self-Assembled Free-Standing Graphite Oxide Membrane," Adv. Mater., (2009). , 21, 3007-3011.

[7] Bai, S, & Shen, X. Graphene-inorganic nanocomposites," RSC Advances, (2012). , 2, 64-98.

[8] Guo, Y, Di, C. -A, Liu, H, Zheng, J, Zhang, L, Yu, G, & Liu, Y. General route toward patterning of graphene oxide by a combination of wettability modulation and spin-coating," ACS Nano (2010). , 4, 5749.

[9] Dong, X, Huang, W, & Chen, P. In Situ Synthesis of Reduced Graphene Oxide and Gold Nanocomposites for Nanoelectronics and Biosensing," anoscale Res Lett., (2011). , 6, 1-6.

[10] Fu, W. Y, Liu, L, Wang, W. L, Wu, M. H, Xu, Z, Bai, X. D, & Wang, E. G. Carbon nanotube transistors with graphene oxide films as gate dielectrics," Science China: Physics, Mechanics and Astronomy, (2010). , 53, 828-833.

[11] Loh, K. P, Bao, Q, Eda, G, & Chhowalla, M. Graphene oxide as a chemically tunable platform for optical applications," Nature Chemistry, (2010). , 2, 1015-1024.

[12] Dreyer, D. R, Park, S, Bielawski, C. W, & Ruoff, R. S. The chemistry of graphene oxide," Chem.Soc.Rev.,(2010). , 39, 228-240.

[13] Lim, G. -K, Chen, Z. -L, Clark, J, Goh, R. G. S, Ng, W. -H, Tan, H. -W, Friend, R. H, Ho, P. K. H, & Chua, L. -L. Giant broadband nonlinear optical absorption response in dispersed graphene single sheets," Nature Photonics, (2011). , 5, 554-560.

[14] Shukla, S, & Saxena, S. Spectroscopic investigation of confinement effects on optical properties of graphene oxide" Appl. Phys. Lett. 98, 073104, (2011).

[15] Johari, P, & Shenoy, V. B. Modulating Optical Properties of Graphene Oxide: Role of Prominent Functional Groups," ACS Nano,(2011). , 5, 7640-7647.

[16] Bao, Q, & Loh, K. P. Graphene Photonics, Plasmonics, and Broadband Optoelectronic Devices," ACS Nano, (2012). , 6, 3677-3694.

[17] Chiu, N. -F, Huang, T. -Y, Kuo, C. -C, Lee, W. -C, Hsieh, M. -H, & Lai, H. -C. Single-Layer Graphene based SPR biochips for tuberculosis bacillus detection," Proc. of SPIE (2012). , 8427, 84273M1-84273M7.

[18] Liu, Y, Dong, X, & Chen, P. Biological and chemical sensors based on graphene materials," Chem. Soc. Rev., (2012). , 41, 2283-2307.

[19] Dong, X, Long, Q, Wang, J, Chan-park, M. B, Huang, b, Y, Huang, W, & Chen, P. A graphene nanoribbon network and its biosensing application," Nanoscale, (2011). , 3, 5156-5160.

[20] Morales-narváez, E, & Merkoçi, A. Graphene Oxide as an Optical Biosensing Platform," Adv. Mater., (2012). , 24, 3298-3308.

[21] Chen, D, Feng, H, & Li, J. Graphene Oxide: Preparation, Functionalization, and Electrochemical Applications" Chem. Rev., (2012). , 112, 6027.

[22] Zhang, Y, Tan, Y. W, Stormer, H. L, & Kim, P. Experimental Observation of Quantum Hall Effect and Berry's Phase in Graphene," Nature., (2005). , 438, 201-204.

[23] Loh, K. P, Bao, Q. L, Ang, P. K, & Yang, J. X. The chemistry of graphene," J. Mater. Chem., (2010). , 20, 2277-2289.

[24] Myung, S, Solanki, A, Kim, C, Park, J, Kim, K. S, & Lee, K. B. Graphene-Encapsulated Nanoparticle-Based Biosensor for the Selective Detection of Cancer Biomarkers," Adv. Mater., (2011). , 23, 2221-2225.

[25] Balapanuru, J, Yang, J. X, Xiao, S, Bao, Q, Jahan, M, Polavarapu, L, Wei, J, Xu, Q. H, & Loh, K. P. A Graphene Oxide-Organic Dye Ionic Complex with DNA-Sensing and Optical-Limiting Properties," Angew. Chem. Int. Ed., (2010). , 49, 6549-6553.

[26] Liu, F, Choi, J. Y, & Seo, T. s. Graphene oxide arrays for detecting specific DNA hybridization by fluorescence resonance energy transfer," Biosens Bioelectron., (2010). , 25, 2361-2365.

[27] Hu, Y, Li, F, Bai, X, Li, D, Hua, S, Wang, K, & Niu, L. Label-free electrochemical impedance sensing of DNA hybridization based on finctionalized graphene sheets," Chem. Commun., (2011). , 47, 1743-1745.

[28] Alwarappan, S, Liu, C, Kumar, A, & Li, C. Z. Enzyme-Doped Graphene Nanosheets for Enhanced Glucose Biosensing,"J. Phys. Chem. C., (2010). , 114, 12920-12924.

[29] Eda, G, & Chhowalla, M. Chemically derived graphene oxide: Towardslarge-area thin-film electronics and optoelectronics",Adv Mater, (2010). , 22, 2392-2415.

[30] Loh, K. P, Bao, Q. L, Eda, G, & Chhowalla, M. Graphene oxide as a chemicallytunable platform for optical applications", Nat Chem, (2010). , 2, 1015-1024.

[31] Stankovich, S, Dikin, D. A, Dommett, G. H. B, Kohlhaas, K. M, Zimney, E. J, Stach, E. A, Piner, R. D, Nguyen, S. T, & Ruoff, R. S. Graphene-based composite materials" nature, (2006). , 442, 282-286.

[32] Liu, Z. B, Zhang, X. L, Yan, X. Q, Chen, Y. S, & Tian, J. G. Nonlinear optical properties of graphene-based materials"Chin Sci Bull, (2012). , 57, 2971-2982.

[33] Liu, Z, Zhao, X, Zhang, X, Yan, X. Q, Wu, Y. P, Chen, Y, & Tian, J. Ultrafast dynamics and nonlinearoptical responses from sp^2- and sp^3-hybridized domains in graphemeoxide", J. Phys. Chem.Lett, (2011). , 2, 1972-1977.

[34] Liu, Z, Wang, Y, Zhang, X, Xu, Y, Chen, Y, & Tian, J. Nonlinear optical properties of graphene oxide in nanosecond and picosecond regimes"Appl. Phys. Lett, (2009). , 94

[35] Palomaki, P. K. B, & Dinolfo, P. H. A Versatile Molecular Layer-by-Layer Thin Film Fabrication TechniqueUtilizing Copper(I)-Catalyzed Azide-Alkyne Cycloaddition"Langmuir., (2010). , 26, 9677-9685.

[36] Raether, H. Surface plasmons on smooth and rough surfaces and on gratings, Springer Tracts Mod. Phys. 111, 1,Verlag, Berlin, (1988). Chapter 2)., 4-39.

[37] Otto, A. Excitation of nonradiative surface plasma waves in silver by the method of frustrated total reflection, Zeitschrift f°ur Physik, (1968). , 216, 398-410.

[38] Kretschmann, E. E, & Raether, H. Radiative decay of non-radiative surface plasmons excited by light, Z. Naturforsch, (1968). , 23, 2135-2136.

[39] Wood, R. W. On a remarkable case of uneven distribution of light in a diffraction grating spectrum, Proc. of the Physical Society of London, (1902). , 18, 269-275.

[40] Chiu, N. -F, Yu, C, Lee, J. -H, Kuan, C. -H, Wu, K. -C, Lee, C. -K, & Lin, C. -W. Enhancement and Tunability of Active Plasmonic by Multilayer Grating Coupled Emission," Optics Express. (2007). , 15

[41] Chiu, N. -F. Jiun-Haw Lee, C.-H. Kuan, K.-C. Wu, C.-K. Lee, C.-W. Lin, "Enhanced luminescence of organic/metal nanostructure for grating coupler active long-range surface plasmonic device," Appl. Phys. Lett., (2007). , 91

[42] Nien, S. -Y, Chiu, N. -F, Ho, Y. -H, Lin, C. -W, Wu, K. -C, Lee, C. -K, Lin, J. -R, Wei, M. -K, & Lee, J. -H. Directional Photoluminescence Enhancement of Organic Emitters via Surface Plasmon Coupling," Appl. Phys. Lett. (2009). , 94

[43] Chang, C. -C, Chiu, N. -F, Lin, D. S, Chu, S. -Y, & Lin, C. -W. High-Sensitivity Detection of Carbohydrate Antigen 15-3 Using a Gold/Zinc Oxide Thin Films Surface Plasmon Resonance-Based Biosensor", Anal. Chem., (2010). , 82, 1207-1212.

[44] Chen, W. P, & Chen, J. M. Use of surface plasma waves for determination of the thickness and optical constants of thin metallic films, J. Opt. Soc. Am., (1981). , 71, 189-191.

[45] Homola, J, Yee, S. S, & Gauglitz, G. Surface plasmon resonance sensors: review, Sens. Actuators B, (1999). , 54, 3-15.

[46] Ebbesen, T. W, Lezec, H. J, Ghaemi, H. F, Thio, T, & Wolff, P. A. Extraordinary optical transmission through subwavelength hole arrays," Nature (London), (1998). , 391, 667-669.

[47] Barnes, W. L, Murray, W. A, Dintinger, J, Devaux, E, & Ebbesen, T. W. Phys. Rev. Lett. 92, 107401 ((2004).

[48] Maier, S. A, & Atwater, H. A. Plasmonics: Localization and guiding of electromagnetic energy in metal/dielectric structures," J. Appl. Phys.,(2005). , 98

[49] Wu, L, Chu, H. S, Koh, W. S, & Li, E. P. Highly sensitive graphene biosensors based on surface plasmon resonance," Optics Express, (2010). , 18, 14395-14400.

[50] Choi, S. H, Kim, Y. L, & Byun, K. M. Graphene-on-silver substrates for sensitive surface plasmon resonance imaging biosensors," Optics Express, (2011). , 19, 458-466.

[51] Salihoglu, O, Balci, S, & Kocabas, C. Plasmon-polaritons on graphene-metal surface and their use in biosensors," Appl. Phys. Lett. 100, 213110, (2012).

[52] Yang, S. L, Xu, B. F, Zhang, J, Huang, X. D, Ye, J. S, & Yu, C. Z. Controllable Adsorption of Reduced Graphene Oxide onto Self-Assembled Alkanethiol Monolayers on Gold Electrodes: Tunable Electrode Dimension and Potential Electrochemical Applications," J. Phys. Chem. C., (2010). , 114, 4389-4393.

[53] Carter, D. C, & Ho, J. X. Structure of serum albumin," Adv. Protein Chem., (1994). , 45, 153-203.

[54] Meziani, M. J, & Sun, Y. P. Protein-conjugated nanoparticles from rapid expansion of supercritical fluid solution into aqueous solution," J Am Chem Soc., (2003). , 125, 8015-8018.

[55] Myszka, D. G. Kinetic analysis of macromolecular interactions using surface plasmon resonance biosensors," Current Opinion in Biotechnology., (1997). , 8, 50-57.

[56] Global tuberculosis control: World Health Organization report 2011. http://whqlibdoc.who.int/publications/2011/9789241564380_eng.pdf).

[57] Notomi, T, Okayama, H, Masubuchi, H, Yonekawa, T, Watanabe, K, Amino, N, & Hase, T. Loop-mediated isothermal amplification of DNA,"Nucleic Acids Res., E63, (2000). , 15

[58] Aryan, E, Makvandi, M, Farajzadeh, A, Huygen, K, Bifani, P, Mousavi, S. -L, Fateh, A, Jelodar, A, Gouya, M. -M, Romano, M, Novelandmoresensitiveloop-mediated, A, & Isothermalamplificationassaytargeting, I. S. for detectionof Mycobacterium tuberculosis complex," Microbiological research, (2010). , 165, 211-220.

[59] Neonakis, I. K, Spandidos, D. A, & Petinaki, E. Use of loop-mediated isothermal amplification of DNAfor the rapid detection of Mycobacterium tuberculosisin clinical specimens,"Eur J Clin Microbiol Infect Dis., (2011). , 30

[60] Lee, S. -Y, Huang, J. -G, Chuang, T. -L, Sheu, J. -C, Chuang, Y. -K, Holl, M, Meldrum, D. R, Lee, C. -N, & Lin, C. -W. Compact optical diagnostic device for isothermal nucleic acids amplification,"Sensors and Actuators B, (2008). , 133, 493-501.

[61] Mohanty, N, & Berry, V. Graphene-Based Single-BacteriumResolution Biodevice and DNATransistor: Interfacing GrapheneDerivatives with Nanoscale andMicroscale Biocomponents,"Nano Lett., (2008). , 8, 4469-4476.

[62] Myung, S, Solanki, A, Kim, C, Park, J, Kim, K. S, & Lee, K. -B. graphene-encapsulated nanoparticle-based biosensor for the selective detection of cancer biomarkers,"Advanced Materials, (2011). , 23, 2221-2225.

[63] Singh, V, Joung, D, Zhai, L, Das, S, Khondaker, S. I, & Seal, S. Graphene based materials: Past, present and future,"Progress in Materials Science, (2011). , 56, 1178-1271.

[64] Wang, Z, Zhou, X, Zhang, J, Boey, F, & Zhang, H. Direct Electrochemical Reduction of Single-Layer Graphene Oxide and SubsequentFunctionalization with Glucose Oxidase,"J. Phys. Chem. C, (2009). , 113, 14071-14075.

[65] Lu, C. -H, Yang, H. -H, Zhu, C. -L, Chen, X, & Chen, G. -N. A Graphene Platform for Sensing Biomolecules,"Angew. Chem. Int. Ed., (2009). , 48, 4785-4787.

Chapter 9

Thermoelectric Power in Graphene

N.S. Sankeshwar, S.S. Kubakaddi and B.G. Mulimani

Additional information is available at the end of the chapter

1. Introduction

Recent years have witnessed considerable interest devoted to the electronic properties of graphene [1-3]. Graphene, a one-atom-thick sheet of carbon atoms arranged in a honeycomb crystal, exhibits unique properties like high thermal conductivity, high electron mobility and optical transparency, and has the potential for use in nano-electronic and optoelectronic devices. With the size of these devices shrinking through integration, thermal management assumes increasingly high priority, prompting the study of thermoelectric effects in graphene systems.

The thermoelectric (TE) effect refers to phenomena by which either a temperature difference creates an electric potential or an electric potential creates a temperature difference. An interesting transport property, thermoelectric power (TEP) has been a source of information to physicists for over a century [4]. TE devices are used as generators and coolers to convert thermal energy into electrical energy or *vice versa*. The potential of a material for TE applications — solid state refrigeration and power generation — generally is determined in large part by a measure of the material's TE figure of merit, $ZT=S^2\sigma T/\kappa$, where S is the thermoelectric power (also called Seebeck coefficient), σ the electrical conductivity, and κ the thermal conductivity of the material. Efficient TE energy conversion, therefore, requires materials that have an enhanced power factor $S^2\sigma$ and reduced κ [4, 5]. The state of art TE materials possess a value ZT ~1, at room temperature [4, 5]. There is no well-defined theoretical limit to ZT. Values of ZT ~ 2-3 would make TE refrigeration competitive with vapour compression refrigeration systems [4, 5]. Even a modest increase in value of ZT would, therefore, provide important opportunities for applications [6]. Recent studies indicate that ZT could be increased nearly fourfold, by optimizing the potential of graphene systems [7].

The interest in the TEP of a material system stems not only from its relation to ZT but also due to its sensitivity to the composition and structure of the system and to the external fields. The TE effect has been able to shed much light on the interaction of electrons and phonons, impurities and other defects. Further, the three transport parameters S, σ and κ are not independent of each other. The Seebeck coefficient, for instance, is partially determined by the electrical conductivity. This provides a challenge for theoreticians and experimentalists alike to search for ways to increase its value. An optimization, say, of the Seebeck coefficient for any material will involve understanding and appropriately modifying its electronic properties. Conventional thermocouples, made from metal or metal alloys, generate Seebeck voltages typically tens of microvolts per Kelvin [8, 9]. Those made from semiconductors with tailored material properties [10] and geometry [11, 12] possess voltages of a few hundreds of microvolts per Kelvin. One of the objectives, therefore, of studies in TEs has been to search for materials with optimized electronic band structures [13] and thermal properties [14, 15]. Much of the recent renewed interest in TEs has been stimulated by the prospect that graphene, with its unique electrical and thermal properties, could provide increased figure of merit.

Ever since its discovery, great interest has been evinced in the electronic properties of graphene [1-3]. Graphene also exhibits interesting TE effects. For instance, compared to elemental semiconductors, it has higher TEP and can be made to change sign by varying the gate bias [16-18]. The unique properties, including high mechanical stiffness and strength, coupled with high electrical and thermal conductivity, make graphene an exciting prospect for a host of future applications in nanoelectronics, thermal management and energy storage devices (For reviews on graphene physics, see [2] and[3]). Technical advances have now made possible the realization of tailor-made 2D graphene systems, such as single-layer graphene (SLG), bilayer graphene (BLG), graphene nanoribbon (GNR), graphene dots, graphene superlattices and defected graphene. Most of the experimental and theoretical work has concerned the electrical and thermal conductivity of such systems. (For a review on recent progress in graphene research, see [19]). However, in the recent past, a good amount of literature has accumulated on the TE properties of graphene systems, and a coherent picture is just emerging into understanding TE effect in graphene.

The present work addresses one of the important components of TE transport in graphene, namely, TEP, also referred to, simply, as thermopower. TEP has been a powerful tool for probing carrier transport in metals and semiconductors [8-12]. Being sensitive to the composition and structure of a system, it is known to provide information complementary to that of resistivity (or conductivity), which alone is inadequate, say, in distinguishing different scattering mechanisms operative in a system.

In this chapter, we review the literature on TEP in graphene systems and present its understanding using the semi-classical Boltzmann transport theory. In Section 1, the electronic structures and phonon dispersion relations for SLG, BLG and GNR systems are described. In the next section, besides a survey of the experimental work, the basic theory of TEP in 2D systems is given, and its relation with other TE transport coefficients is discussed. In Section 3, the diffusion contribution to TEP of graphene systems is discussed. Section 4 deals with the phonon-drag contribution to TEP. An analysis of the experimental data, in terms of the diffusion and drag components, is also presented. This is followed by a summary of the chapter.

1.1. Graphene systems

A single-layer graphene, commonly referred to simply as graphene, is one of the recent nanomaterials. It is a monolayer of graphite with a thickness of 0.34 nm, consisting of carbon atoms in the sp^2 hybridization state, with the three nearest-neighbour carbon atoms in the honeycomb lattice forming σ bonds. The carriers in graphene are confined in this 2D layer [2, 3].

The 2D honeycomb structure of graphene lattice with two equivalent lattice sites, A and B (Figure 1.(a)), can be thought of as a triangular lattice with a basis of two atoms per unit cell, with 2D lattice vectors $\mathbf{a}_1 = (a/2)(3, \sqrt{3})$ and $\mathbf{a}_2 = (a/2)(3, -\sqrt{3})$, where $a = 0.142$ nm is the C-C distance. The inequivalent corners $\mathbf{K} = (2\pi/a)(1/3, 1/3\sqrt{3})$ and $\mathbf{K'} = (2\pi/a)(1/3, -1/3\sqrt{3})$ of the Brillouin zone are called Dirac points. The existence of the two Dirac points, K and K', where the Dirac cones for electrons and holes touch each other in momentum space (Figure 1.(b)), gives rise to a valley degeneracy, $g_v = 2$. Graphene is a zero band-gap semiconductor with linear long-wavelength energy dispersion for both electrons and holes in the conduction and valence bands. The two equivalent lattice sites make carrier transport interesting giving rise to the 'chirality' in its carrier dynamics [3]. The thermoelectric transport properties of graphene, discussed in this chapter, follow from the linear low-energy dispersion and the chiral character of the bands.

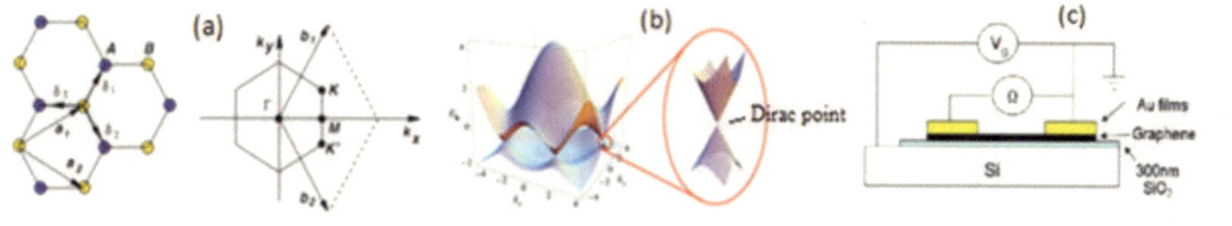

Figure 1. (a) Graphene honeycomb lattice and the Brillouin zone. The two sublattices are shown in different colours. (b) Graphene band structure. An enlargement close to the K and K' point shows Dirac cones. (from [2]) (c) Typical configuration for gated graphene.

Gapless graphene has a charge neutrality point (CNP), that is, the Dirac point, where its character changes from being electron- like to being hole-like. For pure graphene the Fermi surface is at the Dirac point. The system with no free carriers at $T = 0$ K and E_F at the Dirac point is called intrinsic graphene. It has a completely filled valence band and an empty conduction band. However, any infinitesimal doping, as also any finite temperature, with electrons present in the conduction band, makes the system 'extrinsic'. It is possible experimentally, by varying the external gate voltage, to tune the system from being electron-like to being hole-like, with the system going through its intrinsic nature at the CNP [3]. In the case of a gapped system with an insulating region in between, one may not access the intrinsic nature of graphene.

The electronic properties of graphene depend on the number of layers. Generally, the graphene community distinguishes between single-layer, bilayer and few-layer graphene, the latter of which refers to graphene with a layer number less than ten. Bilayer graphene (BLG) consists of two graphene monolayers weakly coupled by interlayer carbon hopping, which depends on the manner of stacking of the two layers with respect to each other; typically they are arranged in A-B stacking arrangement. The bilayer structure, with the various electronic hopping energy parameters γ_i, is shown in Figure 2.(a). The low energy, long wavelength electronic structure of BLG for A-B stacking of the two layers is depicted in Table 1. It may be noted that unbiased BLG is gapless. However, by applying an external voltage, a semiconducting gap can be induced in the otherwise zero-gap band structure [3].

In order to improve applicability, graphene needs to acquire a bandgap. This can be achieved by appropriate patterning of the graphene sheet into nanoribbons. A graphene nanoribbon (GNR) is a quasi-one-dimensional (Q1D) system that confines the graphene electrons in a thin strip of (large) length L and a finite (small, a few nm) width W. Figure 2. (b) shows a honeycomb lattice of a GNR having zigzag edges along the x-direction and armchair edges along the y-direction. The resulting confinement gap, E_g, depends on the chirality of the edges (armchair or zigzag) and the width of the ribbon. Choosing a GNR to be macroscopically large say along the y-direction but finite along the x-direction, gives a GNR with armchair edges (AGNR), and, conversely, a GNR chosen with width along y-direction gives a zigzag terminated GNR (ZGNR). A ZGNR is metallic in nature, whereas an AGNR can be metallic or semiconducting, with E_g inversely proportional to W [3, 20].

One of the strategies adopted to achieve higher mobility in graphene samples is to improve the substrate quality or eliminate the substrate altogether by suspending graphene over a trench. Improved growth techniques have enabled obtaining graphene as a suspended membrane, supported only by a scaffold or bridging micrometer-scale gaps schematic of which is shown in Figure 2.(c). Suspended graphene (SG), shows great promise for use in nanoelectronic devices. With most of the impurities limiting electron transport sticking to the graphene sheet and not buried in the substrate, a large reduction in carrier scattering is reported [21] in current-annealed SG samples. However, unlike supported graphene, only a small gate voltage ($V_g \sim 5$ V) can be applied to a SG sample before it could buckle and bind to the bottom of the trench. Despite limited carrier densities, Bolotin *et al* [22] report a mobility of 1.7×10^5 cm^{2}V^{-1}s^{-1} in ultra-clean SG with $n_s \sim 2\times10^{15}$ m^{-2}. The electronic properties of SG can be affected by strain. The layer(s) may be under strain either due to the electrostatic force arising from the gate or as a result of micro-fabrication, or even by applying strain in a controlled way. Recent studies suggest that strain can be used to engineer graphene electronic states [23] and hence the transport properties.

In the following, the thermoelectric property of TEP will be reviewed with regard to the three systems, namely SLG, BLG and AGNR.

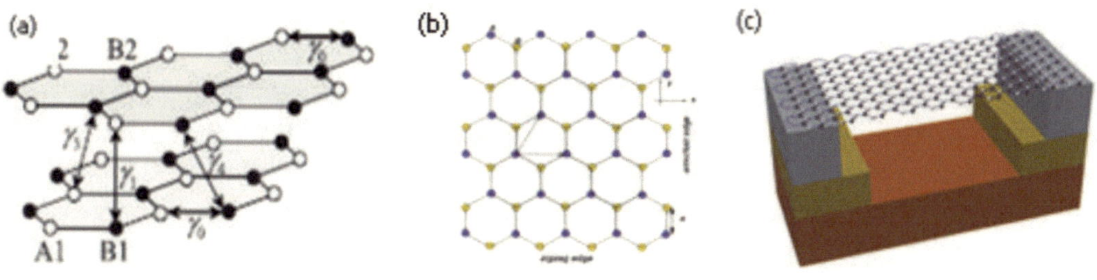

Figure 2. Lattice structure of (a) BLG, and (b) GNR. (from [2]). (c) Schematic of suspended SLG.

1.2. Electronic structures

1.2.1. Single layer graphene

The transport characteristics of a material are intimately related to the energy band structure. The carriers in the graphene lattice are free to move in two dimensions. In the carrier transport of graphene, the carriers — electrons and holes — close to the Dirac points are of importance. Their transport is described by a Dirac-like equation for massless particles [2, 3]:

$$-i\hbar v_F\ \sigma \cdot \nabla \psi(r) = \mathrm{E}\ \psi(r) \tag{1}$$

where $\boldsymbol{\sigma} = (\sigma_x, \sigma_y)$ is the vector of Pauli matrices in 2D and $\psi(\mathbf{r})$ includes a 2D plane wave and a spinor (graphene pseudospin) function. In the continuum limit Eq. (1) corresponds to the effective low energy Dirac Hamiltonian

$$H(k) = \hbar v_F \begin{pmatrix} 0 & k_x - ik_y \\ k_x + ik_y & 0 \end{pmatrix} = \hbar v_F \sigma \cdot \mathrm{k} \tag{2}$$

The electronic band structure of the energy (E) versus wavevector ($\mathbf{k}$) relation for the graphene carriers is given by the solution of (1). The solution of (1) has been calculated in the tight-binding model up to the next-nearest neighbor approximation [24]. The carrier wavefunctions, energy eigenvalues, the density of states and the low-energy (close to the CNP, K) band structure for SLG are given in Table 1 [2].

Being interested mostly in understanding electron transport for small energies and relatively small carrier concentrations, only the low-$\mathbf{k}$, linear dispersion aspects of the band structure are considered close to the K and K′ points where the Dirac cones for electrons and holes touch each other (see Figure 1.b). SLG is thus a zero band-gap semiconductor with a linear, long-wavelength ($k \ll 2\pi/a$) energy dispersion for both electrons (in the conduction band) and holes (in the valence band) with the conduction and valence bands intersecting at $\mathbf{k} = 0$ [2, 3]:

$$E_k = s\,\hbar v_F |k| \tag{3}$$

Here, $s = +1$ (-1) corresponds to the conduction (valence) band, $\mathbf{k} = \mathbf{i}k_x + \mathbf{j}k_y$ denotes the carrier wavevector measured from the relevant Dirac point, and $v_F = 3ta/2\hbar$, a constant, is called the graphene (Fermi) velocity; with the nearest-neighbor hopping amplitude, $t \sim 2.5$ eV, $v_F \sim 10^6$ ms^{-1}.The linearity of the dispersion relation signifies that the effective mass of charge carriers vanishes, and hence the interaction between electrons or holes and the 2D crystalline lattice is weak, and the charge carriers can propagate without collisions through graphene. Graphene can, therefore, be modeled as a 2D gas of massless fermions [3].

1.2.2. Bilayer graphene

The effective Hamiltonian for a BLG, in the low energy, long-wavelength regime is [3]

$$H(k) = -\frac{\hbar^2}{2m}\begin{pmatrix} 0 & (k_x - ik_y)^2 \\ (k_x + ik_y)^2 & 0 \end{pmatrix} \tag{4}$$

where m is the effective mass of the carriers. The electron wavefunctions and energy eigen values around the **K** and **K**′ points at the corners of graphene Brillouin zone are given in Table 1 [25]. The BLG electronic structure, consisting of two branches, depends on the electrostatic potential, V, between the two layers. It controls an effective band-gap opening near Dirac point. For V=0, BLG is a gapless semiconductor with parabolic dispersion. The low-field electron transport is discussed with respect to the lowest branch.

1.2.3. Graphene nanoribbon

The spectrum of GNRs depends on the nature of their edges. The low-energy electronic states of GNRs near the two non-equivalent Dirac points (**K** and **K'**) can be described by the 4x4 Dirac equation and using the appropriate boundary conditions. The effective Hamiltonian for a GNR is [26]

$$H(k) = \gamma\begin{pmatrix} 0 & (k_x - ik_y) & 0 & 0 \\ (k_x + ik_y) & 0 & 0 & 0 \\ 0 & 0 & 0 & (k_x + ik_y) \\ 0 & 0 & (k_x - ik_y) & 0 \end{pmatrix} \tag{5}$$

where γ ($=\sqrt{3}ta/2 = \hbar v_F/\sqrt{3}$) is a band parameter. The expressions for the electron wave functions, energy eigenvalues, density of states and the band structures for a AGNR (the system considered here) derived from admixing the states in the K and K′ valleys and satisfying hard wall

System	Wavefunctions	Eigenvalues	Density of States	Band Structure
SLG	$\psi_{\pm,K}(r)=\frac{1}{\sqrt{2LW}}\begin{pmatrix}e^{-i\theta_k}\\ \pm 1\end{pmatrix}e^{ik.r}$	$E_k=\hbar v_F\lvert k\rvert$	$\rho(E_k)=\frac{2\lvert E_k\rvert}{\pi(\hbar v_F)^2}$	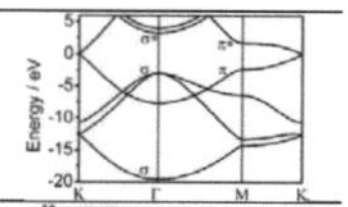
BLG	$\psi_{\pm,K}(r)=\frac{1}{\sqrt{2LW}}\begin{pmatrix}e^{-2i\theta_k}\\ \pm 1\end{pmatrix}e^{ik.r}$	$E_k=\frac{\hbar^2k^2}{2m}$	$\rho(E_k)=\frac{2m}{\pi\hbar^2}$	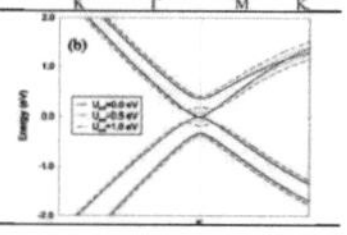
AGNR	$\Psi_{n,k_y}(\mathbf{r})=\left[\psi_{n,k_y}(\mathbf{r})-\psi'_{n,k_y}(\mathbf{r})\right]/\sqrt{2}$ with $\psi_{n,k_y}(\mathbf{r})=\sqrt{1/2LW}\;e^{i[(\Delta K/2)-k_n]x}\varphi_{n,k_y}(y)$ $\psi'_{n,k_y}(\mathbf{r})=\sqrt{1/2LW}\;e^{-i[(\Delta K/2)-k_n]x}\varphi_{n,k_y}(y)$ $\varphi_{n,k_y}(y)=e^{ik_yy}\begin{pmatrix}1\\ -e^{-i\theta_{n,k_y}}\end{pmatrix}$	$E_{n,k_y}=\hbar v_F\sqrt{k_n^2+k_y^2}$	$\rho(E_{n,k_y})=\frac{2}{\pi\hbar v_F}\left[\frac{E_{n,k_y}}{\{E_{n,k_y}^2-(E_n)^2\}^{1/2}}\right]$	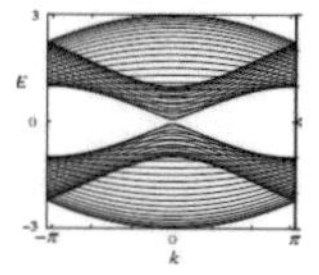

Meanings of symbols (Table 1.):

L : Length of graphene system; W : Width of graphene system; $\mathbf{r}\equiv(x, y)$: 2D electron position vector; $\mathbf{k}\equiv(k_x, k_y)$: 2D electron wavevector

v_F - Fermi velocity; m - Effective mass; $\theta_k=\tan^{-1}(k_x/k_y)$.

$E_n=\pm n\,\pi\hbar v_F/3W$: AGNR subband energy; subband index, n=1, 2, 4, 5, 7, 8....; $E_g=2\pi\hbar v_F/3W$ – Band gap

$k_n=\pm n\pi/3W$: electron transverse wave vector; $\Delta K=4\pi/3a$, a – lattice constant; $\theta_{n,k_y}=\tan^{-1}(k_n/k_y)$.

Table 1. Electron wavefunctions, energy eigenvalues, density of states and band structures of graphene systems [2, 25, 27]

boundary conditions, are given in Table 1 [26, 27]. The confinement of electrons to a Q1D system gives rise to the subband structure with an energy gap at the Dirac point.

1.3. Phonon dispersion relations

Vibrations in the 2D graphene lattice are characterized by two types of acoustic phonons: those vibrating in the plane of layer with linear longitudinal and transverse acoustic branches (LA and TA), and those vibrating out of the plane of the layer – the so-called flexural phonons (ZA) [1-3].

The low-energy in-plane phonons have the usual linear dispersion relation

$$\omega_q=v_sq \tag{6}$$

where, $\mathbf{q}=(q_x, q_y)$ is the 2D phonon wavevector, and v_s denotes the velocity of the in-plane phonons of mode s ($\equiv L, T$). The group velocities v_L = ~13.6x10^{3}ms^{-1} and v_T = ~21.3x10^{3}ms^{-1} are larger than those in silicon.

The acoustic flexural phonons (FPs) are described by an approximately quadratic dispersion relation [1, 28]:

$$\omega_q \approx \alpha |q|^2 \tag{7}$$

with $\alpha = \sqrt{\kappa / \rho}$, κ being bending rigidity and ρ the mass density. Eq.(7) is applicable to the free-standing membranes as in suspended graphene (SG).

The existence and possible modification of the ZA modes, as in the case of SG membrane under tension, are known to lead to the unusual thermal transport in graphene [15]. For slowly varying finite in-plane stresses, the dispersion relation of the FPs is anisotropic. Assuming uniaxial strain $\bar{u}$, the effective dispersion relation of FPs, in the isotropic approximation, may be expressed as as [28]:

$$\omega_q = q\sqrt{\alpha^2 q^2 + \bar{u} v_L^2} \tag{8}$$

The quadratic dispersion relation (8) of FPs becomes linear at long wavelengths [28].

2. Thermoelectric power – Basics

Thomas Johann Seebeck observed that a conductor generates a voltage when subjected to a temperature gradient. This phenomenon is called Seebeck effect, and can be expressed as [4, 5, 8, 9, 11]

$$V = S\Delta T \tag{9}$$

where V is thermoelectric voltage, ΔT is temperature difference, and S is the Seebeck coefficient (see Figure 3). The Seebeck coefficient is also called thermoelectric power. One may define an 'absolute' thermopower characteristic of a particular material.

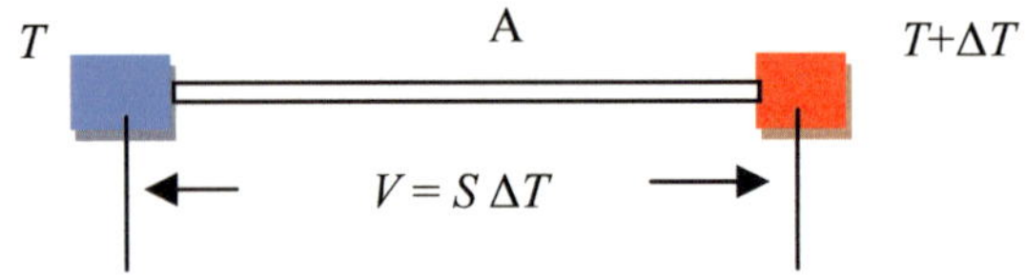

Figure 3. Seebeck effect. Material A cooled at one end (in blue color) with low temperature T and heated at the other end (red color) at higher temperature $T+\Delta T$, results in a voltage difference as a function of temperature difference (ΔT)

On the other hand, Jean Charles Peltier discovered that when an external voltage is applied, the resulting current flow is associated with a heat flow. The Peltier effect is thus the reverse of the Seebeck effect — it refers to the temperature difference induced by voltage gradient. A third thermoelectric phenomenon, called the Thomson effect after its discoverer, William Thomson, is the reversible evolution (or absorption) of heat in a homogeneous conductor that carries an electric current and in which a temperature gradient is also maintained.

The three effects are related to thermal transport, and the coefficients are interrelated. The TEP is relatively easily measured and most of the available results are about this coefficient. Focusing attention, therefore, on TEP, we give below, in brief, the basic theory of TEP which serves as a basis for description of TEP in graphene systems. Also discussed below is the relation of TEP with other transport coefficients.

2.1. Definition and general relations

The thermoelectric effect is due to the interdependence of potential and temperature gradient in a system where no electric current flows. The absolute TEP, *S*, which is a unique physical property of a material is defined by the relation [29, 30]

$$\mathbf{E} = S\nabla T \tag{10}$$

under open-circuit conditions, where **E** is the effective (that is, measured) electric field produced by the temperature gradient ∇T. Since **E** and ∇T are vector quantities, *S* is generally, a tensor.

There are, in general, two contributions to the TEP of the system, namely, the electron-diffusion TEP and the phonon-drag TEP. They will be described later in 2.1.2.

2.1.1. Transport coefficients and thermopower

One can write an expression for the thermoelectric power, *S*, in terms of the fundamental transport tensors. The equations for the electric current density **J** and the heat current density **U**, in the presence of a weak electric field and a small temperature gradient, may be written as [8, 9, 11, 12, 29, 30]

$$\begin{aligned} \mathbf{J} &= L_{11}\mathbf{E} + L_{12}\nabla T \qquad \text{(a)} \\ \mathbf{U} &= L_{21}\mathbf{E} + L_{22}\nabla T \qquad \text{(b)} \end{aligned} \tag{11}$$

where the coefficients, L_{ij}, are, in general, tensors. In order to relate the coefficients to the experimentally measured quantities, such as TEP, it is usual to invert Eq. (11) and write

$$\begin{aligned} \mathbf{E} &= \sigma\mathbf{J} + S\nabla T \qquad \text{(a)} \\ \mathbf{U} &= \Pi\mathbf{J} - \kappa\nabla T \qquad \text{(b)} \end{aligned} \tag{12}$$

Here

$$
\begin{aligned}
\rho &= (L_{11})^{-1} = \sigma^{-1} && \text{(a)} \\
S &= -(L_{11})^{-1} L_{12} && \text{(b)} \\
\Pi &= (L_{11})^{-1} L_{21} && \text{(c)} \\
&\text{and} \\
\kappa &= L_{21}(L_{11})^{-1} L_{12} - L_{22} && \text{(d)}
\end{aligned}
\tag{13}
$$

are the electrical resistivity, the thermopower, the Peltier and the thermal conductivity tensors, respectively; σ is the electrical conductivity tensor. The coefficients in Eq. (11) are not independent. The Onsager relations say that L_{11} and L_{12} are symmetric and [11]

$$L_{12} = -L^{T}{}_{21} / T \tag{14}$$

with the superscript (T) meaning the transpose. The TEP, S, determined under open circuit conditions (that is, with $\mathbf{J} = 0$), is given by Eq. (13b) and can also be expressed as

$$S = (L_{11})^{-1} L_{21} / T = \Pi / T \tag{15}$$

This is known as the second Kelvin relation.

There are two approaches to the evaluation of TEP, S. One can evaluate S directly from the defining relation (10). In this method, usually referred to as the 'Q' approach [9], we need to consider the effect of the electric field and the temperature gradient, simultaneously. One can eliminate the necessity of incorporating a spatially varying temperature in the theory of thermoelectric phenomena by using the Kelvin relation (15), which provides a convenient way of calculating TEP. Instead of a direct calculation of S, we may first determine the Peltier coefficient Π from the solution of an electric conduction problem assuming the temperature gradient to be zero. Imposing the isothermal condition ($\nabla T = 0$), Eq. (11) give

$$
\begin{aligned}
\mathbf{U} &= \Pi \mathbf{J} && \text{(a)} \\
&\text{with} \\
\mathbf{J} &= \sigma \mathbf{E}. && \text{(b)}
\end{aligned}
\tag{16}
$$

S is, then, easily obtained using Kelvin's relation:

$$TS = \sigma^{-1} L_{21} = \Pi = \mathbf{U} / \mathbf{J} = \rho(\mathbf{U} / \mathbf{E}) \tag{17}$$

This method of computing S from Π, advocated by Herring [31] is, usually, known as the Π-approach.

2.1.2. Diffusion and Phonon-drag thermopower

As mentioned earlier, there are two contributions to the thermopower, S, of a system. They are the diffusion TEP and the phonon-drag TEP. In the presence of a temperature gradient the electrons diffuse through the specimen interacting with a random distribution of scattering centres that are assumed to exist in thermal equilibrium at the local temperature T. The heat flux, $\mathbf{U}_{d,}$ carried by the electrons yields the electron-diffusion thermopower, S_d.

When the assumption of the phonon system being in equilibrium is lifted (which is true, especially, at low temperatures) an additional contribution to S appears — the phonon-drag TEP. The principle of phonon-drag is simple. The phonon flux occurring under the temperature gradient will now less readily dissipate its energy and momentum to the lattice than to the electrons *via* the phonon-electron interaction. There is a net additional momentum imparted to the electrons moving down the temperature gradient. The phonon current thus 'drags' electrons with it and extra electrons tend to pile up at the cold end over and above those electrons which are there as a result of the diffusion processes [8, 9, 11, 29, 30]. The heat flux, $\mathbf{U}_g$, carried by the phonon system yields the phonon-drag TEP, S_g.

The total heat current density **U**, therefore, consists of two parts:

$$\mathbf{U} = \mathbf{U}_d + \mathbf{U}_g \tag{18}$$

and, correlatively, the total TEP, S, can be expressed as

$$S = S_d + S_g \tag{19}$$

The treatment presented here is quite general and is applicable to graphene systems.

One can make a simplistic estimate of the magnitude of the diffusion thermopower [11]. It follows from Eq.(16a) that the Peltier coefficient Π, being the ratio of the rate of heat flow to the electrical current, is just the heat per unit charge. For a non-degenerate electron gas, the thermal energy per carrier will be $\sim k_BT$, so that $\Pi \sim \pm k_BT/|e|$ and, from Eq. (15), $S \sim \pm k_B/|e|$ $\sim 86\ \mu V/K$. This suggests that S is a measure of the ratio of the average entropy to the charge. For the degenerate case, the average thermal energy will be reduced by $\sim k_BT/E_F$, so that

$$S \sim \pm \left(k_B/|e|\right)\left(k_BT / E_F\right) \tag{20}$$

Eq.(20) suggests a linear temperature dependence, usually observed in degenerate systems at higher temperatures when the phonon-drag is unimportant.

2.2. Survey of experimental work

Fundamentally related to the electrical conductivity of a material, the TE transport coefficients are also determined by the band structure and scattering mechanisms operative, and can offer unique information complementary to the electrical transport coefficients. The minimal conductivity at the Dirac point is characteristic of graphene [1-3]. Away from the Dirac point, the electron concentration dependence of conductivity depends on the nature of the scatterers. At low temperatures, the conductivity of graphene is limited by scattering off impurities and disorder which depend on the sample preparation. In the absence of extrinsic scattering sources, phonons constitute an intrinsic source of scattering [3].

Measurements of the thermoelectric properties of graphene have helped elucidate details of the unique electronic structure of the ambipolar nature of graphene, which cannot be probed by conductivity measurements alone. Table 2 lists the recent experimental investigations made with regard to the thermoelectric properties of graphene. Here, we primarily review the measurements made in the absence of an applied magnetic field. The presence of a magnetic field is expected to reveal some more interesting important features, as in conventional 2DEG [11, 16-18, 32].

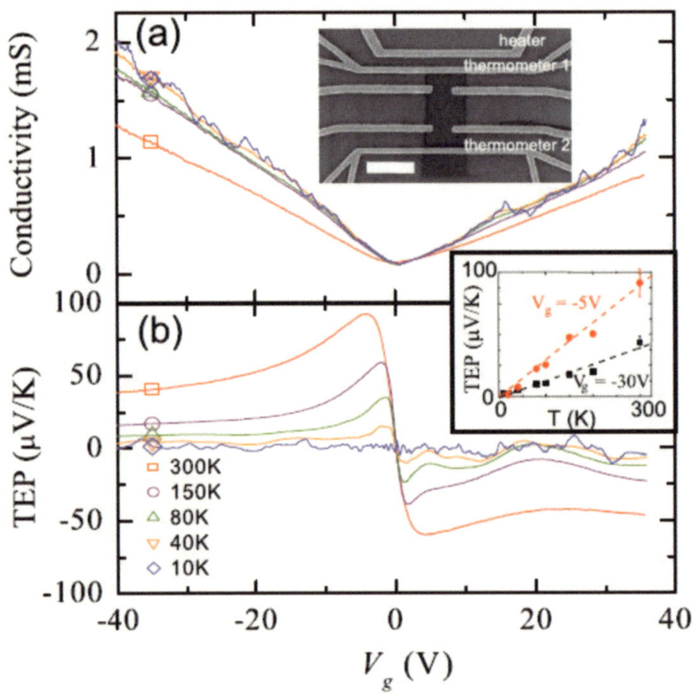

Figure 4. (a) Conductivity and (b) TEP of a graphene sample as function of V_g for T=300 K (square), 150 K (circle), 80 K (up triangle), 40 K (down triangle), and 10 K (diamond). Upper inset: SEM image of a typical device, the scale bar is 2 μm. Lower inset: TEP values taken at V_g = - 30 V (square) and -5 V (circle). Dashed lines are linear fits to the data. (from [16])

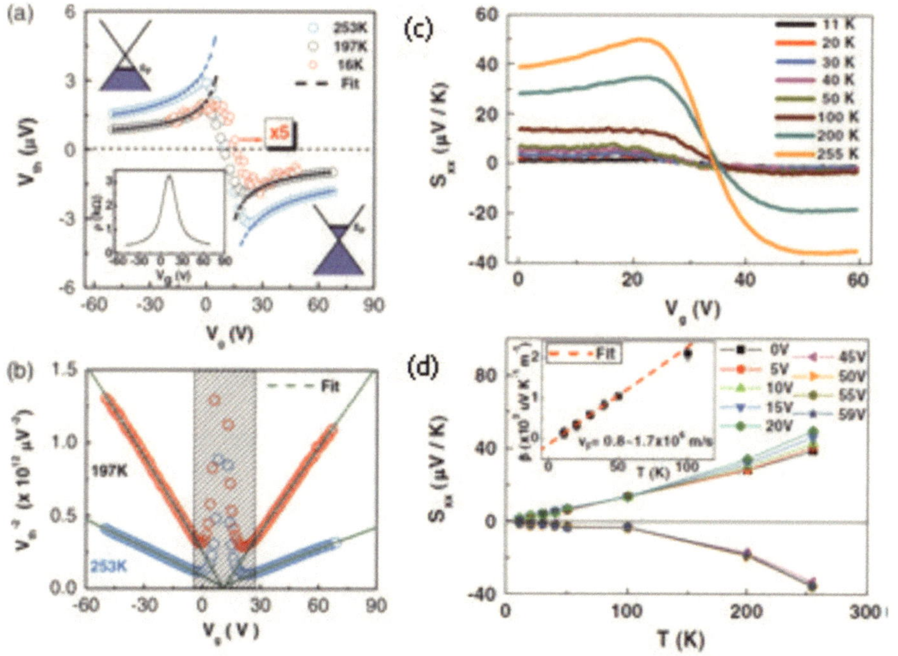

Figure 5. (a) V_{th} vs V_g for three different temperatures. The 16 K data (red circle) were multiplied by a factor of 5. The dashed lines are the fits described by $|S_{xx}| \sim 1/\sqrt{|V_g - V_D|}$. (b) $1/V_{th}^2$ vs V_g plot for the same data shown in (a). The shaded area is for $|V_g - V_D| < 10$ V. Green dashed lines are the best power-law fits with exponent ~0.95. (c) V_g dependence of longitudinal Seebeck coefficient S_{xx} at different temperatures (11–255 K) and zero magnetic field. (d) T dependence of S_{xx} at different gate voltages. The inset is the T dependence of $\beta = S_{xx}\sqrt{|n_{2D}|}$ at V_g= 0 V for low temperatures. (from [17])

The TE effect of Dirac electrons has been initially experimentally investigated in graphene samples mechanically exfoliated on~300 nm SiO_2/Si substrates [16-18]. The number of layers in graphene samples can be identified by optical contrast of the samples cross correlated with scanning probe studies and Raman spectroscopy. A controlled temperature difference ΔT is applied to the sample by a heater and the resulting thermally induced voltage ΔV is measured by the voltage probes to acquire the TEP, $S = -\Delta V/\Delta T$. In a typical set up (Figure 1.c) the underlying degenerately doped silicon substrate acts as a gate electrode for modulating the graphene carrier density. The TEP of graphene can be modulated by the gate voltage, V_g. The nonexistence of a gap in the graphene carrier dispersion as in SLG leads to a direct transition between electron-like transport to hole-like transport as the gate voltage is tuned through the charge neutral Dirac point.

Zuev *et al* [16] measured simultaneously the conductance and TEP of different SLG samples. Figure 4 shows the measured electrical conductivity and TEP as a function of applied gate voltage V_g over a temperature range of 10-300 K. The conductivity becomes minimum at the CNP, corresponding to V_g=V_D, where the Dirac point V_D= 0 V for the device (shown in figure 4). They observe a change in sign of the TEP across the CNP (V_D=0 V) as the majority carrier density changed from electrons to holes. The linear temperature dependence of TEP shown (inset) for two values of V_g, far away from the CNP suggested that the mechanism for thermoelectric generation is diffusive TEP, with the phonon-drag component not present.

Wei *et al* [17] reported similar experimental results and showed that TE transport is uniquely sensitive to the electronic band structure. Away from the Dirac point, the magnitude of the thermovoltage, V_{th}, decreases, scaling approximately with $| V_g - V_D |^{-1/2}$; the dependence is more noticeable in the linear dependence of V_{th}^{-2} on V_g (see Figure 5.(b). The divergence of the Seebeck coefficient S_{xx} as $n_s^{-1/2}$, it may be noted, is a direct manifestation of the linear dispersion of Dirac particles in graphene. Because, assuming the energy dependence as $\sigma \sim E^{\alpha}$, for conductivity of such a highly doped 2D system, the Mott relation [11] yields $S_{xx} \sim - | V_g - V_D |^{-1/2}$. This is in contrast to conventional 2D systems with a quadratic dispersion relation, for which $S_{xx} \sim n_s^{-1}$. Measurements of S_{xx} on a different device with $V_D \sim 33$ V, have indicated (see Figure 5. (c)) electron-hole asymmetry. On the hole side S_{xx} decreases with decreasing V_g, whereas on the electron side S_{xx} remains flat. Further, S_{xx} is found to follow different T dependence for different V_g; S_{xx} is nearly straight on the hole side, whereas, on the electron side, it remains nonlinear in T except at very low temperatures (see Figure 5.(d)). Wei *et al* attribute the departure from the linear T dependence on the electron side to the asymmetric nature of the band of impurity states, which in the impurity scattering model, can be highly asymmetric near Dirac point [33]. This observation brings out the anomalous TE transport in graphene, which may be used as a sensitive probe for impurity bands near the Dirac point.

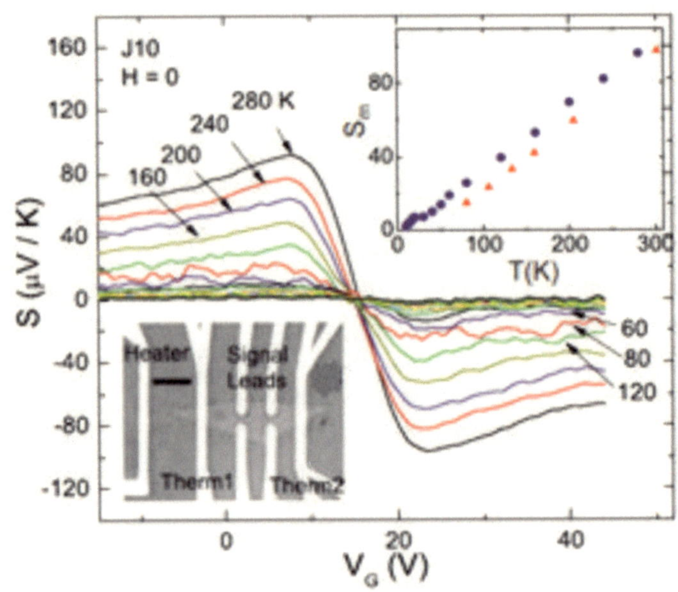

Figure 6. Curves of TEP, $S=-S_{xx}$ vs V_g in sample J10 (left inset) in zero magnetic field at selected T. The curves are antisymmetric about the Dirac point which occurs at the offset voltage V_0=15.5 V. The peak value S_m (right inset) is nominally linear in T from 25 to 300 K. (from [18])

Sample	Measurements	Value	Ref.
	Single Layer Graphene (SLG)		
Mechanical exfoliation on 300nm SiO_2 substrate; $\mu \sim 1–7 \times 10^3$ cm^2/Vs	Gate-dependent conductance and TEP measured simultaneously in zero and non-zero magnetic fields, in linear response regime ($\Delta T << T$); T (10-300K); $S_{xx}(B)$ and $S_{xy}(B)$ quantized	~ 80 μV/K @ RT	[16]
Mechanical exfoliation on 300 nm SiO_2 substrate; $\mu \sim 3 \times 10^3$ cm^2/Vs	Gate voltage (V_g) and temperature dependent TEP measured in zero & applied magnetic fields; $S(T)$ different for different V_g, S_{xx} diverges near CNP as $n_s^{-½}$; Oscillating dependence of both S_{xx} & S_{xy} on n_s @ low T	At T=255 K, S_{xx} ~ 39 μV/K; S_{xy}~ 50 μV/K @ B=8 T, @ CNP	[17]
Exfoliation on 300 nm SiO_2/Si substrate	$V_g$dependent TEP in zero and non-zero magnetic fields; $S_m^{peak}(T)$ is linear S_{xx} & S_{yx} show strong quantum oscillations as functions of V_g.	~ 100 μV/K @ RT	[18]
Exfoliated & supported on SiO_2; W:1.5–3.2; L: 9.5–12.5 μm; G1:3.2 μm parallel to 1.5 μm G2:2.4μm μ =20x10^{3}cm^2/Vs;	Temperature dependence of TEP $S(T)$ curve fitted using theoretical model [35].	G1, G2: ~ 80 μV/K @ RT	[34]
Exfoliated from Kish graphite/ HOPG; $\mu \sim 1.5 – 13 \times 10^3$ cm^2/Vs; n_i=3.3x10^{16} m^{-2}	$V_g$dependence of TEP for 100<T<295 K Effect of charged impurities on the TEP near the Dirac point High μ sample: Mott relation fails near CNP; high T effects obtain agreement Low μ sample: Mott relation holds for all V_g; charged impurities induce high residual n_s	~ 60 μV/K @ 295 K	[36]
Epitaxial on C-face of SiC hole-doped: $n_s \sim 10^{12}$ cm^{-2}; $\mu \sim 20 \times 10^3$ cm^2/Vs @ 4 K	Temperature dependence of TEP $S_{xx}(T)$ nonlinear: $AT+BT^2$; $S_{xx}(B)$ shows quantum oscillations in (1/B); Sign change observed for S_{yx}, suppression of S_{xx} peak	~ 55 μV/K @ 230 K	[41]
Exfoliated on SiO_2/Si using e-beam lithography; $\mu \sim 4.56 – 12.9 \times 10^3$ cm^2/Vs	V_g dependence of TEP of device for three mobility states S_{xx}^{peak} increases with mobility; Effect of carier mobility on $S(B)$	~ 50 – 75 μV/K, @ 150 K	[32]
Fabricated on SiO_2/Si with e-beam lithography; $\mu \sim 12.9 \times 10^3$ cm^2/Vs	Low-T longitudinal and Hall resistivities (R_{xx}, R_{xy}) and Seebeck and Nernst coefficients ($S_{xx}$and S_{xy}) in quantizing magnetic fields, B.	S_{xx}^{peak} ~ 10 μV/K @ 20 K	[39]
Suspended Cu-CVD SLG	The T dependence of TEP for 50 < T < 300 K	9 μV/K @300 K	[45]
Few atomic layer thick, cm size sample CVD grown on Si/ SiO_2/Ni substrates $\rho \sim 3 \times 10^{-5}$ Ωcm	The T dependence of TEP for 75 < T < 300 K	10 μV/K @300 K	[46]
	Bilayer Graphene (BLG)		
Mechanical exfoliation on 300 nm SiO_2/Si substrate;	S_{xx} as function of V_g and T for 15< T <300 K $S_{yx}(T)$ dependence on disorder	$\|S_m\|$ ~ 95 μV/K @ 300 K	[51]

Sample	Measurements	Value	Ref.
$\mu \sim 2.6 \times 10^3$ cm²/Vs			
Mechanical exfoliation on 300 nm SiO_2/Si substrate; $\mu \sim 2-4 \times 10^3$ cm²/Vs	V_g dependence of TEP measured in BLG for various T $S \alpha T$ for large n_s, low T; Deviation from Mott relation for low n_s @ large T; Oscillations in $S_{xx}(B)$ and $S_{xy}(B)$ with n_s observed for high B	~ 100 μV/K @ 250 K	[49]
Mechanical exfoliation on 300 nm SiO_2/Si substrate; $\mu \sim 2-3 \times 10^3$ cm²/Vs	Electric field tuning of TEP in Dual-Gated BLG demonstrated – originates from band-gap opening; Enhanced TEP;	$S_m^{peak} \sim 180$ μV/K @ 250 K	[52]
	Single Layer-Multi Layer Graphene (SLG-MLG)		
S-BLG transistor; Mechanical exfoliation of graphene sheets onto 90 nm SiO_2/Si wafer; SLG/BLG identified by optical contrast & Raman	Optoelectronic response of S-BLG interface junction using photocurrent microscopy as function of V_g (Photocurrent is by photo-TE effect)	~ 6 μV/K @ 12 K	[53]
SLG - TLG epitaxial on 6H-SiC	TEP (over 300 – 550 K) as function of environment composition	p: 10 μV/K n: -20 μV/K (annealed @ 500K)	[54]
SLG–MLG CVD on Cu	Layer-dependence of the graphene Seebeck coefficient is peculiar & unexpected, that exceptionally increases with increasing thickness Gas flow induced voltage in MLG is not proportional to S	~ 30 μV/K (SLG) – 54 μV/K (HLG) @ RT	[55]
	Few Layer Graphene (FLG)		
FLG Pristine: on SiO_2/Si substrate, t ~5nm with possible structural defects; Treated: ACN, TPA attachments	Temperature dependence of TEP Power Factor Enhancement for Few-Layered Graphene Films by Molecular Attachments TEP increased ~ 4.5 times Results supported by simulations based on Kubo's formula	Pristine: ~ 40 μV/K Treated: 180 μV/K; 300< T <575 K	[56]
FLG On SiO_2/Si substrate; SLG & rGO	Temperature dependence of TEP Enhanced TEP of films with Oxygen Plasma Treatment Treatment generates disorders which open the π-π* gap leading to enhancement of TEP and reduction in σ	FLG: Pristine:~80 μV/K Treated:~ 700 μV/K @575 K; SLG: p @low T & n @ high T; S~ - 40 to 50 μV/K	[57]

Table 2. Measurements of thermopower in graphene

Checkelsky and Ong [18] have also reported measurements of TEP, S, and Nernst, S_{yx}, signals in graphene in non-zero and zero magnetic fields. In the absence of a magnetic field, they observe, besides the change in sign of S with V_g, a nominally linear-in-T dependence of the peak value S_m from ~20 K to 300 K (see Figure 6).

Seol *et al* [34], in their study of the thermal properties of graphene, have reported measurements of electrical conductivity, TEP and thermal conductivity of SLG flakes. They obtain room-temperature values of ~ - 80 μV/K for the Seebeck coefficient (S). They fit their data on the temperature dependence of S using the semiclassical theoretical model of Hwang *et al* [35], which assumes screened charged impurity scattering to dominate the electron transport. The temperature-dependent screening is found to produce a T dependence of S not captured in Mott formula.

With a view to investigate the effect of charged impurities on the TEP of graphene near the Dirac point, Wang and Shi [36] have measured both TEP and electrical conductivity of SLG samples with varying degree of disorders as characterized by carrier mobilities ranging from 1.5 – 13.0 x 10^3 cm^2/Vs, and examined the validity of the Mott relation as the low-density region near the Dirac point is approached. The four-point geometry they employed allowed them to measure the graphene resistivity properly by excluding the contact resistance and ensuring that both σ and V_{th} come from the same locations where the local temperatures are measured. They find that at higher temperatures, the Mott relation holds for low-mobility samples but fails in the vicinity of the Dirac point in high-mobility samples; however, below 100 K the deviation is insignificant even in the highest mobility sample. The reason for such a behaviour could be that, in the case of high-mobility graphene, the carrier density near the CNP can be so low that $k_B T \ll E_F$ no longer holds. It may be noted that the Mott relation is only an approximation for degenerate electron systems when T is far below the Fermi temperature. In low-mobility samples, on the other hand, the charged impurities are many and could induce relatively high residual carrier density near CNP so that $T_F \gg T$ and the Mott relation holds at all gate voltages. Wang and Shi suggest that, by properly taking account of the high-temperature effects, the Boltzmann transport theory can satisfactorily explain the experimental data on Seebeck coefficient in low-density electron systems near CNP.

In their recent measurements, shown in Figure (7), Shi and co-workers [32] have investigated the carrier mobility-dependence of TE transport properties of SLG in zero and non-zero magnetic fields. In the absence of magnetic field, they find that, with increase in mobility, the maximum value of S_{xx} increases, and exhibits an increasingly diverging trend accompanied by a sharper peak- to-dip transition around Dirac point. They find that the peak-to-dip width is related to the width of the minimum conductivity plateau, which is broader for the low-mobility state, and is known to be associated with disorder in graphene. Further, S_{xx} is found to converge to the same values at high gate voltages on either side of CNP, for all mobility values. This suggests that the effective carrier density is much greater than the charge density fluctuations induced by charged impurities near the Dirac point.

The magnetic field dependence of TEP has also been studied [16-18, 32]. In a magnetic field, carriers diffusing under the temperature gradient experience a Lorentz force, resulting in a non-zero transverse voltage. In the quantum Hall regime at a high magnetic field, the curves of S vs V_g show pronounced oscillations reflecting Landau quantization of the Dirac states. The peaks in S are aligned with those in conductance. The TEP and Nernst signals, which show quantized behaviour, are in agreement with the generalized Mott relation, except for strong deviations near the CNP. A Nernst signal, S_{xy} ~ 50 μV/K at 8 T is observed near the Dirac point

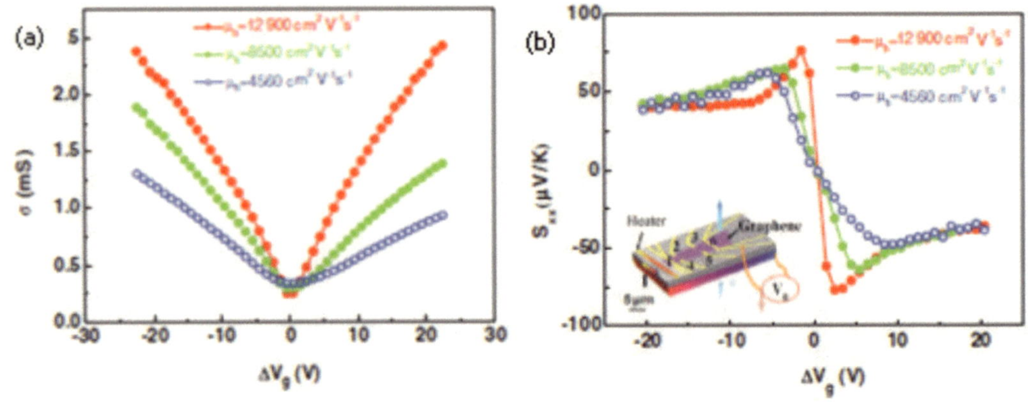

Figure 7. Gate voltage dependence of (a) Electrical conductivity, σ, and (b) Seebeck coefficient, S_{xx}, of device A at 150 K with three hole mobility values 12900, 8500 and 4560 $cm^2V^{-1}s^{-1}$. Inset of (b) shows SEM image of the device. (from [32])

(n = 0 LL). Such an enhancement of the Nernst signal is predicted in conventional 2D electron systems, and depends on the disorder strength [37, 38]. However, the observed large Nernst peak and the strong suppression of *S* at n=0 LL are inconsistent with the flat-top profiles calculated for a 2D system with quadratic dispersion. Shi and co-workers [32, 39] have investigated the dependence of the magneto-TE transport properties of graphene on electrical transport. They find, by independently varying the magnetic field and carrier density, that the derivative relation between S_{xx} and S_{xy}, discovered in conventional 2DEG systems, holds for graphene for high LLs except near the Dirac point, where different mechanisms such as carrier localization may be responsible.

Samples grown by different methods throw light on the different characteristics of TEP in graphene systems. The main graphene production techniques include dry and wet exfoliation, photo-exfoliation, growth on SiC, CVD, MBE and chemical synthesis (for a recent review see [40]). Although initially graphene samples have been mechanically exfoliated, with a view to investigate the TE characteristics further, the samples have been produced by other methods as well.

Wu *et al* [41] have investigated the TE response of relatively high-mobility (~ 20 x 10^3 cm^2/Vs at 4 K) SLG grown epitaxially on SiC substrates. For a carrier (hole) density of 1 x 10^{12} cm^{-2}, away from the Dirac point, the temperature dependence of TEP displays a deviation from the Mott relation. The data is found to obtain a least square fit to $AT+BT^2$, where A and B are temperature independent constants. The additional quadratic dependence instead of the otherwise linear dependence reflects the importance of the screening effect. The dielectric constant of SiC being a factor of 2 higher than that of SiO_2, the effect of screening is expected to be stronger in epitaxial graphene. Hwang *et al* [35] have shown that when the screening effect and its temperature dependence are taken into account, a quadratic correction to the TEP appears. In high magnetic fields, in the quantum Hall regime, a suppression of S_{xx} is observed even for the high mobility sample used. A study [42] finds TEP to be universal in the presence of disorder. The role of disorder needs to be investigated further.

Kim and co-workers [43] report measurement of TEP on graphene samples deposited on hexa-boron nitride substrates where drastic suppression of disorder is achieved. Their results show that at high temperatures where the inelastic scattering rate due to electron-electron (e-e) interactions is higher than the elastic scattering rate by disorders, the measured TEP exhibits an enhancement compared to the expected TEP from the Mott relation.

Graphene structures grown epitaxially on metal surfaces could reach sizes up to a micrometer with few defects. They can also be formed on the surface of SiC with the quality and number of layers in the samples depending on the SiC face used for their growth.

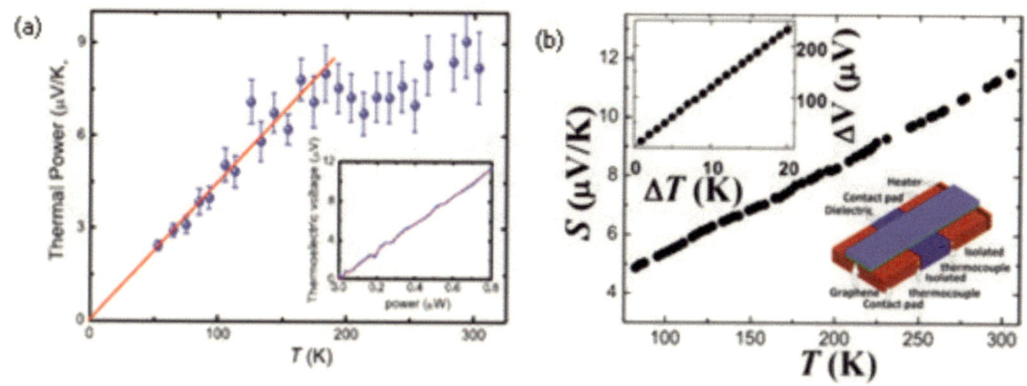

Figure 8. Temperature dependence of TEP measured in CVD-graphene samples of (a) Xu *et al* [45], and (b) Babichev *et al* [46]. Insets: (a) TE voltage as a function of applied Joule heat. (b) Upper-left: dependence of the TEP voltage on the applied temperature difference. Lower-right: a schematic view of device used in TEP measurements.(from [45, 46])

The carbon-terminated surface can produce few layers with low mobility whereas the silicon-terminated surface can give many layers with higher mobility [40]. Chemically exfoliating graphene is another method of preparing good quality and large amount of few-layer graphene sheets [44].

There exist a few reports of measurements of TEP of CVD-grown graphene [45, 46]. Figure 8 shows the observed temperature dependences. Other investigations have demonstrated the TEP of CVD-grown graphene to be a sensitive probe to the surface charge doping from the environment and the device concept promises use in gas/chemical sensing [47]. An initially degassed n-type graphene sample, upon exposure to gases, was found to become p-doped or

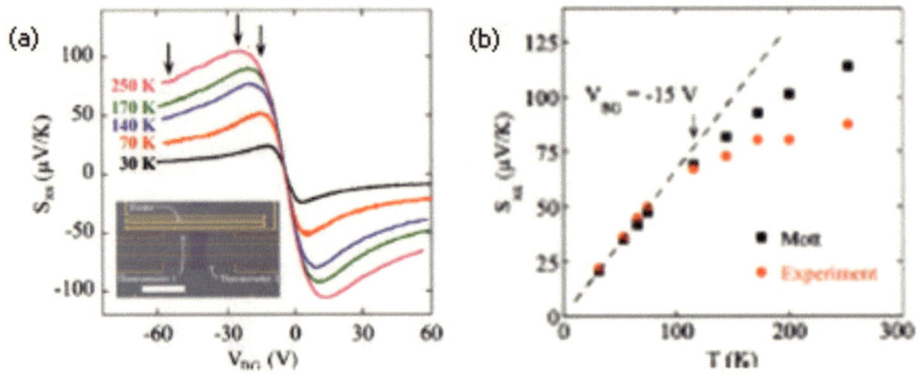

Figure 9. TEP as a function of (a) backgate voltage V_{BG} for T=30, 50, 70, 140, 170 and 250 K, and (b) temperature T for V_{BG}=-15 V. Inset in (a): optical image of a typical device. (from [49])

further n-doped during exposure depending on the properties of the ambient gases as evidenced by a monotonic change in sign of TEP.

In a graphene sample, the substrate on which the graphene layer is exfoliated, affects the morphology of the graphene specimen and is a source of impurities. In a suspended graphene (SG) sample, on the other hand, the substrate is etched away so that the graphene is suspended over a trench approximately 100 nm deep, with most of the impurities sticking to the graphene sheet [48]. Annealed SG samples showed both ballistic and diffusive carrier transport properties with carrier mobilities more than $2x10^4$ cm^2/Vs. The warping of the layers can be avoided with the use of a top gate [21].

In the case of BLG, Nam *et al* [49] have measured the TEP, S_{xx}, for 30<T< 250 K and for different charge-carrier densities. As in SLG, [16-18] the ambipolar nature of the carriers manifests itself as the sign change in Figure 9.(a) at the CNP. Their low-T result of TEP is found to follow the semi-classical Mott formula, and for high carrier densities TEP shows a linear-in-T dependence implying a weak electron-phonon interaction and negligible phonon-drag effect in BLG. For a low carrier density, a deviation from the Mott relation along with a saturating tendency of TEP, is observed at higher temperatures (See Figure 9.(b)) and is attributed to the low Fermi temperature in the BLG.

There do not seem to be till date any reports on measurements of TEP of graphene nanoribbons.

In the following sections, we discuss, based on the Boltzmann formalism, the present theoretical understanding of the observed phenomena, in terms of the diffusion and phonon-drag contributions. An analysis of measured TEP is usually done by separating the two contributions by making use of their characteristic temperature dependences at lower temperatures [11]. Often, in literature, the diffusion component, S_d, for a degenerate system, is assumed to vary linearly with temperature. It may be seen from Eq. (36) that the Mott relation suggests such linear temperature dependence, provided the energy dependence of relaxation time does not vary with temperature, though its magnitude will. S_d, thus, reflects the energy dependence of a scattering mechanism and is determined not only by the magnitude of scattering, but also by details concerning the distribution of the scatterers and their type. The phonon-drag component, S_g, on the other hand, unlike S_d, depends only on the electron-acoustic phonon coupling strength. Its low-temperature dependence is expected, as in conventional 2D systems, to display a characteristic peak, which exhibits the role of phonon scattering mechanisms. Since the effect of electron-phonon coupling in determining the resistivity of graphene is observed to be weak [50], especially at low temperatures, the phonon drag component has often been assumed to be absent, and an analysis of TEP, for the range of temperatures ($10<T<300$ K) investigated in the experiments on graphene, is given based on the diffusion component.

3. Diffusion thermopower

Diffusion thermopower, S_d, is known to provide unique information complementary to the electrical transport coefficients in metals and semiconductors [8-12]. The TE transport coefficients, defined above and related to electrical transport, are determined by the band structure and scattering mechanisms operative in a system. S_d can, therefore, be a sensitive and powerful tool to probe and elucidate details of the thermoelectric transport and understand the carrier transport mechanisms of graphene that cannot be probed by electrical conductivity alone.

In this review, we adopt the Boltzmann approach, found to be robust especially for transport in graphene far away from the Dirac point [58]. We give here, in brief, the basic theory of TEP and the expressions used in the present analysis of S_d in graphene systems.

3.1. Basic formalism – Boltzmann approach

Low field transport in many of the systems is often described by the Boltzmann transport equation (BTE) [59-61]. This semi-classical Boltzmann approach is known to be appropriate for structures in which the potentials vary slowly on both the spatial scale of the electron thermal wavelength and the temporal scale of the scattering processes. The conventional theory of charge carrier transport in 2D semiconductors is based on this formalism, and the TE coefficients are commonly obtained by solving the BTE in the relaxation time approximation [11].

In the regime of large chemical potential, the nature of transport of the massless Dirac fermions through a 2D graphene membrane may be accessed by the Boltzmann formalism [3] and one may write an expression for TEP in graphene systems in terms of the fundamental transport coefficients.

3.1.1. Transport coefficients in graphene systems

We consider a graphene system of length L, and width W to be a 2D homogeneous system of charge carriers, with a density n_s induced by an external gate bias V_g. In the presence of a temperature gradient ∇T applied along the plane of the graphene layer(s), Eq. (12), for the effective electric field, **E**, and the heat current density, **U**, under open circuit conditions (**J** = 0), reduce to

$$\mathbf{E} = S\nabla T \tag{21}$$

and

$$\mathbf{U} = \kappa(-\nabla T) \tag{22}$$

The electric current density, and heat current density, can be evaluated by solving the Boltzmann transport equation in the relaxation time approximation. Assuming the electric field to be weak and the displacement of the distribution function from thermal equilibrium to be small, the electron distribution function $f(E_{\mathbf{k}})$ can be written as [29, 30]

$$f\left(E_k\right) = f^0\left(E_k\right) + \tau(E_k)\mathbf{v_k} \cdot \left[e\mathbf{E} - \left(\frac{E_k - E_F}{T}\right)\nabla T\right]\left(-\frac{\partial f^0\left(E_k\right)}{\partial E_k}\right) \tag{23}$$

where, $\mathbf{v_k}$ is the velocity of the electrons in state $\mathbf{k}$, $\tau(E_{\mathbf{k}})$ is the electron momentum relaxation time, $f^0(E_{\mathbf{k}})$ is the equilibrium Fermi-Dirac distribution function and E_F is the Fermi energy which is determined by the carrier density

$$n_s = \int \rho(E_k)\, f^0(E_k)\, dE_k \tag{24}$$

with $\rho(E_k)$ denoting the density of states of the system.

In Eqs. (11a) and (11b), the current densities **J** and **U**, can be evaluated from the expressions:

$$\mathbf{J} = \left(g_s g_v / At\right) \sum_k e\, \mathbf{v_k}\, f\left(E_k\right) \tag{25}$$

and

$$\mathbf{U} = \left(g_s\, g_v / At\right) \sum_k (E_k - E_F)\, \mathbf{v_k}\, f\left(E_k\right) \tag{26}$$

Here, g_v and g_s denote the valley and spin degeneracies, A=LW is the area of the surface and 't' the layer thickness of the graphene system. Using Eqs.(21) – (23), **J** and **U** can be expressed as [29]

$$\mathrm{J} = e^2 K_{11} E + (e/T) K_{12} (-\nabla T) \tag{27}$$

$$\mathbf{U} = e K_{21} E + (1/T) K_{32} (-\nabla T) \tag{28}$$

where the coefficients $K_{rs,}$ are given by

$$\begin{aligned}
&SLG \quad K_{rs} = \frac{1}{t\pi\hbar^2}\int_0^\infty E_k (E_k - E_F)^{r-1}\tau^s(E_k)\left(-\frac{\partial f^0(E_k)}{\partial E_k}\right) dE_k \quad &(a)\\
&BLG \quad K_{rs} = \frac{m^*}{2t\pi\hbar^2}\int_0^\infty v_k^2 (E_k - E_F)^{r-1}\tau^s(E_k)\left(-\frac{\partial f^0(E_k)}{\partial E_k}\right) dE_k \quad &(b)\\
&\Lambda GNR \quad K_{rs} = \frac{2}{Wt\pi\hbar}\sum_n \int_{E_n}^\infty v_k (E_{n,k_y} - E_F)^{r-1}\tau^s(E_{n,k_y})\left(-\frac{\partial f^0(E_{n,k_y})}{\partial E_{n,k_y}}\right) dE_{n,k_y} \quad &(c)
\end{aligned} \tag{29}$$

In the absence of temperature gradient ($\nabla T = 0$), Eq. (27) gives electrical conductivity as

$$\sigma = e^2 K_{11} \tag{30}$$

From Eqs. (21), (22), (27) and (28), one obtains expressions for the diffusion contribution to thermopower, S_d, and for electronic thermal conductivity, κ_e as

$$S_d = \frac{1}{eT} K_{11}^{-1} K_{21} \tag{31}$$

and

$$\kappa_e = \frac{1}{T}\left[K_{31} - \left((K_{31})^2 / K_{11}\right)\right], \tag{32}$$

respectively. Equation (31) may be expressed as [11]

$$S_d = \frac{1}{eT}\left(\frac{\langle E_k \tau(E_k)\rangle}{\langle \tau(E_k)\rangle} - E_F\right) \tag{33}$$

with $\langle F(E_k)\rangle = \dfrac{\int F(E_k)\, E_k(-\partial f^0/\partial E_k) d E_k}{\int E_k(-\partial f^0/\partial E_k) d E_k}$

Equations (30) – (33) show that evaluation of the transport coefficients requires a knowledge of the relaxation time(s), $\tau(E_k)$. Note that the overall contribution to S_d and κ_e, from the various scattering mechanisms operative in the system, can be evaluated assuming the overall relaxation time $\tau(E_k)$ to be given by Matthiessen's rule [11, 29, 30]:

$$\tau^{-1}(E_k) = \Sigma_j\, \tau_j^{-1}(E_k), \tag{34}$$

where the sum is over all the relevant scattering mechanisms, j.

Often in literature, limiting forms of S_d are used in the analysis of data. In the degenerate limit a good approximation to Eq. (32)is the well-known Mott expression [11, 29, 30]:

$$S_d = \frac{\pi^2 k_B^2 T}{3e} \left[\frac{d \ln \sigma(E_k)}{dE_k} \right]_{E_k = E_F}, \tag{35}$$

where, $\sigma(E_k)$, is the energy dependent conductivity. If the energy dependence of relaxation time is taken as $\tau(E) \sim E^p$, Eq. (35) can be more simply expressed for the three graphene systems, as [35, 62-65]

$$\begin{array}{lll} \text{SLG, BLG} & S_d = \dfrac{\pi^2 k_B^2 T}{3e E_F} \left[p+1 \right] & \text{(a)} \\ \text{AGNR} & S_d = \dfrac{\pi^2 k_B^2 T}{3e E_F} \left[p + \left(\left(\dfrac{E_F}{E_n} \right)^2 - 1 \right)^{-1} \right] & \text{(b)} \end{array} \tag{36}$$

where the first term reflects the scattering mechanisms. The parameter p can also be expressed in the form

$$p = \frac{E_F}{\tau(E_F)} \left[\frac{d\tau(E_k)}{d E_k} \right]_{E_k = E_F}. \tag{37}$$

Eq. (37) brings out the feature that S_d is determined by not only the magnitude of scattering but also by the energy dependence of $\tau(E_k)$ at $E_k = E_F$. It may be noted that provided the energy dependence of $\tau(E_k)$ does not vary with temperature (though its magnitude will), S_d will be a linear function of temperature.

A solution of the Boltzmann equation in the relaxation time approximation may be applied exactly when the important collision processes are all elastic [30, 59-61]. It is also applicable when the inelastic processes include non-polar optic and intervalley phonon scattering. If polar optic phonon scattering is also important, the method is applicable only at high temperatures. Solutions of Boltzmann equation when polar optic phonon scattering is dominant may be obtained by applying variational or numerical methods.

3.2. Scattering mechanisms

Central to understanding the TE transport properties of graphene, are the mechanisms causing the scattering of the charge carriers. A better understanding, therefore, of the relative importance of the operative scattering mechanisms, which varies with temperature and carrier concentrations in graphene, enables useful improvements in the transport properties of graphene for various possible TE applications.

Scattering in graphene which could contribute to carrier transport may result from both intrinsic and extrinsic sources. The extrinsic sources may be vacancies, surface roughness arising from rippling of the graphene sheet, disorder, which can create electron-hole puddles, and charged impurities, known to be the main scattering mechanism in graphene. Apart from the graphene layers, the substrates may also be a source of impurities. Besides, there are additional scattering sources such as neutral point defects [1, 3]. In principle, the limitation due to the extrinsic scattering mechanisms can be reduced by improved growth/fabrication techniques.

In the absence of extrinsic scattering sources, phonons, which constitute an intrinsic source of scattering in a system, limit carrier mobility at finite temperatures [59-61]. Phonon scattering may be due to intravalley acoustic and optical phonons which induce the electronic transitions within a single valley, and intervalley phonon scattering that induces electronic transitions between different valleys [3]. The intravalley acoustic phonon scattering, induced by low energy phonons and considered an elastic process, gives a quantitatively small contribution in graphene even at room temperature due to the high Fermi temperature of graphene. Shishir *et al* [66] in their calculations of mobility and resistivity, respectively, including contributions of optical phonon scattering, found that the effect of optical phonons cannot be neglected. The role of remote interface polar optical phonons in the substrate in graphene transport also seems to be important.

In the case of SG, the intrinsic scattering mechanisms limiting electron transport in SG layers are due to in-plane and out-of-plane (flexural) acoustic phonons. Recent investigations of electron and phonon transport in SG indicate that in the free standing case (absence of strain) the major contribution to resistivity and thermal conductance is from acoustic flexural phonons, and this intrinsic limitation can be reduced by the effect of strain [28, 67].

Evaluation of the transport coefficients requires the knowledge of the relaxation times of the scattering mechanisms. In the following, we give the expressions for the momentum relaxation times of the extrinsic and intrinsic scattering mechanisms.

3.2.1. Relaxation times

The expressions for the momentum relaxation times for the various scatterings in graphene systems may be expressed as [3]

$$\frac{1}{\tau_j(E_k)} = \frac{2\pi}{\hbar}\sum_{k'}\sum_{q} |C_j(\boldsymbol{q})|^2 F(\theta)(1 - cos(\theta))\Delta(E_k, E_{k'}) \tag{38}$$

where 'θ' is the angle between initial (k) and final states (k'), and $|C_j(q)|^2$ is the matrix element for the interaction of electrons with the scattering source, 'j'. The form factor, $F(\theta)$, due to the overlap of the wavefunctions and arising due to the chiral nature of the graphene carriers, is given by [3, 27] $F(\theta)=\frac{1}{2}(1+\cos(\theta))$SLG, $F(\theta)=\frac{1}{2}(1+\cos(2\theta))$BLG, $F(\theta)=\frac{1}{2}(1+\cos(\theta_{n,k_y} - \theta_{n',k'_y}))$ AGNR

The overall momentum relaxation time $\tau(E_k)$ can be taken to be given by Matthiessen's rule (Eq.(34)). The factor $\Delta(E_k, E_{k'})$ in Eq. (38) is given, for elastic scatterings, by

$$\Delta(E_k, E_{k'}) = \delta(E_k - E_{k'}) \tag{a}$$

In the case of phonons,

$$\Delta(E_k, E_{k'}) = \frac{N_q\left[1-f^0\left(E_k+\hbar\omega_q\right)\right]\delta\left(E_k-E_{k'}+\hbar\omega_q\right)+(N_q+1)\left[1-f^0\left(E_k-\hbar\omega_q\right)\right]\delta\left(E_k-E_{k'}-\hbar\omega_q\right)\theta\left(E-\hbar\omega_q\right)}{\left[1-f^0\left(E_k\right)\right]} \tag{b}$$

$$\Delta(E_k, E_{k'}) = \frac{2k_BT}{\hbar\omega_q} \tag{c}$$

(39)

In eq.(39b), $\theta(x)$ is the unit-step function and the first (second) term corresponds to the absorption (emission) of a phonon of wavevector $\mathbf{q}$ and energy $\hbar\omega_{\mathbf{q}}$. The phonon distribution N_q is, for low fields, given by the equilibrium Bose distribution, $N_q^0 = (\exp(\hbar\omega_q/k_BT)-1)^{-1}$. At high temperatures, in the equipartition (EP) regime, where $\hbar\omega_q << k_BT$, one may consider scattering of electrons from acoustic phonons to be quasi-elastic and with $N_q \sim k_BT/\hbar\omega_q$, the factor $\Delta(E, E')$ in Eq.(38) becomes (39c). Often, in literature, expressions for electron–acoustic phonon relaxation times obtained in the EP approximation are used in the analysis of transport properties [3].

In the 2D material of graphene at low temperatures, an understanding of electron-phonon interaction is important both from basic physics and technology points of view [3]. In typical conductors, electrons are scattered by phonons producing a finite temperature-dependent resistivity $\rho(T)$ [59-61]. Recent investigations [68] of resistivity of graphene show that there is a change in temperature dependence of resistivity from $\rho(T) \sim T$, in the high temperature limit, to $\rho(T) \sim T^4$, at low temperatures, reflecting the 2D nature of the electrons and the acoustic phonons in graphene. At low enough temperatures, recent theoretical studies [69] of the phonon-induced graphene resistivity have shown that, the resistivity eventually goes as $\rho(T) \sim T^6$, for screened phonon scattering. At such low temperatures the major contribution to this resistivity is from in-plane acoustic phonons [50, 70]. The crossover between the two distinct regimes can be described by the characteristic Bloch-Gruneisen (BG) temperature defined as $T_{BG} = 2\hbar v_s k_F / k_B$, where v_s is the velocity of sound and $k_F = (\pi n_s)^{1/2}$ is the Fermi wavevector [68]. The electrostatic tunability of the chemical potential, $E_F = \hbar v_F k_F$, in graphene allows for a wide range of control of T_{BG}; for graphene with $n_s = 1 \times 10^{15}$ m^{-2}, $T_{BG} \sim 17$ K for LA modes.

The expressions for the momentum relaxation rates for in-plane and flexural acoustic phonon, non-polar optical phonon, surface polar optical phonon and for roughness, impurity and vacancy scatterings in the graphene systems are given in Table 3.

Suspended graphene (SG) allows for the investigation of the intrinsic properties of the material, unperturbed by the presence of a substrate. It has been realized that mechanical deformations of graphene sheets affect the electronic properties. This is of special relevance for strain-engineering aimed at controlling the electronic properties of graphene by suitably engineering the deformations ([67] and references therein). Employing the semi-classical Boltzmann transport formalism, Mariani and Oppen [67] and Ochoa *et al* [28], have studied electron-phonon contribution to resistivity in free-standing SLG and BLG and discussed the role of strain in SG. Restricting to the regime of $T \gg T_{BG}$, and assuming the phonons to couple to electrons through a screened scalar deformation potential (constant g) and a vector potential (β) associated with the changes in bond length between carbon atoms, they have given expressions for the momentum relaxation rates in SG for both free-standing and strained layer cases (see Table 3).

3.3. Diffusion thermopower in graphene systems

Besides the experimental investigations (see Table 1.), the TEP of graphene has also attracted much theoretical attention. The theoretical investigations made to understand the experimental results have so far been mostly on the basis of the diffusion TEP, ignoring the drag component.

The following features of TEP, first observed in SLG samples by Zuev *et al* [16], Wei *et al* [17] and Chekelsky and Ong [18], are found to be characteristic of graphene. The measured TEP reaches a value ~ 100 μV/K at room temperature. The sign of TEP changes across the charge neutrality point (CNP) as the majority carriers change from electrons to holes. Away from the CNP, the TEP shows a $n_s^{-1/2}$ dependence on the carrier density n_s. At low temperatures, it exhibits a linear temperature dependence, in consonance with the Mott formula Eq.(36). For higher temperatures, a deviation from Mott formula is observed. Measurements on high mobility graphene samples [51], show that Mott relation fails near the Dirac point. In the case of FLG samples, Li *et al* [55] find that TEP depends on the number of layers, increasing with increase in thickness and reaching a peak value for six layers.

In the presence of a quantizing magnetic field, the TEP of graphene exhibits additional interesting effects [16-18, 32]; however, these do not form the content of the present review. The theoretical studies of S_d of graphene systems, in the absence of magnetic fields, based on the semi-classical Boltzmann formalism are briefed below; the drag component is discussed in the next section.

3.3.1. *Diffusion thermopower in SLG*

Much of the theoretical investigations of diffusion TEP in graphene has been made on SLG with interest being devoted mostly to the TEP at higher temperatures ($10<T<300$K).

Using a phenomenological theory for transport in graphene, close to Dirac point, based on the semiclassical Boltzman approach Peres and coworkers [72, 76] have obtained an expression for diffusion TEP including the scattering mechanism involving midgap states arising from local point defects in the form of vacancies, cracks, boundaries, impurities in the substrate or in corrugated graphene. They find that this mechanism leads to a similar k dependence as charged impurities, and show that the dependence of S_d on the particle density is different from that of the conventional 2DEG and of graphene sheet with only charged impurities in the substrate.

Lofwander and Fogelstrom [33], have presented calculations for the linear response to electrical and thermal forces in graphene for the case of strong impurity scattering in the self-consistent t-matrix approximation. At low temperatures, the electronic contribution to TEP is found to be linear in T with slope proportional to the inverse of the impurity density and the impurity strength, so that TEP could provide information about impurities in graphene. Further, for moderately large impurity strengths, a non-linear temperature dependence is obtained and $S \sim 100$ μV/K.

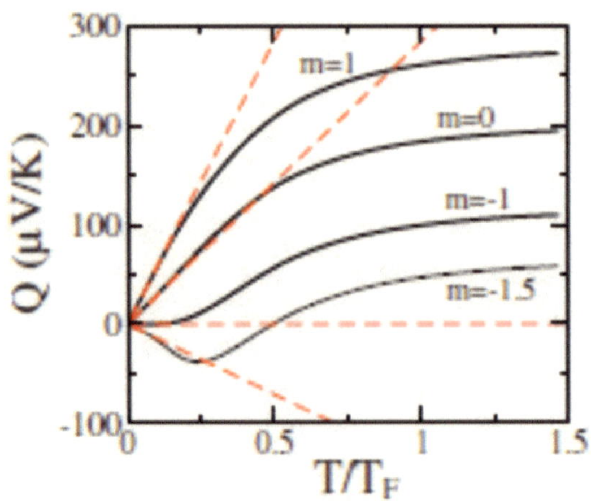

Figure 10. Hole TEP for different energy dependent scattering times, $\tau \sim E^m$. Dashed lines show corresponding Mott variations.(from [35])

Scattering Mechanism	Relaxation Rate, $\tau^{-1}(E)$	Ref.
	Single Layer Graphene (SLG)	
Acoustic Phonons	$\frac{2\pi}{\hbar}\int_{0}^{2\pi}\int_{0}^{\infty}\frac{D_{ac}^{2}\hbar q}{2A\rho v_s}\left(1-\left(\frac{q}{2k}\right)^2\right)\Delta(E_k, E_{k'})k'dk'd\theta$	[71]
Optical Phonons	$\frac{D_{op}{}^{2}}{\rho\,\omega_{LO}(\hbar v_F)^2}[(E-\hbar\omega_{LO})(N_q+1)\,\theta(E-\hbar\omega_{LO})+(E+\hbar\omega_{LO})N_q]$	[66]
Impurities	$\frac{\sqrt{n_i}\,Z e^2}{\hbar^2 v_F k\,(4\varepsilon\varepsilon_o(1+(\gamma_i/k_F)))}$	[72]
Surface roughness	$\frac{2k}{v_F}\left(\frac{\Lambda n_s e^2\Delta}{8\pi\hbar\varepsilon\varepsilon_o}\right)^2\frac{1}{\sqrt{1+k^2\Lambda^2}}\exp\left(\frac{k\Lambda}{\sqrt{1+k^2\Lambda^2}}\right)$	[66]
Vacancies	$\frac{\pi^2 v_F n_{vac}}{k(\ln(kR_o))^2}$	[72]
	Suspended Single Layer Graphene (SG) *	
Acoustic Phonons (In-plane)	$\left[\frac{g^2}{2}+\frac{\hbar^2 v_F^2\beta^2}{4a^2}\left(\frac{1}{v_L^2}+\frac{1}{v_T^2}\right)\right]\frac{E}{2\rho\hbar^3 v_F^2}k_B T$	[28]
Acoustic Phonons (Flexural)	$\left(\frac{g^2}{2}+\frac{\hbar^2 v_F^2\beta^2}{4a^2}\right)\left[\frac{(k_B T)^2}{64\pi\hbar\kappa^2 E}+\frac{k_B T E}{32\pi\hbar^2 v_F^2\kappa\sqrt{\rho\kappa}}\right]\ln\left(\frac{k_B T}{\hbar\omega_c}\right)$	[28]
Acoustic Phonons (Flexural) (strained)	$\left[\frac{g^2}{2}+\frac{\hbar^2 v_F^2\beta^2}{4a^2}\right]\frac{E(k_B T)^4}{16\pi\rho^2\hbar^5 v_F^2 v_L^6\bar{u}^3}\times\left[\Re_2\left(\frac{\alpha k_B T}{\hbar v_L^2\bar{u}}\right)+\Re_1\left(\frac{\alpha k_B T}{\hbar v_L^2\bar{u}}\right)\right]$	[28]
	Bilayer Graphene (BLG)	
Acoustic phonons	$\frac{4mD_{ac}^2\sqrt{2mE_k}}{\pi\rho\hbar^3 v_s}\int_0^{2k}\frac{(q/2k)^3}{\sqrt{1-(q/2k)^2}}\frac{g(q,k)}{1-f(E_k)}x\left\{N_q[1-f(E_k+\hbar\omega_q)]+(N_q+1)[1-f(E_k-\hbar\omega_q)]\right\}dq$	[73]
Surface polar optical phonons	$\frac{2mAe^2F^2}{\pi\hbar^3}\left\{\left[(n_q+1)\int_0^{\varphi_{max}}\frac{1}{\left\lvert\varepsilon(q_\pm^e)\right\rvert^2}\left[\frac{e^{-2q_\pm^e d}+e^{-2q_\pm^e(d+c)}}{2}\right]\frac{g(q_\pm^e)d\varphi}{\left\lvert 2q_\pm^e-2k\right\rvert}\right]+\left[n_q\int_0^{\pi}\frac{1}{\left\lvert\varepsilon(q_\pm^a)\right\rvert^2}\left[\frac{e^{-2q_\pm^a d}+e^{-2q_\pm^a(d+c)}}{2}\right]\frac{g(q_\pm^a)d\varphi}{\left\lvert 2q_\pm^a-2k\right\rvert}\right]\right\}$	[74]
Coulomb impurities	$\frac{2mn_i}{\pi\hbar^3}\left(\frac{2\pi e^2}{\varepsilon}\right)^2\int_0^{2k}\left(\frac{q}{2k}\right)^2\frac{g(q,k)}{k[1-(q/2k)^2]^{1/2}}\frac{dq}{(q+q_{TF})^2}$	[75]
Short range disorder	$\frac{n_d V_0^2 m}{\pi\hbar^3}2\int_0^{2k}\left(\frac{q}{2k}\right)^2\frac{g(q,k)}{k[1-(q/2k)^2]^{1/2}}dq$	[75]
	Armchair Graphene Nanoribbon (AGNR)	
Acoustic phonons	$\frac{2\pi}{\hbar}\sum_{n',k'_y}\sum_{q_y}\frac{D_{ac}^2\hbar\lvert q_y\rvert}{\rho L\,W^3 v_s}\left\lvert F_{n,n'}\right\rvert^2(1+\cos(\theta_{n,k_y}-\theta_{n',k'_y}))(1-\cos(\theta))$ $\times\Delta(E_{n,k_y},E_{n',k'_y})\delta_{q_y,k_y-k'_y}$	[65]
Optical phonons	$\frac{2\pi}{\hbar}\sum_{n',k'_y}\sum_{q_y}\frac{D_{op}^2\hbar}{\rho L\,W^3\omega_{LO}}\left\lvert F_{n,n'}\right\rvert^2(1+\cos(\theta_{n,k_y}-\theta_{n',k'_y}))(1-\cos(\theta))$ $\times\Delta(E_{n,k_y},E_{n',k'_y})\delta_{q_y,k_y-k'_y}$	[65]

Scattering Mechanism	Relaxation Rate, $\tau^{-1}(E)$	Ref.
Edge roughness	$\frac{2\pi}{\hbar}\sum_{n',k'_y}\sum_{q_y}\frac{2E_{n'}^2}{L}\frac{H^2}{W^4}\frac{\Lambda}{1+(q_y\Lambda)^2}\lvert F_{n,n'}\rvert^2(1+\cos(\theta_{n,k_y}-\theta_{n',k'_y}))(1-\cos(\theta)) \times\Delta(E_{n,k_y},E_{n',k'_y})\delta_{q_y,k_y-k'_y}$	[65]
Impurities	$\int dx_0dy_0dz_0N_I(x_0,y_0,z_0)\tau_{IMP}^{-1}(E_{n,k_y})$ $\tau_{IMP}^{-1}(E_{n,k_y})=\frac{2\pi}{\hbar}\sum_{n',k'_y}\sum_{q_y}\lvert C_{IMP}(q_y)\rvert^2(1+\cos(\theta_{n,k_y}-\theta_{n',k'_y}))(1-\cos(\theta))\Delta(E_{n,k_y},E_{n',k'_y})\delta_{q_y,k_y-k'_y}$, with $\lvert C_{IMP}(q_y)\rvert^2=\frac{2}{L^2W^2}\left(\frac{e^2}{4\pi\varepsilon_0\kappa}\right)^2\left\lvert\int_{-W/2}^{W/2}\left[\cos(k_{n'}-k_n)x-\cos(\Delta K-k_{n'}-k_n)x\right]K_0(\lvert q_y\rvert b)dx\right\rvert^2$	[65]

*Expressions for SG are for $T>>T_{BG}$

Meanings of symbols (Table 3.):

v_L – Longitudinal sound velocity; v_T – Transverse sound velocity; ρ - mass density

D_{ac} - Acoustic phonon deformation potential constant; D_{op} - Optical phonon deformation potential constant

$N_{\mathbf{q}}$ – Bose-Einstein distribution function with wave vector **q**

Δ - rms height; Λ - Correlation length for interface roughness; n_i - Density of charged impurities in the sample;

n_{vac} - Number of vacancies of radius R_o; $\gamma_i = D(E_F)e^2 / 2\varepsilon\varepsilon_o$; ε – permeability of substrate

g - Screened deformation potential constant; $\beta=\partial\log t/\partial\log a$, a - Distance between nearest carbon atoms

$\omega_c = a\, q_c^2$ - Infrared cutoff frequency; $\Re_n(\gamma)=\int_0^\infty dx\, x^3/(\gamma^2x^2+1)\left[\exp(\sqrt{\gamma^2x^4+x^2})-1\right]^n$

d - distance between graphene and substrate ; $F^2=(\hbar\omega_s/2A\varepsilon_0)[1/(\varepsilon_s^\infty+1)-1/(\varepsilon_s^0+1)]$; $\hbar\omega_s$ - energy of SPPs

ε_s^∞ (ε_s^0) - high (low) frequency substrate dielectric constant,

$g(q,k)=(1-2(q/2k)^2)^2$;$\varphi_{max}=a\cos\sqrt{\hbar\omega_s/E_k}$;$q_\pm^e=k\cos\phi\pm\sqrt{k^2\cos^2\phi-\frac{2m\hbar\omega_s}{\hbar^2}}$; $q_\pm^a=-k\cos\phi\pm\sqrt{k^2\cos^2\phi+\frac{2m\hbar\omega_s}{\hbar^2}}$

$g(q_\pm^e)=\frac{\sqrt{E_k}-\sqrt{E_{q_\pm^e}}\cos\phi}{\sqrt{E_k-\hbar\omega_s}}$; $g(q_\pm^a)=\frac{\sqrt{E_k}+\sqrt{E_{q_\pm^a}}\cos\phi}{\sqrt{E_k+\hbar\omega_s}}$; $E_{q_\pm^{e(a)}}=\frac{\hbar^2q_\pm^{e(a)2}}{2m}$;

$q_{TF}=4me^2/\varepsilon\hbar^2$: Thomas-Fermi screening wavevector ; n_d (V_0) - density (strength of potential) of short range impurities.

W – Width of AGNR; L – Length of AGNR;

$\lvert F_{n,n'}\rvert^2=\left\lvert\left[\frac{\sin(k_{n'}-k_n)(W/2)}{k_{n'}-k_n}-\frac{\sin(\Delta K-k_{n'}-k_n)(W/2)}{\Delta K-k_{n'}-k_n}\right]\right\rvert^2$

Ionized impurity atom located at distance $b=[(x-x_0)^2+(y-y_0)^2+(z-z_0)^2]^{1/2}$ away from the center of the ribbon

(x_0, y_0, z_0) – Impurity atom position; $N_I(x_0, y_0, z_0)$- Density of impurities in GNR layer and the substrate.

Table 3. Relaxation rates for various scattering mechanisms in graphene systems

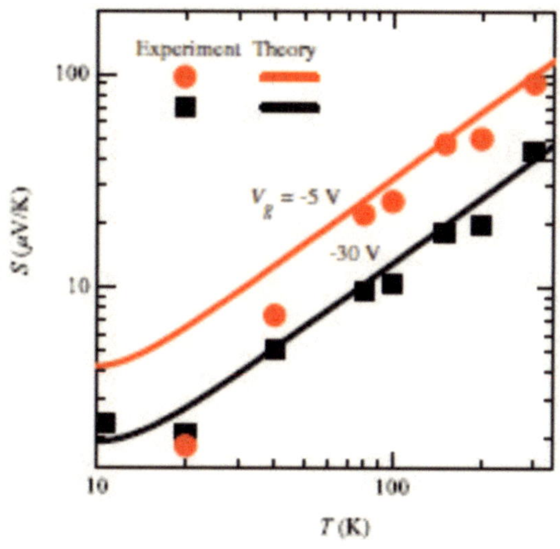

Figure 11. Temperature variation of TEP, S (= S_d + S_g) calculated using balance equation method for n_s=2.16×10^{16} m^{-2} (black curve) and 3.6×10^{15} m^{-2} (red curve), corresponding, respectively, to V_g = −30 V (black squares) and −5 V (red dots) of the experimental data of [16]. (from [77])

Kubakaddi [62] in his study of TEP at low temperatures, has given Mott formula Eq.(36a) for S_d, and predicted the $n_s^{-1/2}$ and linear-in-T dependences. S_d is found to be relatively dominant for $T < 2$ K and $T > 10$ K (see Figure (19)).

Motivated by the experiments of [16], [17] and [18], Hwang *et al* [35] have developed, in the linear response approximation, a theory for the thermopower of graphene, ignoring the drag component. Incorporating the energy dependence of various transport times, they elucidate the comparative importance of scattering mechanisms in graphene. They find that a reasonable explanation for the measured TEP [16-18] can be given by the scattering by random screened charged impurities located in the graphene environment. The density dependence shows the expected $n_s^{-1/2}$ behaviour at high densities. Figure 10 shows the temperature dependence of their calculated TEP for different scattering-time energy-dependence exponents (m). At high temperatures, the TEP is independent of T and approaches a limiting value. Further, the Mott formula is found to apply well, for $T \leq 0.2T_F$ and high carrier densities, but fail in the low-density limit, where electron-hole puddles may dominate and an effective-medium theory may be required.

Considering scattering of electrons by both impurities and phonons, Bao *et al* [77] have studied the behaviour of S_d away from CNP. Within the framework of balance equation approach, they obtain an expression for S_d, which in the limit of large concentrations is shown to reduce to Mott formula. Comparing their calculations of TEP, S (=S_d+S_g) with the data of Zuev *et al* [16] for two values of gate voltage (see Figure 11), they find that for $T > 10$ K, S_d plays an essential role and the TEP shows a linear dependence on T.

Vaidya *et al* [63] have studied S_d for 30<T<300 K, using Eq.(33) and considering the electrons to be scattered not only by phonons [50, 66] but also other disorder-related scattering mechanisms namely, charged impurities *via* long-range Coulomb interaction [72], the vacancies [72] in the system and surface roughness [66] arising from the rippling of the graphene sheet deposited on oxidized Si substrate (see Figure (12)). S_d is found to increase almost linearly with temperature, determined mainly by vacancy and impurity scatterings. In the case of the scattering of electrons by non-polar optical phonons *via* deformation potential coupling, although the emission of optical phonons is possible only if the kinetic energy of carriers exceeds $\hbar\omega_{LO}$, this mechanism could become important for higher energy carriers. For n_s = 2×10^{16} m^{-2}, a value realized in the sample of Zuev *et al* [16], E_F=135 meV, and the onset of optical phonon emission is only a few k_BT times greater than E_F. A departure from linear T-dependence due to optical phonons is noticed (Figure 12a) and, as a function of carrier concentration, a change in the sign of $|S_d|$ is observed. Their analysis, assuming S_g to be negligible, obtains a fit with *S(T)* data of Zuev *et al* (Figure 12b).

The effect of electron-phonon scattering processes on TEP of extrinsic graphene has been studied by Munoz [78]. From a variational solution of the Boltzmann equation, he obtained analytical expressions for the transport coefficients and the leading contribution to phonon-limited TEP. Figure 13 represents his results of temperature dependence of TEP at different electronic densities. At lower temperatures (T<~60 K), TEP is found to show a linear in T dependence (see inset of Figure 13) and to reproduce the $n_s^{-1/2}$ dependence with carrier density, in agreement with reported experimental data [17].

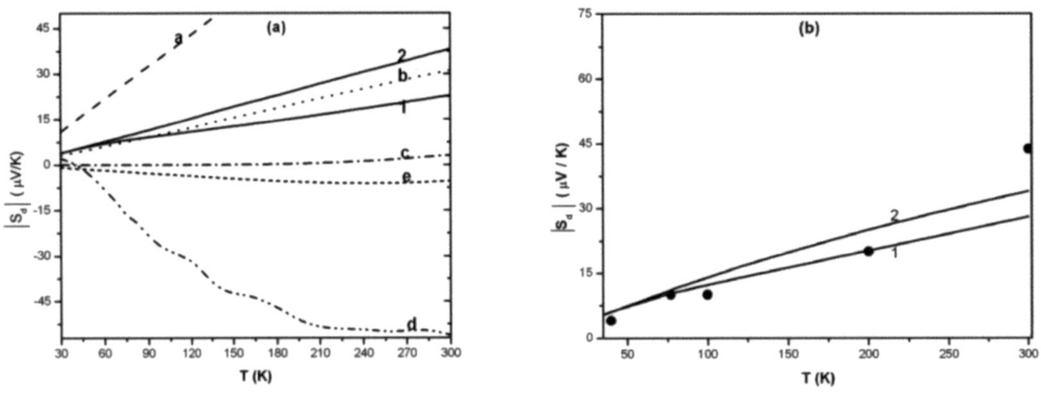

Figure 12. Temperature dependence of the S_d calculated using Eq. (33) for SLG with n_s = 2x 10^{16} m^{-2}. (a) Dashed, dotted, dash-dotted, short-dashed and dash-double-dotted curves represent individual contributions from impurities, vacancies, acoustic phonons, surface roughness and optical phonons, respectively. Curve 1 represents the overall contribution. Curve 2 represents the variation of S_d according to the Mott expression (36a). (b) Comparison of calculations (curves) with the experimental data of [16] (dots). Curves 1 and 2 represent calculations using Eqs. (33) and (36a) respectively. (from [63])

The distinctive features observed in the energy dependence of the relaxation times $\tau(E_k)$ of the scattering mechanisms in graphene are clearly exhibited in the case of the intrinsic acoustic phonon scattering in the Bloch-Gruneissen (BG) regime, where acoustic phonon scattering dominates and phonon energies are approximately of the order of thermal energy ($\hbar\omega_q \sim k_BT$). In their experimental study of the resistivity of graphene, Efetov and Kim [68] have demon-

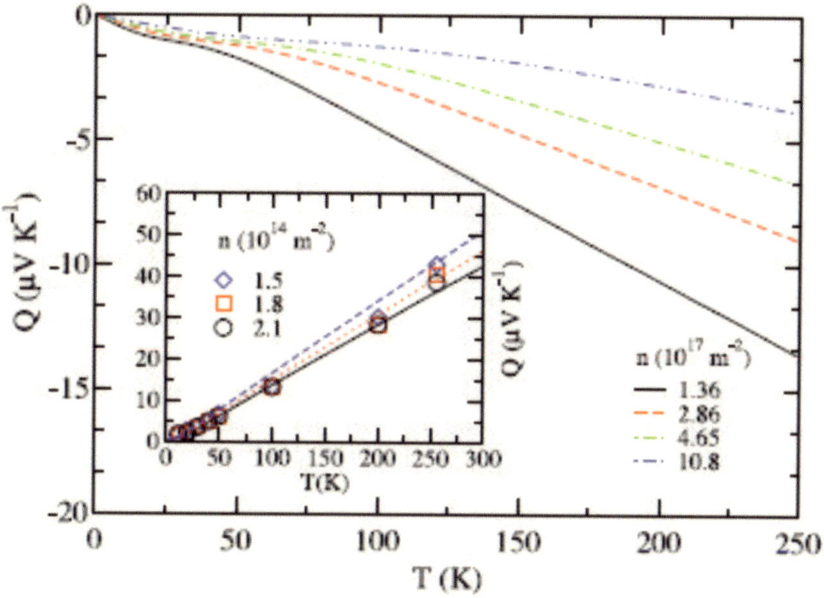

Figure 13. Temperature dependence of TEP at different electronic densities, *n*. Inset: Comparison with data of [17] at corresponding hole densities.(from [78]).

strated that the transition from higher temperature into the BG regime is reflected by a change in the character of electron scattering by acoustic lattice vibrations. With a view to examine the influence of the changed character of electron scattering on S_d, Sankeshwar *et al* [71] first study the behaviour of acoustic phonon limited scattering rates, $\tau^{-1}(E_k)$ in SLG for $T < T_{BG}$. For a graphene sample with n_s=10^{16}m^{-2}, T_{BG} ~60 K. Figure 14.(a) shows the strong influence this change has on the energy dependence of the unscreened electron-phonon interaction momentum relaxation times, τ_{BG}^{-1} around $E = E_F$. The sharp decrease in τ_{BG}^{-1} at $E = E_F$ for T = 4.2K is due to the factor $\Delta(E_k, E_{k'})$ in Eq. (38), governing probability occupation factors of electrons and phonons. With increase in the temperature from 4.2K to 77K to 300 K, the extent of the dip in τ_{BG}^{-1} at E_F is seen to decrease and gradually disappear at higher temperatures, as also the difference between the magnitudes of the rates τ_{BG}^{-1} and τ_{EP}^{-1}, the momentum relaxation time in the EP approximation.

In their investigation of the temperature dependence of S_d in the BG regime, Sankeshwar *et al* have shown that, corresponding to the transition into BG regime, S_d exhibits a non-linear temperature dependence (dotted curve in Figure 14. (b)) before becoming linear at very low temperatures [71]. Such dependence is also exhibited in conventional 2D systems [79]. The minimum value of $|S_d|$ occurs around T= T_{BG} and a change in the sign of $|S_d|$ is also noticed (dotted curve in Figure 14. (b)). On the other hand, the variation of $|S_d|$ in EP regime, calculated using Eq. (31), (38) and (39c) is found to be almost linear for $T > T_{BG}$, eventually merging with $|S_d|$ at higher temperatures [71].

The influence that S_d can have on the behavior of total TEP, S (= S_d+S_g), is illustrated in Figure 14.(b). The dashed curve depicts S_g, calculated in phonon-boundary scattering limit with τ_p= Λ / v_s for $n_s = 1 \times 10^{16}$ m^{-2} and phonon mean free path, Λ=100 nm, and curve 1 represents S. Here, S_g being small for $T < 25$ K, the overall TEP, S, exhibits the nonlinear temperature dependence of S_d in the BG regime ($T < 60$K), whereas for higher temperatures, is influenced by S_g. A comparison of curve 1 with curves 2 (Λ =500 nm) and 3 (Λ =1000 nm), shows that an increase in Λ leads to not only an increase in magnitude of S but also a change in its behavior in the BG regime. It is seen from Figure 14.(b) that the non-linear structure of S, exhibited due to S_d for smaller Λ, is masked by S_g for larger Λ. The electrostatic tunability of the chemical potential, $E_F = \hbar v_F k_F$, in graphene allows for a wide range of control of T_{BG} and could therefore, with proper control of S_g, allow the non-linear behavior of S_d to be observed. Figure 14.(c) shows comparison of the calculations for Λ =200 nm, with experimental data of Wei *et al* [17]. The deviation from data for higher temperatures suggests possible influence of other electron and phonon scattering mechanisms operative in real systems [33, 80], and calls for more detailed investigations.

Mariani and Oppen [67] and Ochoa *et al* [28], in their investigations of the resistivity of suspended SLG and BLG, have studied the effect of strain. They show that in the absence of strain, the FPs dominate the phonon contribution to resistivity, whereas in the presence of strain, the contribution due to FPs is suppressed and the in-plane phonon modes become dominant. These features seen in the scattering rates get reflected in the TE properties [81, 82]. Vaidya *et al* [81] have investigated the acoustic phonon limited S_d of suspended SLG, considering contributions from both in-plane and flexural acoustic phonons. Restricting their analysis to scattering in SG with high electron densities, in EP regime ($60 < T < 300$ K), they study the role of strain in influencing the total TEP, S.Figure 15.(a) shows the temperature dependence of TEP of a SG with $n_s = 1 \times 10^{16}$ m^{-2} [81]. The dominant contribution to S is from the phonon drag component, S_g for $T < 90$K, in the absence as well as presence of strain. However, for higher temperatures ($T > 150$ K) the contribution from S_d becomes important. The effect of strain, which is known to suppress the electron-FP interaction [28], is found to suppress S_d and to alter its behavior, the effect being larger at higher temperatures. The total diffusion TEP due to both in-plane (I) and flexural (F) phonons is calculated taking the overall momentum relaxation rates to be given by Mattheissen's rule: $\tau_{ns}^{-1} = \tau_I^{-1} + \tau_F^{-1}$, for non-strained SG, and $\tau_s^{-1} = \tau_I^{-1} + \tau_{F,\,str}^{-1}$, for strained SG. Figure 15.(b) shows the fit obtained with the calculations of S_d (curves 1 and 2) by Vaidya *et al* [82] for the data (dots) of suspended unstrained Cu-CVD SLG sample of Xu *et al* [45], for which, as suggested by thermal conductivity measurements, most of the heat is carried by FPs in the absence of strain. The almost linear temperature dependence suggests that the major contribution to TEP for $T < 170$ K is due to the diffusion component, as observed by Xu *et al* [45]. The deviation for $T > 170$ K, from experimental data, suggests the role of additional scattering mechanisms.

Apart from those mentioned above, there have been reports of other related TEP studies. The investigations of Sharapov and Varlamov [83] and Patel and Mukerjee [84], with regard to the effect of opening a gap in the graphene spectrum, find that the TEP is found to be proportional to the band gap. Zhou *et al* [85] have studied dependence of TEP on the polarization strength

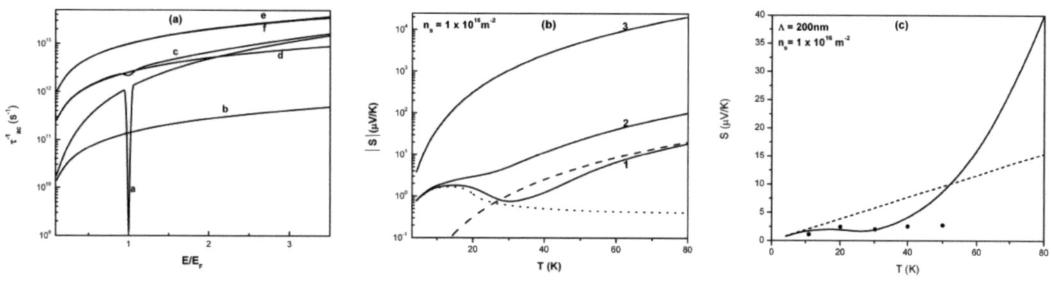

Figure 14. (a) Variation of acoustic phonon limited rates, τ_{BG}^{-1} and τ_{EP}^{-1}, as function of carrier energy E in SLG with n_s = 10^{16} m^{-2}. Curves a, c, and e depict variation of τ_{BG}^{-1} for T= 4.2, 77 and 300 K respectively. Curves b, d and f depict the temperature variation of τ_{EP}^{-1}. (b) Temperature dependence of acoustic phonon limited S_d (dotted curve), S_g(dashed curve, with Λ = 100 nm) and overall TEP, S (curve 1) of SLG with n_s = 1 x 10^{16} m^{-2}. Curves 2 and 3 depict S for S_g with Λ = 500 and 1000 nm, respectively. (c): Comparison of calculated S for Λ = 200 nm (full curve) and Mott S_d (dashed curve) with data of [17] for V_g= 50 V (dots). (from [71])

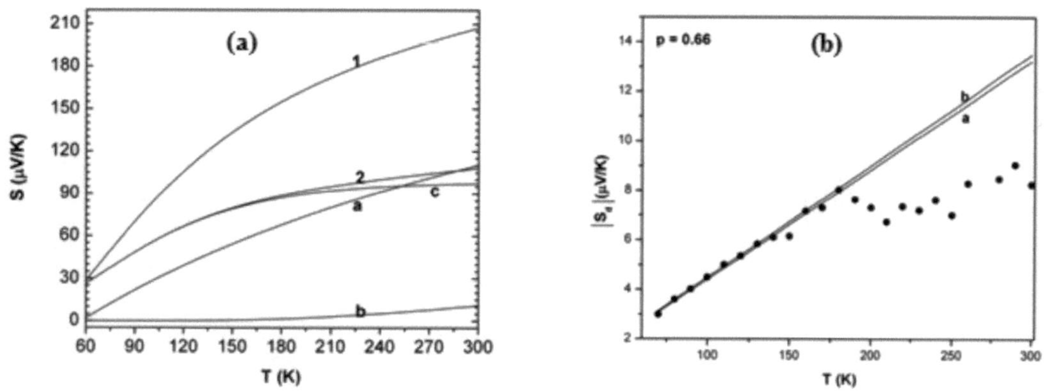

Figure 15. Temperature dependence of acoustic phonon limited TEP of suspended SLG. (a). Curves a and b represent S_d calculated using Eq. (33) using n_s – 1x10^{16} m^{-2} for non-strained and strained samples, respectively. Curve c depicts S_g. Curves 1 and 2 represent $S(= S_d + S_g)$ in non-strained and strained samples, respectively. (b) Comparison of calculations of S_d for n_s – 1x10^{17} m^{-2} with data(dots)of [45]. Curves 1 and 2 depicts, S_d due to FPs calculated using Eq. (32) and (36a) (from [81], [82])

of ferromagnetic leads, and find TEP to depend weakly on the strength and the magnetic configuration of the system.

3.3.2. Diffusion thermopower in BLG

In the last few years, TEP in BLG has been studied both theoretically [64, 86] and experimentally [49]. Nam *et al* [49] find the low-T TEP of BLG with large n_s, to follow the Mott relation with a linear-in-T dependence, suggesting a weak electron-phonon interaction. Hao and Lee [86], in their theoretical investigation, have studied the temperature and carrier density dependence of TEP of gapped relatively clean BLG in terms of the Kubo formula, with impurity scattering treated in the self-consistent Born approximation. They find that introducing a gap enhances the TEP.

Kubakaddi and Bhargavi [64, 73] give an expression for S_d, and compare their calculations of S_d with the data of Nam *et al* [49]. Figure 16.(a) shows the T dependence of S_d calculated taking into account the contributions from acoustic phonons (APs) *via* deformation potential coupling [73], surface polar phonons (SPPs) [74], charged impurity (CI) and short range disorder (SD) [75]. Although the contributions from SPPs (for $T<\sim200$ K) and the APs (for $T<\sim20$ K) exhibit non-monotonic behavior, the overall contribution is found to show almost linear temperature dependence. The non-linear T variation due to APs is attributed to the dip arising around $E=E_F$ in the energy dependence of the carrier relaxation time (as in Figure 14.(a) for SLG). Their calculations of S_d for parameters characteristic of the samples of Nam *et al* [49] obtain a qualitative agreement with data (see inset in Figure 16.(a)). Figure 16.(b) shows the carrier density dependence of S_d at T = 300 K. Although the dependence for SPPs is $S_d \sim n_s^{-0.75}$, the overall S_d is found to show almost $S_d \sim n_s^{-1}$ behavior. This inverse dependence of S_d on n_s, is unlike $\sim n_s^{-1/2}$ dependence observed for SLG but similar to that for degenerate conventional 2DEG [11]. Detailed investigations of TEP in BLG over wider temperature ranges are required to better understand the contribution of S_d relative to that of S_g; they are found to be similar (see Figure (22)).

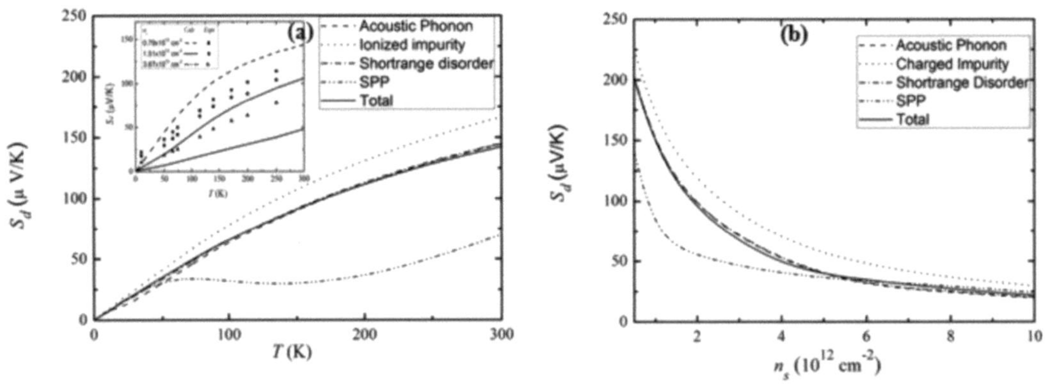

Figure 16. Variation of S_d of BLG as a function (a) T with n_s = 1x10^{12} cm^{-2}, and (b) n_s, at T=300 K. Inset in (a) shows T dependence of S^d calculated for sample of [49]. (from [73])

In the case of suspended BLG, S_d is found to be enhanced [73]. However, the behavior of S_d considering contributions from APs and CI, is found to be same as for supported BLG.

3.3.3. *Diffusion thermopower in AGNR*

The theoretical efforts, to understand the diffusion contribution to TEP of GNRs have been based on different techniques. Divari and Kliros [87] have studied TEP of ballistic wide graphene ribbons with aspect ratio (W/L ≥ 3) using linear response theory and the Landauer formalism. Xing *et al* [88] have studied TEP of GNRs in zero and non-zero magnetic field using non-equilibrium Green's function technique and shown that TEP depends on the chirality of the GNR. The TE properties have also been investigated by solving the electron and phonon transport equations in the nonequilibrium Green's function formalism [7, 89-91]. With a view

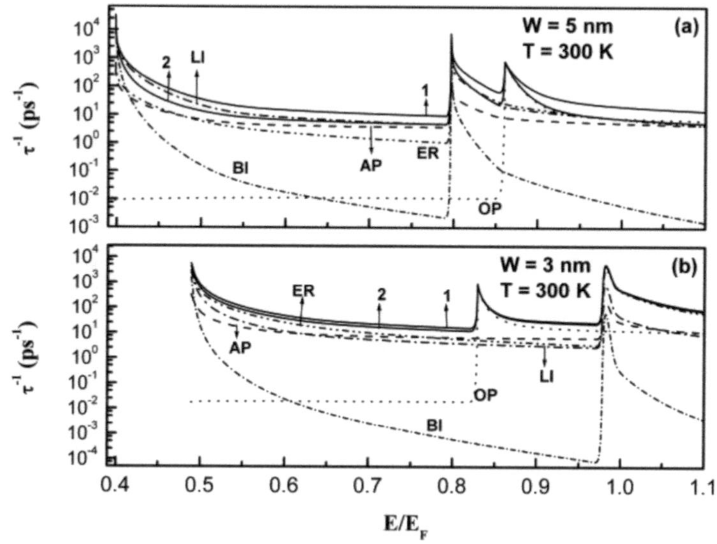

Figure 17. Energy dependence of relaxation rates for AGNRs of widths (a) 5 nm, and (b) 3 nm, with $n_l = 5\text{x}10^8 m^{-1}$ at T = 300 K. Dash-double-dotted, dash-dotted, short-dash dotted, dashed and dotted curves, respectively, represent contributions from ER, LI, BI, AP and OP scatterings. Curves 1 and 2, respectively denote overall relaxation rates τ_1^{-1} and τ_2^{-1}, given in the text. (from [65])

to achieve a high TE figure of merit, Mazzamoto *et al.* [92] in their recent study of the TE properties of GNRs have proposed optimized patterning of the ribbons with regard to their width and edge orientations.

Recently, in their systematic study of S_d of semiconducting AGNR systems employing Boltzmann formalism, Nissimagoudar and Sankeshwar [65] have illustrated the relative importance of various electron scattering mechanisms, and studied the influence of GNR width and subband structure on the behavior of S_d. Considering the electrons to be scattered by edge roughness (ER), impurities (IMPs) and deformation-potential coupled acoustic phonons (APs) and optical phonons (OPs), they obtain expressions for the relaxation times, $\tau_j^{-1}(E_k)$, for $j \equiv$ AP, OP, ER, IMP (see Table 3). Owing to the peculiar nature of the density of states, the energy dependences of the momentum relaxation times, unlike as in the case of SLG (see Figure 14.(a)), exhibit distinctive features as depicted in Figure 17 for AGNRs with $n_l = 5 \times 10^8$ m^{-1}, for $T < 300$ K. Two peaks, one corresponding to the subband and the other due to OP emission, are observed. The dominant contribution to the overall rate $\tau_1^{-1} = \tau_{ER}^{-1} + \tau_{LI}^{-1} + \tau_{AP}^{-1} + \tau_{OP}^{-1}$ (curve 1) is from LI scattering (impurities in the graphene layers only), the contribution from AP scattering being less. Assuming the impurities to be distributed uniformly not only in the layer but also in the substrate, the overall rate $\tau_2^{-1} = \tau_{ER}^{-1} + \tau_{BI}^{-1} + \tau_{AP}^{-1} + \tau_{OP}^{-1}$ (curve 2) is found to be dominated by ER scattering indicating that the contribution from the IMPs in the substrate is small [27]. The rates for a AGNR with $W = 3$ nm (shown in Figure 17.(b)) are found to exhibit the same features as seen for $W = 5$ nm (Figure 17.(a)), however, with a corresponding shift in the positions of the two peaks. In this case, it may be noted that the onset of OP emission occurs before the occupation of the second subband.

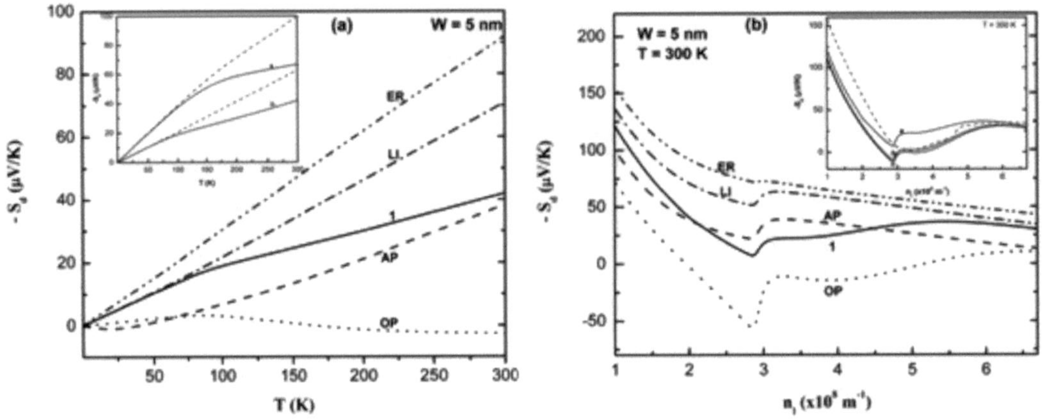

Figure 18. Variation of S_d of an AGNR of W= 5 nm, with (a) temperature for $n_l = 2\text{x}10^8\text{m}^{-1}$, and (b) linear carrier density at T=300 K. Curve 1 denotes overall contribution to S_d, calculated using τ_1^{-1} given in text, from scatterings due to ER (dash-double-dotted curve), LIs (dash-dotted curve), APs (dashed curve) and OPs (dotted curve). The inset in (a) shows overall S_d calculated using τ_1^{-1} for two AGNR widths: 3 nm (curve a) and 5 nm (curve b). Dashed curves show the overall S_d according to Mott expression. The inset in (b) shows overall S_d for an AGNR of width 5 nm for three values of LI concentrations, n_s: 10^{14}m^{-2} (curve a), 10^{13} m^{-2} (curve b), and 10^{12}m^{-2} (curve c). Dashed curve shows overall S_d for AGNR of W= 3 nm, with $n_s = 10^{14}$ m^{-2}. (from [65])

The above mentioned changed energy dependences of the relaxation times are found to influence the behavior of and lead to distinctive features in S_d as shown in Figure 18. The temperature dependence of the overall S_d (curve 1 in Figure 18.(a)), for an AGNR of width 5 nm, supporting an electron density 2 x10^8 m^{-1}, is found to be almost linear up to $T \sim 75$ K, the dominant contribution being from LI scattering. At higher temperatures, the contribution due to OP scattering becomes increasingly important and S_d is found to exhibit room temperature value of 42 μV/K. A decrease in the GNR width (to W= 3 nm), is found not only to enhance the magnitude of S_d but also alter its behavior (depicted by curve a in inset of Figure 18.(a)) due to the changed temperature and energy dependences of the individual contributions. Figure 18.(b) illustrates the influence of subband structure on room temperature S_d for an AGNR of width 5 nm. As a function of carrier density, S_d shows a step-like behavior, increasing in magnitude, when Fermi energy moves into the second subband, and reflects the singular nature of the AGNR density of states.

With a proper choice of parameters characterizing the extrinsic scattering mechanisms, and the possibility of modulating the Fermi level with a control on gate and bias voltages [93], the behavior of overall S_d in AGNRs could, therefore, be tuned and the subband structure in S_d be detected. Further, the enhancement in S_d with a decrease in the ribbon width, portends promising applications in thermoelectric devices.

4. Phonon-drag thermopower in graphene systems

As mentioned in section 2, in the presence of temperature gradient ∇T, the flow of phonons carries a momentum current, a fraction of which is transferred to electrons via phonon-electron interaction, giving rise to the phonon-drag component, S_g. S_g, unlike the diffusion contribution, depends only on the electron-acoustic phonon coupling strength. It has been extensively studied in conventional semiconductor 2DEG [11, 94, 95]. As recent observations of very large thermal conductivity in graphene [96-98] suggest the phonon mean free path Λ to be large, S_g, which also depends on the phonon mean free path Λ, is expected to be large in graphene systems. Λ in a system is decided by the various phonon-scattering mechanisms operative in it.

The formal theory of S_g, based on the semi-classical Boltzmann formalism, has been developed for a conventional 2DEG coupled to 3D phonons in semiconductor heterostructures [99]. With appropriate modifications, Kubakaddi [62, 64, 100] have applied it to graphene systems. The coupled Boltzmann equations for electrons and phonons are solved and the current density due to phonon-drag is obtained. The 2D electrons are assumed to interact with the in-plane 2D LA phonons of energy $\hbar\omega_q$, with $\omega_{\mathbf{q}}=v_s q$, and $\mathbf{q}=(q_x, q_y)$ is the 2D phonon wave vector. For the electron-acoustic phonon interaction *via* unscreened deformation potential coupling, S_g is given by

$$S_g = \frac{2e}{A\sigma k_B T^2} \sum_{\mathbf{k},\mathbf{k}',\mathbf{q}} \hbar\omega_{\mathbf{q}} f(E_{\mathbf{k}})\left[(1-f(E_{\mathbf{k}'})\right] P_{\mathbf{q}}^{a}(\mathbf{k},\mathbf{k}')\tau_p \mathbf{v}_p \cdot (\mathbf{v}_{\mathbf{k}}\tau_{\mathbf{k}} - \mathbf{v}_{\mathbf{k}'}\tau_{\mathbf{k}'}) \tag{40}$$

where $\mathbf{v}_p = v_s\mathbf{q}/|\mathbf{q}|$ is the phonon group velocity, $\tau_{\mathbf{k}}$ ($\tau_{\mathbf{q}}$) is the electron (phonon) momentum relaxation time. The transition rate, $P_{\mathbf{q}}^{a}(\mathbf{k}, \mathbf{k}')$, at which the electron in state $\mathbf{k}$ makes transition to state $\mathbf{k}'$ by absorbing a phonon, is given by

$$P_{\mathbf{q}}^{a}(\mathbf{k},\mathbf{k}') = \frac{2\pi}{\hbar} |C(q)|^2 F(\theta) N_q \delta(E_{k'} - E_{\mathbf{k}} - \hbar\omega_q)\delta_{\mathbf{k}',k+q}. \tag{41}$$

The phonon-drag is known to be important at low temperatures [8, 9, 11, 94, 95]. At these temperatures the phonon scattering is dominated by boundary scattering and $\tau_{\mathbf{p}} = \Lambda / v_s$, with Λ corresponding to smallest linear dimension of the sample.

4.1. Phonon-drag thermopower in SLG

Kubakaddi has studied in detail the low-temperature behavior of S_g. The low temperature S_g, in the boundary scattering regime, can be expressed as [62]

$$S_g = -\frac{D_{ac}^2 \Lambda}{2\pi\rho e E_F k_B T^2 \hbar^2 v_s{}^3 v_F} \int_0^\infty dq \int_\gamma^\infty dE_{\mathbf{k}} (\hbar\omega_{\mathbf{q}})^3 \sqrt{1-(\gamma / E_{\mathbf{k}})^2} N_{\mathbf{q}} f(E_{\mathbf{k}})\left[1 - f(E_{\mathbf{k}} + \hbar\omega_{\mathbf{q}})\right], \tag{42}$$

where $\gamma = \hbar v_F q/2$. Figure 19 shows that the low temperature dependence of S_g calculated using Eq.(42), with D_{ac} = 19 eV and Λ= 10 μm [101] for three values of n_s, is similar to that of conventional 2DEG [11]. Also shown (dotted curve) is the temperature dependence of S_d obtained from Mott formula (Eq. 36a) with p=1 for n_s=1.0x10^{12}cm^{-2}. The relative magnitudes of S_g and S_d depend upon the values of D_{ac}, Λ, n_s, T and p. S_g is found to be not negligible, contrary to the earlier qualitative remarks [35]. A system, for which S_d could be made to vanish (say, with p = -1) will be suitable for study of S_g and to estimate D_{ac}.

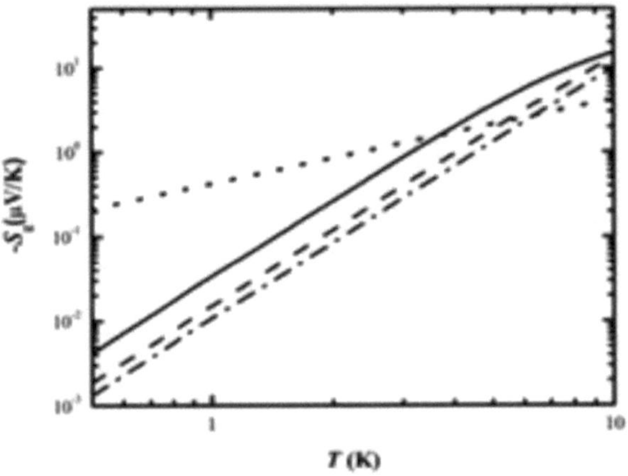

Figure 19. S_g as a function of T for n_s = 1.0x10^{12} cm^{-2}(solid curve), n_s = 5.0x10^{12} cm^{-2}(dashed curve) and n_s = 10.0x10^{12} cm^{-2}(dot-dashed curve). Dotted curve represents S_d, with p=1, for n_s = 1.0x10^{12} cm^{-2}. (from [62])

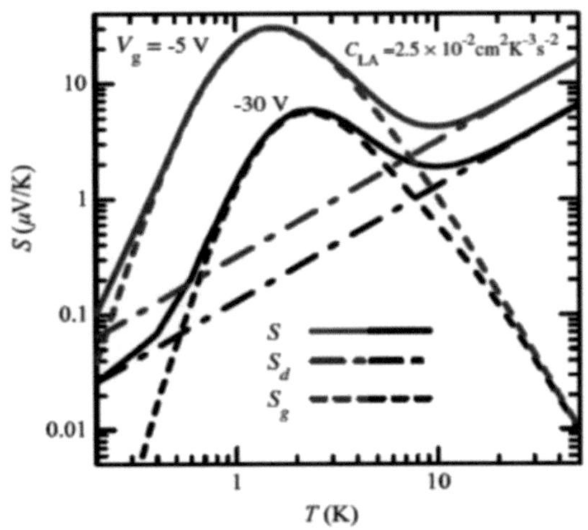

Figure 20. Temperature dependence of S_g, S_d and S (= S_g +S_d) of SLG for n_s = 2.16x10^{12} cm^{-2}(V_g= -5 V), n_s = 3.6x10^{11} cm^{-2}(V_g= -30 V). (from [77]).

Taking account of boundary scattering as well as phonon-phonon interaction in the phonon relaxation processes, Bao *et al* [77] have given the theory of S_g, using balance equation approach. Covering a larger range of temperature, they find (Figure 20) that S_g is important for $T< 10$ K and that the phonon-phonon interaction leads to a peak in its T dependence.The dash-dotted curve represents S_d considering scattering of electrons by impurities and phonons. The main contribution to TEP is mainly due to S_d for T >10 K.

In the BG regime, Kubakaddi gives a simple power law for S_g [62]:

$$S_g = S_{g0} T^3 \tag{43}$$

where S_{g0} (= $S_{g0}^{(SLG)}$) = $-[D_{ac}^2 \Lambda k_B^4 4!\zeta(4)]/[2\pi e\rho E_F \hbar^3 v_s^4 v_F]$ and $\zeta(n)$ is the Riemann zeta function. The cubic T dependence of S_g is a characteristic of 2D nature of phonons and is in contrast to the $S_g \sim T^4$ dependence of unscreened deformation potential scattering in conventional 2DEG [11, 94, 99, 102]. The inverse dependence of S_g on E_F suggests $S_g \sim n_s^{-1/2}$ in contrast to $n_s^{-3/2}$ in conventional 2DEG [11, 94]. This n_s dependence of S_g may be verified experimentally in graphene as it is possible to control n_s experimentally, say through the applied gate voltage.

In conventional 2DEG, in the BG regime, S_g and phonon limited mobility μ_p are known to be related by Herring's formula: $S_g\mu_p \sim T^{-1}$, first given for bulk semiconductors [9, 94, 103, 104]. Since in 2D graphene $\mu_p \sim T^{-4}$ [50], Eq.(43) gives $S_g\mu_p = - v_s \Lambda T^{-1}$ [62] so that Herring's law is validated even in 2D graphene, in which 2D electrons with linear dispersion interact with 2D phonons with $\omega_q \sim q$. This relation can, therefore, be used to determine, as in 2D GaAs system [105], a value for μ_p, from the measured S_g. It may be mentioned here that for the 2D phonons with $\omega_q \sim q^2$ (flexural modes) in semiconducting thin films, Herring's law is shown to be invalidated [106].

A useful and simple approach to calculate S_g is from the force balance argument that $S_g \,\alpha\, -fC_v/n_s e$ [11], where C_v is the lattice specific and f is the fraction of momentum lost by the phonons to the carriers. At very low T, $C_v \sim T^2$ for 2D phonons in graphene giving approximately $S_g \sim T^2$ [62] in contrast to $S_g \sim T^3$ for 3D phonons [11].

4.2. Phonon-drag thermopower in BLG

The theory of S_g in BLG has been developed by Kubakaddi and Bhargavi [64] in the phonon-boundary scattering regime, at low T. Its expression is given by

$$S_g = -\frac{m^{3/2} D_{ac}^2 \Lambda}{2^{3/2} n_s e k_B T^2 \rho \pi^2 (\hbar v_s)^4} \int_0^{\infty} d(\hbar\omega_{\mathbf{q}})(\hbar\omega_{\mathbf{q}})^3 \int_{E_{\mathbf{q}}}^{\infty} dE_{\mathbf{k}} F(E_{\mathbf{k}}, \omega_{\mathbf{q}}) N_{\mathbf{q}} \frac{f(E_{\mathbf{k}})[1 - f(E_{\mathbf{k}} + \hbar\omega_{\mathbf{q}})]}{\sqrt{E_{\mathbf{k}} - E_{\mathbf{q}}}}, \tag{44}$$

where $E_q=(\hbar^2/2m)[(q/2)-(mv_s/\hbar)]^2$ and $F(E_k, \omega_q)=[1-(\hbar\omega_q)^2/4mv_s E_k]^2$ is the function due to the chiral character of the carriers. Setting $F(E_k, \omega_q)=1$, Eq.(44) gives $S_g = S_{g,\,I2D}$ for the ideal 2D system (with no chiral character of electrons).

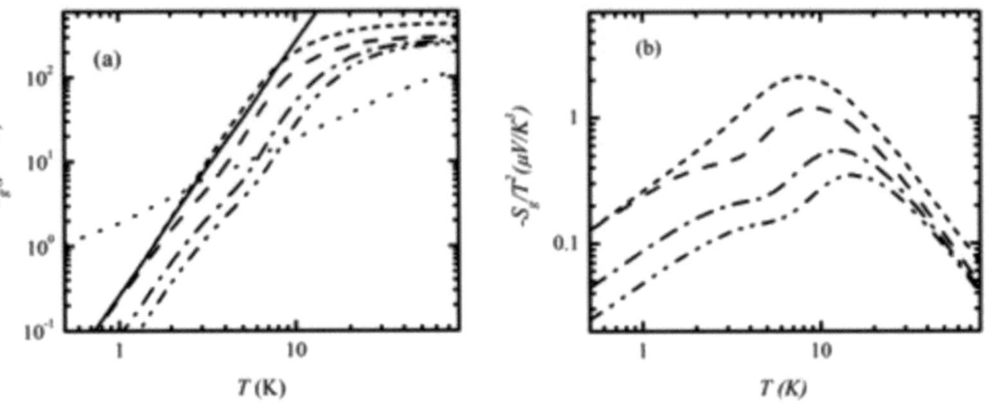

Figure 21. Temperature dependence of (a) S_g and (b) S_g/T^2 for Λ=10 µm and D=20 eV. Dashed curve is for n_s=0.5x10^{12} cm^{-2}; dash-dotted curve is for n_s=1.0x10^{12} cm^{-2}; and dot-dot-dashed curve is for n_s=1.5x10^{12} cm^{-2}. Dotted curve is due to S^d for n_s=0.5x10^{12} cm^{-2}, n_i=1x10^{11} cm^{-2} and $n_dV_o^2$=2 (eVÅ)^2. Short dashed curve is due to S_g with $F(E_{\mathbf{k}}, \omega_{\mathbf{q}})$=1 for n_s=0.5x10^{12} cm^{-2} and solid curve is due to S_g for n_s=0.5x10^{12} cm^{-2} in the BG regime.(from [64])

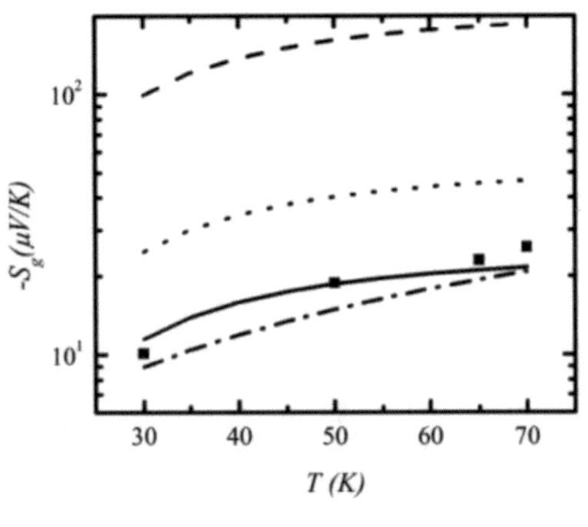

Figure 22. S_g as a function of T for BLG sample of [49] with Λ=8.8 µm and n$_s$=3.67x10^{12} cm^{-2}for D_{ac}= 6.8 eV (solid curve). D_{ac}=20 eV (dashed curve), D_{ac}=10 eV(dash-dotted curve). Dotted curve represents S_d for n_i=1x10^{11} cm^{-2}, $n_dV_o^2$=2 (eVÅ)^2. Closed squares are experimental data of [49]. (from [64])

Figure 21 shows the temperature dependence of S_g calculated using Eq.(44) [64]. Starting with T^3 at very low T, S_g gradually changes in to sublinear behavior in higher T region, and then flattens, thereby producing a knee. Inclusion of scattering due to Umklapp processes and point defects in the phonon relaxation time is expected to induce a peak at larger T, as found in the behavior of S_g of carbon nanotube [107] and of thermal conductivity of SLG [97]. Such an inclusion obtained good fits with data in the case of conventional 2D systems [108]. The dotted curve in Figure 21.(a) represents S_d, calculated using Eq.(36a), for n_s=0.5x10^{12} cm^{-2} assuming scattering by ionized impurities (with concentration n_i) and short-range disorder (of strength $n_d V_o^2$).

Kubakaddi and Bhargavi [64] have studied the influence on S_g of the form factor $F(E_k, \omega_q)$ in Eq.(44), arising from the chiral nature of the electrons, It is shown to produce a kink in the curves of S_g/T^2 *vs* T around 6 K (Figure 21(b)). The dashed curve corresponding to $F(E_k, \omega_q) = 1$, for ideal 2D, does not show such a kink. The effect of chiral character is shown to vanish in the BG regime and reduce the range of T for validity of BG regime. In the higher T regime the chiral nature is shown to reduce the magnitude of S_g, as compared to that of ideal 2DEG, S_{gI2D}, and the reduction is T dependent.

Nam *et al* [49], from their experimental investigations of TEP of a BLG sample for different n_s, for 30<T<250 K, find that the low-T TEP can be explained by Mott formula. However, Kubakaddi and Bhargavi [64] show that the low-T (30<T<70 K) data of [49] can be explained by S_g alone by varying D_{ac}. Figure 22 shows the fit to the data for n_s=3.67x10^{12} cm^{-2}.

In BG regime, S_g can be described by Eq.(44), but with S_{g0} (= $S_{g0}^{(BLG)}$) = $-[m^2 D_{ac}^2 \Lambda k_B^4 4!\zeta(4)]/[2\pi^{5/2} e\rho n_s^{3/2} \hbar^2 (\hbar v_s)^4]$. It may be noted that $S_g \sim T^3$ is a manifestation of the 2D phonons, as in SLG, in contrast with the T^4 dependence in conventional 2D systems [102]. It is inferred that the power of T is determined by the dimensionality and dispersion of phonons. In BLG, since $\mu_p \sim T^{-4}$ in BG regime [101], Herring's formula $S_g \mu_p \sim T^{-1}$ is shown to be valid [64].

With regard to n_s dependence in BLG [64], $S_g \sim n_s^{-3/2}$ in the BG regime. This is in contrast with $S_g \sim n_s^{-1/2}$ in SLG [62]. This can be attributed to the different dispersion relation of the electron spectrum in BLG as in conventional 2DEG [102]. It may be inferred that, n_s dependence of S_g is determined both by the dimensionality of the electron gas and its dispersion relation. The n_s dependence of S_g suggests the possibility of tuning the magnitude of S_g in BLG by tuning n_s.

4.3. Phonon-drag thermopower in AGNR

As shown in section 3.3.3, with regard to the diffusion TEP, the geometry and edge roughness can greatly influence the TE properties of GNRs [7, 65, 88-90, 109]. A dramatic reduction in phonon transport in ZGNR [7] indicates small value for Λ. However, in an AGNR, the phonon conductance is shown to be at least one order of magnitude higher than the electronic contribution indicating a larger value for Λ in this system [110]. The role of quasi-one-dimensionality, temperature, Fermi energy and ribbon width on S_g of a semiconducting n-type AGNR is investigated by Bhargavi and Kubakaddi [100]. The Q1D electrons are assumed to interact, *via* unscreened deformation potential coupling, with the 1D acoustic phonons of frequency $\omega_q = v_s q$, wave vector $\mathbf{q} = (0, q_y)$ with the velocity v_s. The acoustic phonon energy being small, S_g in AGNR, for intrasubband ($n = n'$) transitions, can be expressed as [100]

$$S_g = -\frac{2g_s g_v |e| v_s}{L\sigma k_B T^2 \hbar^2} \sum_{n,k_y} \sum_{q_y} \hbar\omega_q |C(q)|^2 F(\theta)[\tau(E)/\rho(E)] \ \tau_p N_q f(E_{n,k_y})[1 - f(E_{n,k_y} + \hbar\omega_q)]\delta(E_{n',k'_y} - E_{n,k_y} - \hbar\omega_q), \quad (45)$$

where, the various quantities are already defined. In the low-T boundary scattering regime, S_g is shown to be given by [100]

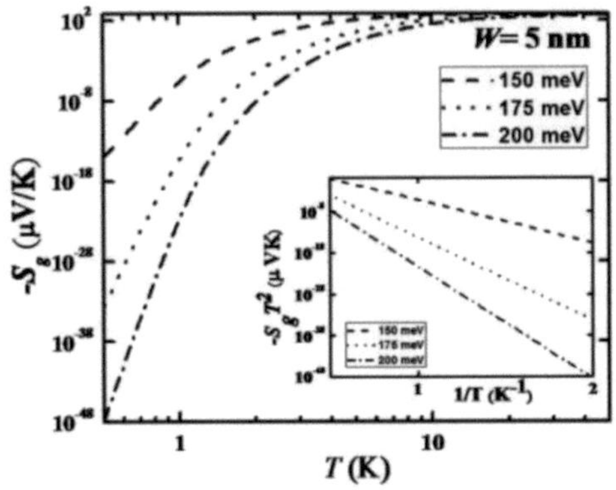

Figure 23. Phonon-drag thermopower S_g as a function of temperature T for an AGNR of width W=5nm for E_F = 150, 175 and 200 meV. Inset: S_gT^2 vs $1/T$ showing an activated behavior at very low T. (from [100])

$$S_g = -\frac{D_{ac}^2 E_f \Lambda}{\pi\rho W |e| k_B T^2 (\hbar v_f)^3 (E_f^2 - E_n^2)^{1/2} v_s^2} \int_{E_n}^{\infty} \frac{g(E)E}{\rho(E)G(E)\sqrt{E^2 - E_n^2}} f(E)\{1 - f(E + \hbar\omega_{q_o y})\}(\hbar\omega_{q_o y})^2 N_{q_o y} dE \tag{46}$$

where, $E_{n,k_y} = E$, $|q_y| = |2[(k_n^2 + k_y^2)^{1/2}(v_s/v_f) - k_y]| = |q_{oy}(k_n, k_y)|$, $g(E) = [1 + [\{E^2 + (E^2 - E_n^2)^{1/2}\hbar v_f q_{oy}\} / \{E[E^2 + 2q_{oy}(E^2 - E_n^2)^{1/2}\hbar v_f + q_{oy}^2(\hbar v_f)^2]^{1/2}\}]/2$, and $G(E) = |[2E(v_s/v_f) - (E^2 - E_n^2)^{1/2}] / [E_n^2 + [2E_n(v_s/v_f) - (E^2 - E_n^2)^{1/2}]^2]^{1/2}|$.

In the BG regime, the energy integration in Eq.(46) gives

$$S_g \sim (1/T^2) exp(-2\hbar v_s k_F / k_B T) \tag{47}$$

Figure 23 illustrates the temperature dependence of S_g calculated in the 1D quantum limit (n = 1) for D_{ac}=20 eV [62] and Λ=10 μm [97, 98]. In the inset, a plot of $S_g T^2$ vs $1/T$ (for T < 2K) shows an activated behavior (Eq. (47)). At very low T, S_g is found to be exponentially suppressed, which is a new feature. This is attributed to the peculiar nature of the one-dimensional Fermi-surface consisting of discrete points. It may be mentioned here that such suppression is similar to that observed in semiconducting single wall carbon nanotubes (SWCNTs) [95, 111], but is in contrast to the well-known algebraic power laws $S_g \sim T^3$ in SLG [62] and BLG [64] and $S_g \sim T^4$ in conventional 1DEG [112]. In the high temperature region, S_g deviates from exponential behavior and finally levels off as observed in SWCNT [107, 113].

The dependence of S_g on the Fermi energy is shown in Figure 24.(a). As seen from Eq.(47), this dependence is strong unlike the case of SLG ($S_g \sim E_F^{-1}$) [62] and BLG ($S_g \sim E_F^{-3/2}$) [64]. The sensitive dependence of S_g on W is shown in Figure 24.(b). The edge roughness of AGNR modulates the phonon mean free path, which, as in SLG and BLG, significantly influences S_g. Since the ribbon edge geometry determines the electronic structure of the system, S_g in ZGNR is expected to be different from that in AGNR.

The above results in AGNR, the TEP measurements and S_g calculations in a SWNT [95, 107, 113-116], and S_g calculations in conventional 1D electron gas in semiconductor heterostructures [112, 117, 118] clearly show the importance of S_g in 1D systems.

From Eqs.(42)-(44) and (46), it may be seen that $S_g \sim \Lambda$, so that S_g may be tuned with Λ, through the sample size. Further, defining the effective phonon mean free path as $\Lambda_{eff} = \Lambda(1+p)/(1-p)$, and with specularity parameter p=0.9 [97], S_g is shown to enhance by about an order of magnitude. For example, in BLG with n_s= 0.5x10^{12} cm^{-2} at T=10 K, S_g (p=0)=0.118 mV/K and S_g (p=0.9)≈1.2 mV/K [64].

In all the three graphene systems considered above, $S_g \sim D_{ac}^2$. In literature, the value of D_{ac}, being uncertain, is taken in the range 10-50 eV [101]. Recent measurements of resistivity [68] and hot electron energy loss rates [119] in SLG indicate the electron-acoustic phonon interaction to be unscreened. Further, a value of 19 eV for D_{ac} is found to explain well the low-T energy-loss rates, which depend only on the electron-acoustic phonon interactions. As S_g depends only upon electron-acoustic phonon coupling, unlike mobility which depends upon scattering due to other sources, detailed experimental and theoretical studies of S_g in graphene systems, particularly at low T, may also be used to determine D_{ac}.

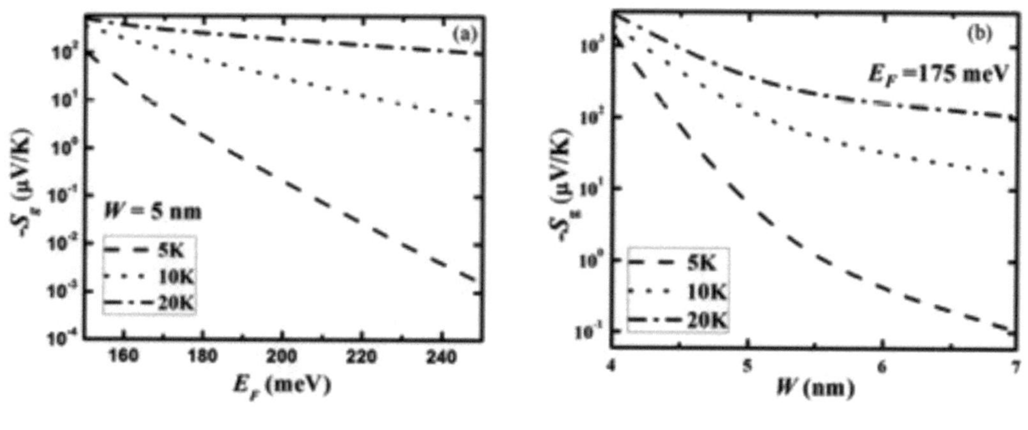

Figure 24. (a) S_g of AGNR as a function of (a) E_F with W = 5 nm, and (b) W with E_F = 175 meV, for T = 5, 10 and 20 K. (from [100])

5. Summary

In this chapter, we have reviewed the current status of the experimental investigations of the important and interesting transport property, namely TEP in graphene and described a theoretical treatment of the diffusion and phonon-drag components of TEP, in graphene systems. The treatment presented, employing the conventional Boltzmann formalism in the relaxation time approximation, gives a basic understanding of TEP in graphene systems, namely, SLG, BLG and AGNR. It gives a description of the dependences of TEP on temperature and gate bias. This understanding is expected to provide a useful guideline for improvement and optimization of performances of graphene-based TE modules.

Measurements of TEP of graphene reveal unique features not observed in metals [9] and conventional 2D semiconductor systems [10]. The graphene systems exhibit a range of TEP values up to 100 μV/K, at room temperature. The TEP changes sign across the CNP as the gate bias is varied. Away from the CNP, the TEP shows a $n_s^{-1/2}$ dependence on the carrier density n_s. At low temperatures, it exhibits a linear temperature dependence, in consonance with the Mott relation. However, a deviation from Mott formula is observed for higher temperatures.

Future experimental endeavours may aid not only in improving applicability in TE devices but also in understanding better the TE processes in graphene. Graphene may be a suitable system to realize a large range of BG regime. A detailed investigation of low-temperature (say, T<10 K) TEP may enable a better analysis of the relative contributions of S_d and S_g. Exclusive data for S_g may help estimate D_{ac}. On the other hand, with proper control of S_g, the non-linear structure of S and a change in sign, exhibited in the BG regime due to S_d for smaller phonon mean free paths, may be observed. A non-linear behavior of S with a change of sign has been observed in recent experiments [120] on exfoliated SPS processed graphene. There are not many reports of measurements of TEP in BLG. Measurements of TEP of GNRs, which are awaited, can reveal the role of quasi-one-dimensionality and of the structure patterning (say, width and edge chirality) of graphene sheets. The effect of physical modifications, such as suspending and/or straining of the graphene structures, is to alter the magnitude as well as the behavior of TEP. We note that a systematic investigation of TEP as well as thermal conductivity of graphene systems is needed to describe their TE figure of merit and to get a much better understanding of scattering mechanisms operative.

Conventional low-dimensional systems, such as quantum wires and superlattices, are known to provide not only new approaches for achieving higher ZT, but also new applications such as thermal management of integrated circuits [4, 5]. The possibility of increasing ZT through engineering the electron and phonon transport, therefore, makes graphene systems attractive, in future, for applications in efficient thermoelectric devices.

Acknowledgements

This work was supported by UGC (India). The assistance of Mr. A.S. Nissimagoudar in the preparation of the manuscript is acknowledged.

Author details

N.S. Sankeshwar[1*], S.S. Kubakaddi[1] and B.G. Mulimani[2]

*Address all correspondence to: n_s_sankeshwar@hotmail.com

1 Department of Physics, Karnatak University, Dharwad, Karnataka, India

2 B.L.D.E. University, Bijapur, Karnataka, India

References

[1] Katsnelson MI. Graphene: Carbon in Two Dimensions. Cambridge: Cambridge University Press; 2012

[2] Castro Neto AH, Guinea F, Peres NMR, Novoselov KS, Geim AK. The electronic properties of graphene. Rev Mod Phys 2009; 81(1):109-62

[3] Das Sarma S, Adam S, Hwang EH, Rossi E. Electronic transport in two-dimensional graphene. Rev Mod Phys 2011; 83(2):407-70.

[4] Rowe DM., editor. Thermoelectrics Handbook - Macro to Nano. Boca Raton: Taylor and Francis; 2006

[5] Tritt TM., editor. Recent Trends in Thermoelectric Materials Research III. Semiconductors and Semimetals Volume 71. San Diego: Academic Press; 2001

[6] Mahan GD. Good thermoelectrics. In: Ehrenreich H. and Spaefen F. (eds.) Solid State Physics, vol 51, New York; Academic Press, 1998. p 81

[7] Sevincli H, Cuniberti G. Enhanced thermoelectric of merit in edge-disordered zigzag graphene nanoribbons. Phys Rev B 2010; 81(11):113401-1-4..

[8] Barnard RD. Thermoelectricity in metals and alloys. London: Taylor and Francis; 1972.

[9] Blatt FJ, Schroeder PA, Foiles CL, Greig D. Thermoelectric power of metals. New York: Plenum Press; (1976).

[10] Ure RW. Thermoelectric effects in III – V Compounds. In: Willardson RK and Beer AC. (eds.) Semiconductors and Semimetals, vol 8, New York; Academic Press, 1972. p 67-102

[11] Gallagher BL, Butcher PN. Classical transport and thermoelectric effects in low dimensional and mesoscopic semiconductor structures. In: Landsberg PT. (ed.) Handbook on semiconductors, vol 1, Amsterdam; Elsevier; 1992. p.721-816.

[12] Fletcher R. Magnetothermoelectric effects in semiconductor systems. Semicond Sci Tech 1999; 14(4):R1-R15

[13] Dresselhaus MS. Quantum wells and quantum wires for potential thermoelectric applications. In: Willardson RK and Weber ER. (eds.) Semiconductors and Semimetals, vol 71, New York; Academic Press, 2001. p 1

[14] Chen G. Phonon transport in low-dimensional structures. In: Willardson RK and Weber ER. (eds.) Semiconductors and Semimetals, vol 71, New York; Academic Press, 2001. p 203

[15] Balandin AA. Thermal properties of graphene and nanostructured carbon materials. Nat Mat 2011;10(8):569–81

[16] Zuev YM, Chang W, Kim P. Thermoelectric and magnetothermoelectric transport measurements of graphene. Phy Rev Lett 2009; 102(9): 096807-1-4.

[17] Wei P, Bao W, Pu Y, Lau CN, Shi J. Anomalous thermoelectric transport of Dirac particles in graphene. Phy Rev Lett 2009; 102(16): 166808-1-4.

[18] Checkelsky JG, Ong NP. Thermopower and Nernst effect in graphene in a magnetic field. Phy Rev B 2009; 80(20): 081413(R)-1-4.

[19] Novoselov KS, Falko VI, Colombo L, Gellert PR, Schwab MG, Kim K. A roadmap for graphene. Nature 2012;490(7419):192-200

[20] Palacios JJ, Fernandez-Rossier J, Brey L, Fertig HA. Electronic and magnetic structure of graphene nanoribbons. Semicond Sci Technol 2010;25(3):033003-1–10.

[21] Bolotin KI, Sikes KJ, Hone J, Stormer HL, Kim P.Temperature-dependent transport in suspended graphene. Phys Rev Lett 2009; 101(9):096802-1-4

[22] Bolotin KI, Sikes KJ, Jiang J, Klima M, Fundenberg G, Hone J, Stormer HL, Kim P. Ultrahigh electron mobility in suspended graphene. Solid State Commun 2011; 146(9): 351-55

[23] Levy N, Burke SA, Meaker KL, Panlasigui M, Zettl A, Guinea F, Castro Neto AH, Crommie MF. Strain-induced pseudo– magnetic fields greater than 300 tesla in graphene nanobubbles. Science 2010;329 (5991): 544-47

[24] Wallace PR. The band theory of graphite. Phys Rev 1947; 71(9);622-34.

[25] Min H, Sahu B, Banerjee SK, MacDonald AH. Ab initio theory of gate induced gaps in graphene bilayers. Phys Rev B 2009; 75(15):155115-1-7.

[26] Wakabayashi K, Takane Y, Yamamoto M, Sigrist M. Electronic transport properties of graphene nanoribbons. New J of Phys 2009; 11(9): 095016-1-21.

[27] Fang T, Konar A, Xing H, Jena D. Mobility in semiconducting graphene nanoribbons: phonon, impurity, and edge roughness scattering. Phys Rev B 2008; 78(20): 205403-1-8.

[28] Ochoa H, Castro E, Katsnelson MI, Guinea F. Scattering by flexural phonons in suspended graphene under back gate induced strain. Physica E 2012;44 (6):963-66..

[29] Ziman JM. Electrons and phonons. Oxford: Claredon Press; 1960.

[30] Ashcroft NW, Mermin ND. Solid state physics. New York: Brooks/Cole; 1976.

[31] Herring C, Geballe TH, Kunzler JE. Phonon-drag thermomagnetic effects in n-type germanium. I. general survey. Phys Rev 1958; 111(1):36-57.

[32] Liu X, Wang D, Wei P, Zhu L, Shi J. Effect of carrier mobility on magnetothermoelectric transport properties of graphene. Phy Rev B 2012;86(15):155414-1-7.

[33] Löfwander T, Fogelström M. Impurity scattering and Mott's formula in graphene. Phy Rev B 2007; 76(19):193401-1-4

[34] Seol JH, Jo I, Moore AL, Lindsay L, Aitken ZH, Pettes MT, Li X, Yao Z, Huang R, Broido D, Mingo N, Ruoff RS, Shi L. Two-dimensional phonon transport in supported graphene. Science 2010; 328(5975):213-16

[35] Hwang EH, Rossi E, Das Sarma S. Theory of thermopower in two-dimensional graphene. Phy Rev B 2009; 80(23): 235415-1-5.

[36] Wang D, Shi J. Effect of charged impurities on the thermoelectric power of graphene near the Dirac point. Phy Rev B 2011; 83(11): 113403-1-4.

[37] Jonson M, Girvin SM. Thermoelectric effect in a weakly disordered inversion layer subject to a quantizing magnetic field. Phys Rev B 1984; 29(4):1939-46.

[38] Oji H, Thermomagnetic effects in two-dimensional electron systems. J Phys C: Solid State Phys. 1984;17(17):3059-66

[39] Liu X, Ma Z, Shi J. Derivative relations between electrical and thermoelectric quantum transport coefficients in graphene. Solid State Commun 2012; 152(6):469-72.

[40] Bonaccorso F, Lombardo A, Hasan T, Sun Z, Colombo L, Ferrari AC. Production and processing of graphene and 2D crystals. Materials Today 2012; 15(12):564-89.

[41] Wu X, Hu Y, Ruan M, Madiomanana NK, Berger C, de Heer WA. Thermoelectric effect in high mobility single layer epitaxial graphene. Appl Phys Lett 2011; 99(13): 133102:1-3

[42] Bergman DL, Oganesyan V. Theory of dissipationless Nernst effects. Phys Rev Lett 2010;104(6):066601-1-4.

[43] Ghahari F, Zuev Y, Watanabe K, Taniguchi T, Kim P. Effect of electron-electron interactions in thermoelectric power in graphene. Bull.American Phys Soc; 57(1) Mar2012

[44] Puneet P, Podila R, Oleveira L, Tritt T, Rao A. Transport properties of pristine and doped graphene. Bull.American Phys Soc; 57(1)Mar2012

[45] Xu X, Wang Y, Zhang K, Zhao X, Bae S, Heinrich M, Bui CT, Xie R, Thong JTL, Hong BH, Loh KP, Li B, Oezyilmaz B. Phonon transport in suspended single layer graphene. arXiv:1012.2937v1 [cond-mat.mes-hall] 14Dec2010.

[46] Babichev AV, Gasumyants VE, Butko VY. Resistivity and thermopower of graphene made by chemical vapor deposition technique. J Appl Phys 2013;113(7): 076101-1-3.

[47] Sidorov AN, Sherehiy A, Jayasinghe R, Stallard R, Benjamin DK, Yu Q, Liu Z. , Wu W, Cao H, Chen YP, Jiang Z, Sumanasekera GU. Thermoelectric power of graphene as surface charge doping indicator. Appl Phys Lett 2011; 99(1): 013115:1-3.

[48] Fogler MM, Guinea F, Katsnelson MI. Pseudomagnetic fields and ballistic transport in a suspended graphen sheet. Phys Rev Lett 2008; 101(22):226804-1-4.

[49] Nam SG, Ki DK, Lee HJ. Thermoelectric transport of massive Dirac fermions in bilayer graphene. Phys Rev B 2010; 82(24): 245416-1-5.

[50] Hwang EH, Das Sarma S. Acoustic phonon scattering limited carrier mobility in two-dimensional extrinsic graphene. Phys B 2008; 77(11):115449-1-6.

[51] Wang CR, Lu WS, Lee WL. Transverse thermoelectric conductivity of bilayer graphene in the quantum Hall regime. Phys Rev B 2010; 82(12): 121406(R)-1-4.

[52] Wang CR, Lu WS, Hao L, Lee WL, Lee TK, Lin F, Cheng IC, Chen JZ. Enhanced thermoelectric power in dual-gated bilayer graphene. Phy Rev Lett 2011; 107(18): 186602-1-4.

[53] Xu X, Gabor NM, Alden JS, van der Zande AM, McEuen PL. Photo-thermoelectric effect at a graphene interface junction. Nano Lett 2010; 10(2):562-66.

[54] Sidorov AN, Gaskill K, Nardelli MB, Tedesco JL, Myers-Ward RL, Eddy CR Jr., Jayasekera T, Kim KW, Jayasingha R, Sherehiy A, Stallard R, and Sumanasekera GU. Charge transfer equilibria in ambient-exposed epitaxial graphene on (0001) 6 H-SiC. J Appl Phys 2012; 111(11): 113706:1-6.

[55] Li X, Yin J, Zhou J, Wang Q, Guo W. Exceptional high Seebeck coefficient and gas-flow-induced voltage in multilayer graphene. Appl Phys Lett 2012; 100(18): 183108:1-3

[56] Sim D, Liu D, Dong X, Xiao N, Li S, Zhao Y, Li L, Yan Q, Hng HH. Power factor enhancement for few layered graphene films by molecular attachments. J Phys Chem C 2011; 115(5):1780-1785

[57] Xiao Ni, Dong X, Song L, Liu D, Tay YY, Wu S, Li L, Zhao Y , Yu T, Zhang H, Huang W, Hng HH, Ajayan PM, Yan Q. Enhanced thermopower of graphene films with oxygen plasma treatment. ACS Nano 2011; 5(4):2749-2755.

[58] Shaffique A, Hwang EH, Galitski VM, Das Sarma S. A self-consistent theory for graphene transport. PNAS 2007;104(47): 18392–18397.

[59] Nag BR. Electron transport in compound semiconductors. Berlin: Springer-Verlag; 1980.

[60] Ferry DK, Goodnick SM, Bird J. Transport in nanostructures. Cambridge: Cambridge University Press; 2009.

[61] Ridley BK. Quantum processes in semiconductors. 2nd ed. Oxford: Clarendon Press; 1988.

[62] Kubakaddi SS. Interaction of massless Dirac electrons with acoustic phonons in graphene at low temperatures. Phy Rev B 2009;79(7): 075417-1-6.

[63] Vaidya RG, Kamatagi MD, Sankeshwar NS, Mulimani BG. Diffusion thermopower in graphene. Semicond Sci Technol 2010; 25(9): 092001-1-6.

[64] Kubakaddi SS, Bhargavi KS. Enhancement of phonon-drag thermopower in bilayer graphene. Phys Rev B 2010;82(15) 155410-1-7

[65] Nissimagoudar AS, Sankeshwar NS. Electronic thermal conductivity and thermopower of armchair graphene nanoribbons. Carbon 2013; 52:201-08.

[66] Shishir RS, Chen F, Xia J, Tao NJ, Ferry DK. Room temperature carrier transport in graphene. J Comput Electron 2009; 8(2): 43–50 [54]

[67] Mariani E, Oppen F. Temperature-dependent resistivity of suspended graphene. Phys Rev B 2010;82(19):195403-1-11.

[68] Efetov DK, Kim P. Controlling electron-phonon interactions in graphene at ultrahigh carrier densities. Phys Rev Lett 2010; 105(25):256805-1-4.

[69] Min H. Hwang EH, Das sarma S. Chirality-dependent phonon-limited resistivity in multiple layers of graphene. Phys Rev B 2012; 83(16):161404(R)-1-4.

[70] Castro EV, Ochoa H, Katsnelson MI, Gorbachev RV, Elias DC, Novoselov KS, Geim AK. Limits on charge carrier mobility in suspended graphene due to flexural phonons. Phys Rev Lett 2010; 105(26):266601-1-4.

[71] Sankeshwar NS, Vaidya RG, Mulimani BG. Behavior of thermopower of graphene in Bloch–Gruneisen regime. Physica Status Solidi B 2013; 250(7): 1356-62.

[72] Stauber T, Peres NMR, Guinea F. Electronic transport in graphene: A semiclassical approach including midgap states. Phy Rev B 2007;76(20): 205423-1-10.

[73] Bharaghavi KS, Kubakaddi SS. Scattering mechanisms and diffusion thermopower in a bilayer graphene.Physica E 2013; 116-21.

[74] Li X, Borysenko KM, Nardelli MB, Kim KW. Electron transport properties of bilayer graphene. Phys Rev B 2011; 84(19)195453-1-5

[75] Das Sarma S, Hwang EH, Rossi E, Theory of carrier transport in bilayer graphene, Phys Rev B 2010; 81(16):161407-1-4.

[76] Peres NMR, Lopes dos Santos JMB, Stauber T. Phenomenological study of the electronic transport coefficients of graphene. Phys. Rev. B 2007; 76(7): 073412-1-4.

[77] Bao WS, Liu SY, Lei XL. Thermoelectric power in graphene. J Phys Condens Matter 2010;22(31):315502-1-7.

[78] Munoz E. Phonon-limited transport coefficients in extrinsic graphene. J Phys Condens Matter 2012; 24(19): 195302-1-8.

[79] Sankeshwar NS, Kamatagi MD, Mulimani BG. Behaviour of diffusion thermopower in Bloch–Grüneisen regime in AlGaAs/GaAs and AlGaN/GaN heterostructures. Physica Status Solidi B 2005;242(14):2892-2901.

[80] Chen W, Clerk AA. Electron-phonon mediated heat flow in disordered graphene. Phys Rev B 2012; 86(12): 125443-1--14

[81] Vaidya RG, Sankeshwar NS, Mulimani BG. Diffusion thermopower in suspended graphene: effect of strain. J Appl Phys 2012;112(9):093711-1-6.

[82] Vaidya RG, Sankeshwar NS, Mulimani BG. Diffusion thermopower in suspended graphene. In: Singh V, Katiyar M, Mazharier SS, Das U, Dutta A, Sodhi R, Anantharamakrishna S. (eds.) : SPIE Proceedings Vol. 8549: proceedings of the 16th national Workshop on Physics of Semiconductor Devices, IWPSD2011, 19-22 December 2011, IIT-Kanpur, India. Wahington: SPIE; 2012. doi: 10.1117/12.925343

[83] Sharapov SG, Varlamov AA. Anomalous growth of thermoelectric power in gapped graphene. Phs Rev B 2012; 86(3): 035430-1-5.

[84] Patel AA, Mukerjee S. Thermoelectricity in graphene: effects of a gap and magnetic fields. Phy Rev B 2012;86(7): 075411-1- 5.

[85] Zhou B, Zhou B. Liu Z, Zhou G. Thermoelectric effect in a graphene sheet connected to ferromagnetic leads. J Appl. Phys 2012; 112(7):073712-1-4.

[86] Hao L, Lee TK, Thermopower of gapped bilayer graphene, Phys Rev B 2010; 81(16): 165445-1-8

[87] Divari PC, Kliros GS. Modeling the thermopower of ballistic graphene ribbons. Physica E 2010; 42(9):2431-2435.

[88] Xing Y, Sun Q, Wang J. Nernst and Seebeck effects in a graphene nanoribbon. Phys Rev B 2009; 80(23): 235411-1-8.

[89] Ouyang Y, Guo J. A theoretical study on thermoelectric properties of graphene nanoribbons. Appl Phys Lett 2009; 94(26):263107-1-3.

[90] Karamitaheri H, Neophytou N, Pourfath M, Faez R, Kosina H. Engineering enhanced thermoelectric properties in zigzag graphene nanoribbons. J Appl Phys 2012; 111(5): 054501-1-9.

[91] Zheng H, Liu HJ, Tan XJ, Lv HY, Pan L, Shi J. Enhanced thermoelectric performance of graphene nanoribbons. Appl Phys Lett 2012; 100(9):093104-1-5.

[92] Mazzamuto F, Hung-Nguyen V, Apertet Y, Caer C, Chassat C, Saint-Martin J, Dollfus P. Enhanced thermoelectric properties in graphene nanoribbons by resonant tunneling of electrons. Phys Rev B 2011; 83(23):235426-1-7.

[93] Lin Y, Perebeinos V, Chen Z, Avouris P. Electrical observation of subband formation in graphene nanoribbons. Phys Rev B 2008; 78(16):161409(R)-1-4.

[94] Fletcher R, Zaremba E, Zeitler U. Electron-Phonon interactions in low dimensional structures. In: Challis L. (ed.) Oxford: Clarendon; 2003. p149

[95] Tsaousidou M. Thermopower of low dimensional structures: the effect of electron-phonon coupling. In: Narlikar AV and Fu YY. (eds.) Frontiers in Nanoscience and nanotechnology vol 2. Oxford: Oxford University Press; 2010. p477.

[96] Balandin AA, Ghosh S, Bao W, Calizo I, Teweldebrhan D, Miao F, Lau CN. Superior thermal conductivity of single-layer graphene. Nano Lett 2008;8(3): 902-07.

[97] Nika DL, Pokatilov EP, Askerov AS, Balandin AA. Phonon thermal conduction in graphene: role of umklapp and edge roughness scattering. Phys Rev B 2009;79(12) 155413-1-12.

[98] Ghosh S, Nika DL, Pokatilov EP, Balandin AA. Heat conduction in graphene: experimental study and theoretical interpretation. New Journal of Physics 2009;11(09) : 095012-1-19.

[99] Cantrell DG, Butcher PN. A calculation of the phonon-drag contribution to the thermopower of quasi-2D electrons coupled to 3D phonons. J Phys C: Solid State Physics 1987; 20(13): 1985-2003.

[100] Bhargavi KS, Kubakaddi SS. Phonon-drag thermopower in an armchair graphene nanoribbon. J Phys Condens Matter 2011;23(27):275303-1-5

[101] Viljas JK, Heikkila TT. Electron-phonon heat transfer in monolayer and bilayer graphene.Phys Rev B 2010;81(24):245404- 1-9.

[102] Fletcher R, Pudalov VM, Feng Y, Tsaousidou M, Butcher P N. Thermoelectric and hot-electron properties of a silicon inversion layer. Phys Rev B 1997;56(19) 12422-28

[103] Herring C. Theory of thermoelectric power of semiconductors. Phys Rev 1954;96(11): 1163-1-25.

[104] Tsaousidou M, Butcher PN, Triberis GP. Fundamental relationship between the Herring and Cantrell-Butcher formulas for the phonon-drag thermopower of two-dimensional electron and hole gases. Phys Rev B 2001;64(16): 165304-1-10

[105] Tieke B, Fletcher R, Zeitler U, Henini M, Maan JC. Thermopower measurements of the coupling of phonons to electrons and composite fermions. Phys Rev B 1998;58(4): 2017-25

[106] Kubakaddi SS. Effect of acoustic-phonon confinement on the phonon-drag thermopower of a two-dimensional electron gas in a semiconductor thin film. Phys Rev B 2004;69(03):035317-1-5

[107] Scarola VW, Mahan GD. Phonon drag effect in single-walled carbon nanotubes. Phys Rev B 2002;66(20):205405-1-7

[108] Kamatagi MD, Sankeshwar NS, Mulimani BG. Wide temperature range thermopower in GaAs/AlGaAs heterojunctions. AIP Conf Proc.2009; 1147:514-20.

[109] Chen Y, Jayasekera T, Calzolari A, Kim KW, Nardelli MB. Thermoelectric properties of graphene nanoribbons, junctions and superlattices. J Phys Condens Matter 2010;22(37):372202-1-6

[110] Mazzamuto F, Hung Nguyen V, Nam DV, Caer C, Chassat C, Saint-Martin J, Dollfus P. In: proceedingsof 14th International workshop on computational electronics p 1-4, IWCE, Pisa, IEEE Xplore; 2010.

[111] Tsaousidou M. Theory of phonon-drag thermopower of extrinsic semiconducting single-wall carbon nanotubes and comparison with previous experimental data. Phys Rev B 2010;81(23):235425-1-9

[112] Kubakaddi SS. Electron-phonon interaction in a quantum wire in the Bloch-Gruneisen regime. Phys Rev B 2007;75(7) 075309-1-7

[113] Vavro J, Llaguno MC, Fischer JE, Ramesh S, Saini RK, Ericson LM, Davis VA, Hauge RH, Pasquali M, Smalley RE. Thermoelectric power of p-doped single-wall carbon nanotubes and the role of phonon drag. Phys Rev Lett 2003;90(06):065503-1-4

[114] Zhou W, Vavro J, Nemes NM, Fischer JE, Borondics F, Kamarás K, Tanner DB. Charge transfer and Fermi level shift in p-doped single-walled carbon nanotubes. Phy Rev B 2005;71(20): 205423-1-7.

[115] Yu C, Shi L, Yao Z, Li D, Majumdar A. Thermal conductance and thermopower of an individual single-wall carbon nanotube. Nano Lett 2005;5(9):1842-46

[116] Tsaousidou M. Phonon-drag thermopower of ballistic semiconducting single-wall carbon nanotubes and comparison with the semiclassical result. Europhysics Lett 2011;93(4) 47010-1-8

[117] Kubakaddi SS, Butcher PN. A calculation of the phonon-drag thermopower of a 1D electron gas. J Phys Condens Matter 1989;1(19):3939-46

[118] Tsaousidou M, Butcher PN. Phonon-drag thermopower of a ballistic quantum wire. Phys Rev B 1997;56(16):R10044-10047

[119] Baker AMR, Alexander-Webber JA, Altebaeumer T, McMullan SD, Janssen TJBM, Tzalenchuk A, Lara- Avila S, Kubatkin S, Yakimova R, Lin C-T, Li L-J, Nicholas RJ. Energy loss rates of hot Dirac fermions in epitaxial, exfoliated and CVD graphene. Phys Rev B 2013;87(04):0454014-1-6

[120] Rao A. Private communication 2012.